BIOLOGY

AN APPRECIATION OF LIFE

CONTRIBUTING CONSULTANTS

Advisory Board

Wilfred J. Wilson, Ph.D., California State University, San Diego
Theodore Friedmann, M.D., University of California, San Diego
Paul D. Saltman, Ph.D., University of California, San Diego
Gail R. Patt, Ph.D., Boston University

Susan Bryant, Ph.D., University of California, Irvine
Leland N. Edmunds, Jr., Ph.D., State University of New York, Stony Brook
Peter H. Hartline, Ph.D., University of California, San Diego
Jonathan Hodge, Ph.D., University of California, Berkeley
Yashuo Hotta, Ph.D., University of California, San Diego
William A. Jensen, Ph.D., University of California, Berkeley
Lee H. Kronenberg, Ph.D., candidate, University of California, San Diego
Robert D. Lisk, Ph.D., Princeton University
Vincent T. Marchesi, M.D., Ph.D., National Institute of Health
David M. Phillips, Ph.D., Washington University
David M. Prescott, Ph.D., University of Colorado
Roberts Rugh, Ph.D., Columbia University
Howard A. Schneiderman, Ph.D., University of California, Irvine
Michael Soulé, Ph.D., University of California, San Diego
Albert Szent-Györgyi, M.D., Ph.D., Nobel Prize, Marine Biological Laboratory
J. Herbert Taylor, Ph.D., Florida State University
Robert H. Whittaker, Ph.D., Cornell University

BIOLOGY

AN APPRECIATION OF LIFE

CRM BOOKS

Del Mar, California

Many people who have stumbled across an unusual or beautiful creature of the earth take it to a biology instructor and ask, "What is it?" In *The Forest and the Sea*, Marston Bates tells of some who have gone on to ask him, "But what good is it?" His response invariably has been, "What good are you?" We don't presume that this level of naïveté is common among students of biology, especially when groups of college students are cleaning petroleum from the wings of sea birds and young people of all sorts are campaigning for better ecological behavior. But the assumption that a man is superior to a 1,000 year old tree or a humpback whale often goes unchallenged. Many people also assume that man's place on earth is guaranteed without wondering why the guarantee for the dinosaur ran out. So it seems right for our biology book to encourage an appreciation of life.

Of all the sciences, biology is the one that attempts to establish the relationship of man to all of life and to the part life plays in the universe. In a biology course, this attempt involves a process of observation and analysis that can and should be conducted on many levels. Whether this examination be undertaken in terms of atoms, of individuals, of communities, or of habits and habitats, however, a remarkable story begins to unfold.

This book describes the fascinating diversity of all living things, as well as the unifying threads that bring order to this diversity. It tells of the beginning of life and the evolution of interlocked dependence of one creature on another and on the world in which they have evolved. It also reveals that small but important part about what we are, where we have been, and where we might be going.

John H Painter Jr.

Publisher
Del Mar, California

Preface

Contents

**Unit II
The Continuity
of Life**

Unit III
The Ascending Complexity of Life

Contents

BIOLOGY

AN APPRECIATION OF LIFE

Unit I

The Origin and Evolution of Living Systems

Biology is, by etymology, the study of life. Life is difficult to define, and the easiest solution is to decide, as so many people have done, that its definition is impossible. In his book The Nature of Life, [Albert] Szent-Györgyi writes: "Life as such does not exist; nobody has ever seen it. . . . The noun 'life' has no sense, there being no such thing."

Yet, it is well known that definition is among the methods for discovery. It is, as a matter of fact, an excellent heuristic method. For it obliges one to condense the essential of a category or of a phenomenon into a formula—the formula containing everything it has to contain, and excluding everything it has to exclude. To cast a good definition is therefore useful, for this exercise compels critical consideration of all the terms or aspects of a problem.

Life may be considered either as a property, or as a manifestation, or as a state of organisms. This might or might not satisfy the biologist. The physicist will immediately ask two questions: (1) What is an organism? (2) What is the specific property of living organisms that does not exist in the inanimate world and is therefore characteristic of life?

— Andre Lwoff (1962)

3

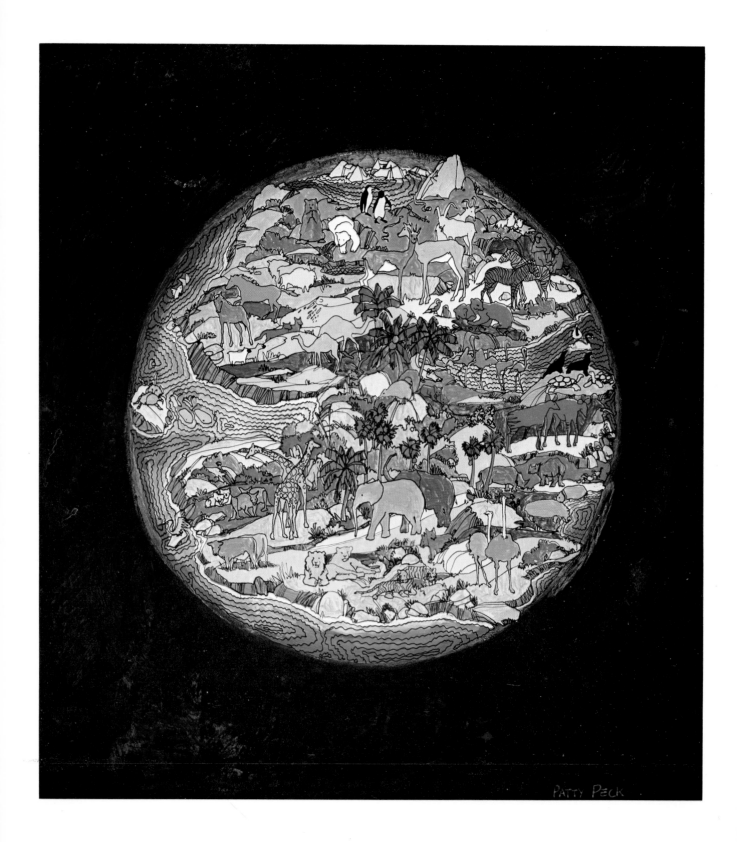

We live, most of us, in a remarkable environment called a city. It is our own creation, our own artifact, just as our chairs and forks and television sets are artifacts shaped in our cultural image out of the materials of the natural world. A city has form and structure; it functions to house and feed and keep occupied all the creatures of our kind who live together there. We can be born into this environment, grow up in it, and die there without seeing much else of the world.

Far out in the African savanna another artificial environment rises above the parched earth. It has form, and an architect would marvel at its structural integrity and its functional air-cooling system. Many of the creatures who live there will never know what it is like outside its splendidly designed walls. They will be born there, interact with creatures of their own kind, and eventually will die there.

Both artifacts were erected with purposeful behavior; both provide a setting for an organized way of living that ensures a tolerable existence for generation after generation of its inhabitants. But in spite of the similarities, even the inexperienced eye can see differences—far beyond the obvious difference in scale between a human city and a termite tower. We humans know at once that we are the more advanced, for instance. We can think and talk about what they are doing, but it is safe to say that not one of them is capable of thinking very much about what we are up to; even if they were, they could not do much about it.

This remarkably distinctive capacity we have for conceptual thought sets us apart from every other creature on the earth. We are our creatures' keeper.

But every now and then we stumble across something that has edged slightly over the line that keeps us apart. Perhaps at some time in our lives we happen to peer through a microscope—and see that the bits of matter composing different creatures do not appear to be all that different from the bits of matter composing human beings. Perhaps we read a book or watch a documentary on the life habits of the baboon or the chimpanzee. And we think about juvenile primates playing tag; about mothers caring attentively for their offspring and letting other females take turns at holding their infant; about a group of adult males carrying out what undeniably is an organized effort to go after not a leafy branch or a banana but another living creature. Whales "sing," and warblers migrating at night may use the stars for navigation as much as any human traveler.

The lines we draw separating ourselves from other creatures grow dim at times, and shift about, depending on how deeply we are willing to explore the phenomenon called life. There is a unity to life everywhere you look; atom joins with atom in ways that are basic to all living matter, and cell joins with cell in similar ways even as they achieve complex levels of integration. Aggregations

1

The Unity and Diversity of Life

Figure 1.1 Patterns of organization in biological systems. All areas in biology can be grouped within this organizational framework, and biologists can be specifically identified according to their fields of interest in a number of ways, including the following:

1. *Level of study.* Examples are cell biologists, organismic biologists, and population biologists.
2. *Kind and type of organism studied.* Examples are botanists, invertebrate zoologists, and protozoologists.
3. *Specific feature or process that is common to a large group of organisms.* Examples are physiologists, geneticists, morphologists, and developmental biologists.
4. *Where an organism lives.* Examples are aquatic or marine biologists and terrestrial biologists.
5. *Combination of two or more of the above categories.* Examples are marine invertebrate physiologists, population ecologists, and fruit fly cytogeneticists.
6. *Combination of one or more biological fields with other disciplines, such as chemistry, physics, or one of the social sciences.* Examples are biochemists, quantitative biologists, and social biologists.

of cells we call organisms share fundamental patterns of response to events that occur in their environment. And no one can deny that life is linked with life in the phenomenon of reproduction, the multistranded thread that has carried living systems out of the dim past into the present.

But the thread is not totally intact; the terminals of many strands lie buried in the past as shattered fragments of life that could not continue under some unrelenting test of strength. Nor are the delicate strands that now exist of a singular configuration. Life appears today in numberless variations, and perched precariously on the end of individual strands are such creatures as the kangaroo, the woolly aphid, the salmon, the carrot, the sponge, the bowerbird, and man.

Suppose that the thread of life began from some fundamental combination of materials—carbon, hydrogen, oxygen, and similar elements—as you will read in the next chapter. Assuming this

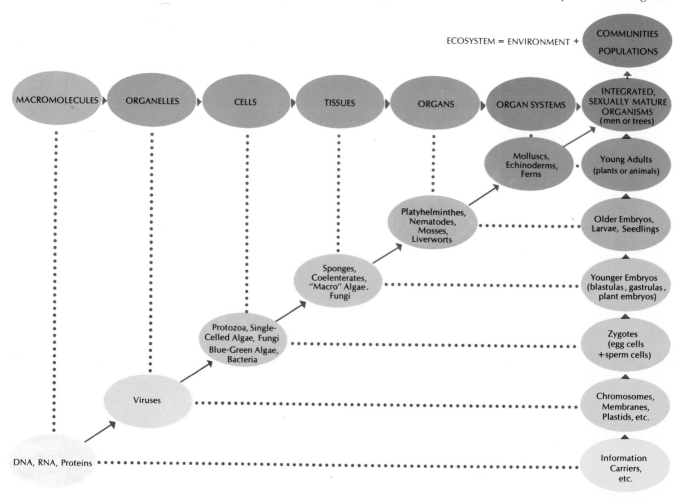

Figure 1.2 The French naturalist Jean-Baptiste Lamarck (1744–1829). His belief in spontaneous generation, where the simplest organisms are formed directly from inorganic matter, led him to the conclusion that the more complex plants and animals are produced indirectly by the complexification of these simplest organisms over eons of time.

common beginning, what explanation could possibly be advanced that would account for the transition of these fundamental materials into the diversity we see today?

THE THEORY OF NATURAL SELECTION

An early suggestion was made in the 1790s by the French naturalist Jean-Baptiste Lamarck, who knew nothing of the unifying material substrate of life but who was intrigued with the characteristics of the earth's creatures. In undertaking a classification of animals, he came to the conclusion that animals fit into a scheme of progression that began with the simplest creature and culminated in man, the most complex.

Lamarck was among the first to hold that the earth had been habitable for many millions of years, not just a few thousand. He argued that the motions of active vital fluids, which are the essential properties of all living bodies, can increase the organization present in a plant or animal, and any gains in organization are conserved in reproduction. According to Lamarck, however, these progressive increases in the degrees of organization have been modified by subsidiary causes. Vital motions in plants are speeded up by the action of heat and are therefore affected by climatic variations around the world. When responding to changes in their environment, vertebrate animals may adopt new habits, leading to the increased or decreased use of certain organs. The vital fluids in these organs will thus be accelerated, and the development of the organs will be enhanced or suppressed accordingly. Lamarck also believed that all such changes produced by these new habits will be inherited by the descendants of future generations. The doctrine that characters due to the environment's influence on an individual will be passed on to its offspring is often called the Lamarckian theory of inheritance, but this view had been a common one since the time of the Greeks. For Lamarck, this claim was simply a special case of his general thesis that the vital fluids of living bodies can produce permanent gains in the degree of organization.

Lamarck's theory is not a hypothesis of common descent, which ascribes the common characteristics of a particular species to their common descent from a single species. He claimed that although all mammals are produced by the gradual complexification of reptiles they are not descended from the *same* reptiles. He assumed that all mammals share certain characteristics because they have all reached the same general degree of organization.

A contrary suggestion was advanced by Charles Darwin, who in the 1830s had sailed around the world as a naturalist aboard the *H.M.S. Beagle*. During his travels he had been particularly impressed with the diverse characteristics of the fossils and the living creatures of the remote and isolated Galápagos Islands, some 600 miles off the coast of Ecuador. He perceived that there was a range

Figure 1.3 Charles Darwin (1809–1882), the English naturalist who formulated the theory of evolution by natural selection.

of variation in the characteristics of one *species*—a group of related individuals that resemble one another and are capable of interbreeding. He also perceived that different species apparently could rise from a common ancestral stock. He combined his detailed research with Thomas Malthus' earlier idea that population growth can outpace available food supplies, which creates a state of *competition* for nutritional resources. Darwin synthesized his ideas of a common origin of different species with the Malthusian doctrine. It was not that species passed on the characteristics they had worked hard to acquire; only the species that had *already* acquired competitively advantageous characteristics would survive to produce more of their kind.

Darwin's theory of natural selection comprises the following arguments: First, there is *variation* in all living species. Second, all individuals tend to *reproduce geometrically*—that is, they produce more offspring than will survive. Third, a changing environment will, by means of *natural selection*, select those individuals that are the best adapted, or most fit (*survival of the fittest*). Finally, Darwin stated that, as part of the selection process, all of life is locked in a *struggle for existence*. Modern evolutionary biologists have added that the best-adapted characteristics to the changing environment will be passed on to future generations, according to the laws of genetics.

Since the time of Darwin, the theory of evolution by natural selection has become a central tenet in biology. Little in science remains totally unchanged; new information can bring about modifications to an idea, and new ideas can supplant earlier ones. The core of Darwinian theory, however, has remained remarkably intact with the passing of time. Advances in the study of geology and in the study of heredity have served to strengthen the theory by answering questions that Darwin knew existed but could not answer with the knowledge of his time.

The first of the major post-Darwinian additions to the theory concerned the age of the earth. Natural selection would have to be a slow, gradual process, and an immense span of time would be

Figure 1.4 Some species of animals on the Galápagos, such as the famed giant tortoises (left), are very different from species on the mainland and presumably have been evolving independently on the islands for a long time. Other species are quite similar to mainland species and are assumed to have reached the islands relatively recently. Another inhabitant, the marine iguana (right), is one of the five kinds of reptiles found on the islands.

required to bring about the diversity that existed. Most eighteenth-century geologists reckoned the age of the habitable earth in terms of thousands of years or, at most, a few million years. The sorts of changes that Darwin envisioned—the descent and divergence of all species from a common ancestor—could not have taken place within that period of time. As geology advanced into the twentieth century, increasingly accurate dating methods confirmed that the earth was on the order of 4 or 5 billion years old. There was time enough, then, for natural selection to have occurred.

Darwin maintained that the theory of natural selection could stand alone, even if the underlying mechanisms of hereditary variation remained an enigma. Without that knowledge, however, acceptance of his theory would be slow in coming. If differences in heredity accounted for differences in the forms of life, what were the *units* of heredity? Where were they located? It was only with the explosive growth and discoveries of the field of genetics that Darwin's perception of the link between natural selection and hereditary variation would be proved essentially correct.

PATTERNS OF EVOLUTION

In retrospect, it can now be said that evolution itself has been a continuous process through which species become better adapted to

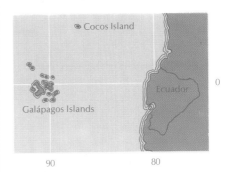

Figure 1.5 The Galápagos Archipelago, shown in relation to Ecuador (600 miles to the east) and Cocos Island (400 miles to the northwest).

Figure 1.6 This ship, the H. M. S. *Beagle*, took Darwin around the world on his surveying expedition between 1831–1836.

Figure 1.7 Comparison of the ideas of Lamarck and Darwin regarding the evolutionary process. The origin of the giraffe's long neck is illustrated according to Lamarck's theory of inherited acquired characteristics (above) and Darwin's theory of natural selection (below).

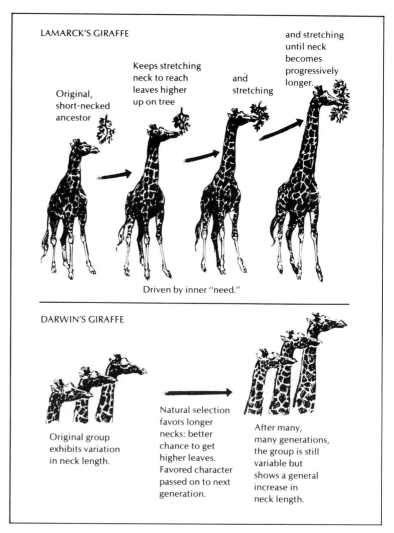

environmental conditions. The theory of natural selection implies only that certain creatures are more successful in reproduction and thereby cause certain units of heredity to become more widely represented in the population. Other variables are at work, including the outcome of apparently random or chance events.

Consider how a variety of one species might be converted into a new species. To begin with, some barrier may arise between one part of a species and another. Such barriers may be geographic— part of a population may cross a land bridge, for example, which might later become submerged under the sea. Following reproductive isolation, genetic changes may then occur that will yield an increasing number of differences between the isolated groups. Some of the newly acquired differences may be attributed to new and accidental genetic events that will be of selective advantage in the new environment. Other differences may result from previously existing characteristics that suddenly become much more or less favorable in a new environment and that are thus treated differently by the winnowing effects of natural selection. Finally, a point is reached at which divergence is so great that the two populations no longer can interbreed, which means *speciation* has occurred.

In this example, two processes are implicit. The first is the acquisition of genetic variation, which can come about in a number of ways. The second process is that of selection, which acts on the variants and results in a group of organisms that are most closely adapted to their environment. But the combination of variation and selection need not necessarily lead to evolutionary change. If a species has become well adapted to its environment, and if the environment remains relatively constant, selection will tend to be a stabilizing force for the species, acting to remove the extremes of variation and preserving the "average" type of organism. Only when the environment changes or when the organisms living there are no longer in balance will the average type not be the best-adapted representative of the species. In such cases, preservation of the status quo ceases and selection becomes directional. Under conditions of *directional selection*, a variant that was once considered an undesirable extreme may become the new "average" type for the species because it represents the best adaptation at that particular time. When a balance is again achieved between the environment and the average organism, selection will revert to the stabilizing sort.

Fossil records show that the first appearance of a particular major kind of organism in an area is usually followed by a relatively rapid process of speciation and divergence. Many distinct populations become established. Interbreeding among the populations eventually ceases and the various populations may be regarded as separate species. Each species becomes more diversified as it adapts to a unique set of environmental conditions, a process known as *adaptive radiation*.

One classic example of adaptive radiation was first observed by Darwin among the finches of the Galápagos Islands. Thirteen different species of finches have been identified on the Galápagos Islands (Figure 1.8). Often referred to as Darwin's finches, they are sufficiently different from other finches of the world that only one related species is found outside the Galápagos. They differ in size, in feeding habits and other behavioral patterns, and in many minor characteristics. The various species have specialized in obtaining certain kinds of food from their environment, and major differences in beak structure are closely related to the kind of food eaten.

Modern evolutionists agree with Darwin's speculation that these different species descended from a single, ancestral species that reached the islands long ago. The original population probably spread throughout the islands and gradually developed some variation of characteristics in the differing environments available.

Figure 1.8 The fourteen species of Darwin's finches arranged on an "evolutionary tree of development." Thirteen of the species are found on the Galápagos; one is found only on Cocos Island, 400 miles away. The different species live in different environments and eat different types of foods so there is little competition. Each species falls into one of two broad categories: ground finches (the most primitive) and tree finches (which evolved later). Although all ground finches share the same type of environment, each species has a different size of beak and tends to feed on different sizes of seeds. One ground finch feeds mainly on prickly pear cactus. The tree finches exhibit similar variety although they live mostly in the moister forests and feed chiefly on insects. One species has a chisel-shaped bill like that of a woodpecker; this finch carries a cactus spine or twig to pry insects out of crevices—one of the few known examples of the use of a tool by an animal other than a primate. Another has a parrotlike bill and lives on fruits and buds.

The relatively isolated populations on the small, outer islands probably tended to change most drastically. Eventually some of these isolated populations may have been sufficiently different from the interbreeding populations on the main islands that they were no longer capable of interbreeding. Thus, when members of the isolated populations happened to reach the main islands, they formed the nucleus of new populations there. Individuals in these new populations would be most likely to survive if they were able to find food sources and places to live overlooked by the more numerous main population. The effects of competition between populations on the main islands would force the new species to diverge still more in their characteristics. Members of these new species would eventually find their way to the outer islands, where further divergence due to isolation could occur.

In opposition to speciation is the evolutionary process of *extinction*. As Darwin thoughtfully observed, "Certainly no fact in the long history of the world is so startling as the wide and repeated exterminations of its inhabitants." Perhaps in competing for a limited supply of resources one species will completely crowd out another. Perhaps a species is well adapted to an existing environment but may not have the genetic flexibility for coping with change. And finally, a catastrophic change can obliterate even those species that have acquired genetic flexibility.

FINDING ORDER IN DIVERSITY

As a consequence of Darwin's work, it is possible to look upon all organisms as some part of a unified thread of life. In fact, the system of classifying living systems is now based primarily on theories of evolution. Beginning with the broadest of categories to indicate degrees of likeness, the scheme is made up of the general groups in Table 1.1. The species is the fundamental unit in this scheme, for

Figure 1.9 The Galápagos Islands, showing the five relatively large islands and nearby smaller islands. It is now generally agreed that these islands have never been connected with the mainland—apparently they were pushed up out of the sea by volcanoes over a million years ago.

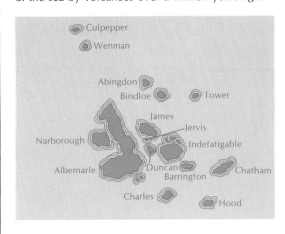

Figure 1.10 Examples of fossil animals. The fossil eurypterid shown at left is from an ancient group of Arachnids that were abundant in the seas about 400 million years ago. Some species reached gigantic proportions of over 6 feet in length. The fossil trilobite (middle) is thought to have been the basic prototype giving rise to contemporary arthropod lines. At right is a fossil nautiloid cephalopod.

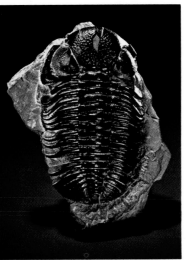

Table 1.1
Partial Classification Scheme of a Plant and an Animal

Taxonomic Rank	Marijuana	Man
Kingdom	Plantae: multicellular plant; autotroph; cells with chloroplasts, well-developed vacuoles, and cellulose cell walls	Animalia: multicellular animal; heterotroph; cells with well-developed centrioles and ingesting organelles
Phylum	Tracheophyta: plants that have well-developed vascular tissues	Chordata: animals with a rodlike notochord for support; dorsal tubular nerve cord; gill slits in pharynx
Subphylum	Pteropsida: plants with large leaves	Vertebrata: animals with a spinal column of segmented vertebrae
Class	Angiospermae: flowering plants with seeds inside fruits	Mammalia: animals with body hair and mammary glands
Subclass	Dicotyledoneae: plants with net-veined leaves; flower parts in fours and fives; embryos with two cotyledons	Eutheria: female members carry developing offspring (nourished by placenta)
Order	Urticales: elms, nettles, and mulberries	Primates: animals with four generalized, five-digit limbs bearing flat nails
Family	Moraceae: mulberry family (herbs to trees); flowers small and inconspicuous; leaves simple	Hominidae: animals with upright posture: brain size of 1,000cc or more; prolonged infancy and skeletal maturation; flat face
Genus	*Cannabis*: hemp, marijuana; annual herbs; axillary, greenish flowers with five drooping stamens; five to eleven divided leaves	*Homo*: one living member of species; double-curved spine
Species	*Cannabis sativa*: plants 40–120cm tall; leaves 5–10cm long with coarsely-toothed edges	*Homo sapiens*: well-developed chin; high forehead; thin skull bones

it is in a sense an indicator of the evolutionary divergence that has occurred in the past. The further back in time the divergence, the closer to some ancient population a species was. Species are actual populations, and the paths leading up from a species (through genus, family, order, class, and phylum) to a kingdom indicate evolutionary crossroads when one group of organisms diverged from another.

Modern species (such as man) have not evolved from modern species (such as apes); the myriad forms of life that exist today are not ranked against one another on some evolutionary ladder, with one "higher" or "lower" than the other. For example, *Homo sapiens* (the "wise" animal) is a modern species and has no counterpart alive today. But some fossil remains are thought to represent other species of the genus *Homo*. This genus is grouped in the family Hominidae with some other genera of fossil creatures similar to man but not similar enough to be considered part of the genus *Homo*. The family Hominidae is part of the order Primates —distinguished by members with flat nails instead of claws, forward-facing eyes, and large brains. This order has roots in the class Mammalia, which includes all animals that suckle their young. The Mammalia are categorized under the phylum Chordata, which includes all organisms having a tubular nerve cord, gill slits in their pharynx, and a supporting structure called a "notochord." Finally, the Chordata are representatives of the most inclusive category, the kingdom Animalia.

Figure 1.11 Modified version of the five-kingdom system of classification proposed by Robert Whittaker (listing common examples of organisms instead of specific phyla). The kingdoms include Monera, Protista, Fungi, Plantae, and Animalia. Whittaker regards plants, fungi, and animals as three groups of organisms that, through evolution, have come to specialize in three ways of food-getting: photosynthesis, absorption, and ingestion. Protista include a variety of diverging lines of evolution, all composed of unicellular (single-celled), nucleated organisms. Monera are the simplest known organisms without a well-developed membrane system within their cells.

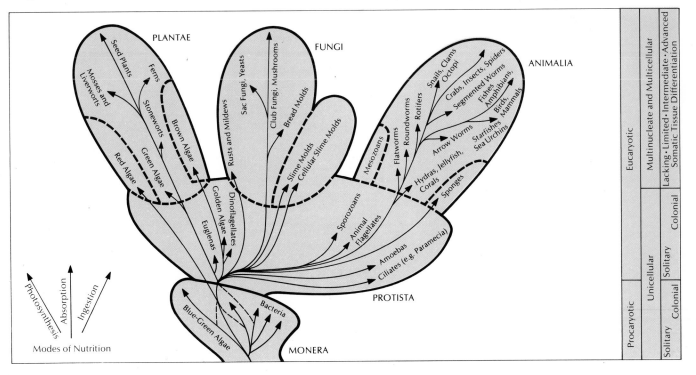

Monera

Figure 1.12   The Kingdom Monera is composed of two phyla: the blue-green algae (Phylum Cyanophyta) and the bacteria (Phylum Schizophyta). The cells of the Monera are the smallest and simplest systems known to be classified as living organisms. All Monera are procaryotic organisms. They lack mitochondria, plastids, centrioles, (9 + 2) flagella or cilia, and nuclear membranes. All of the blue-green algae are photosynthetic autotrophs. Some utilize gaseous nitrogen in forming amino acids, a property shared only with the nitrogen-fixing species of bacteria. All blue-green algae contain chlorophyll a as the major photosynthetic pigment; an accessory pigment, phycocyanin, gives the cell a blue tinge. Bacteria may either be heterotrophic or autotrophic. Many species play important roles in the cycling of nitrogen through the biosphere. Most bacteria secrete substances that inhibit the growth of other organisms.

(a) The common blue-green alga *Anabaena*, noted for its "chain of beads" aggregation. Note the larger cells (called heterocysts) that are found at intervals along the filaments.

(b) Colonial blue-green algae. Some of these organisms may play a role as nitrogen-fixers.

(c) Large masses of blue-green algae often form scumlike layers on the surfaces of polluted lakes and ponds.

(d) A blue-green alga of the genus *Gloeotrichia*. Note the clusters of brown basal cells toward the center of the photograph. (*Walter Dawn*)

(e) Filamentous forms of blue-green algae. The upper filament represents the species *Spirulina versicolor*, and the lower two filaments belong to the genus *Arthrospira*. *Arthrospira* directly contributes to the flamingos' pink color; the pigment comes from carotenes in the blue-green algae that make up part of the birds' diet.

(f) The blue-green alga *Nostoc*. (*Courtesy Carolina Biological Supply Company*)

(g) Bacteria culture of *Clostridium botulinus* grown on an egg yolk medium. Botulus toxin in exceedingly small amounts can produce the deadly food poisoning called botulism.

(h) Stained anthrax bacteria.

(i) Early micrograph of anthrax bacillus, first isolated by the German physician Robert Koch in 1876. After studying bacteria for a few years, Koch found several dyes that can be used to stain bacteria, increasing their visibility under the microscope. He also developed a method for the preparation of pure cultures of a single type of bacteria. Variations of these techniques are still used.

The most commonly-known members of the Kingdom Monera are the *bacteria*, for they have played an influential role in the affairs of mankind. Most of the major disease epidemics that have swept the civilized world can be attributed to the disease-causing (pathogenic) bacteria, whereas the fertility of the soil and the decomposition of waste products of living matter are accomplished in large measure by nonpathogenic bacteria. Most bacteria, particularly the actinomycetes, secrete substances that inhibit the growth of other organisms. Many antibiotics, used to destroy disease-causing bacteria in plants and animals and to prevent bacterial decay of food, have been obtained by purification of these substances. Bacteria are able to survive unfavorable conditions, to go into a kind of "suspended animation" when their food supply dwindles, and then to reproduce rapidly when conditions are favorable. These abilities have apparently been major factors in their evolutionary success. Practically all of the bacteria are of three forms: bacillus (rod-shaped), coccus (spherical), or spirillum (corkscrew-shaped).

Less well-known members of the Kingdom Monera are the *blue-green algae*, which are primitive life forms found throughout the world wherever there is water—including such hostile environments as the rocks lining hot springs. They are somewhat like bacteria but are enveloped in a film of slime and are different in other structural and chemical ways. They do not have any notable means of moving about.

Monera frequently form colonies, or colonial aggregations, as shown in Figure 1.12. There is little doubt that they are the most representative of the ancestral stock from which all life has evolved. Their divergence from some ancestral species took place in the dim past, and they may be considered the most remote "relatives" of life forms that now exist.

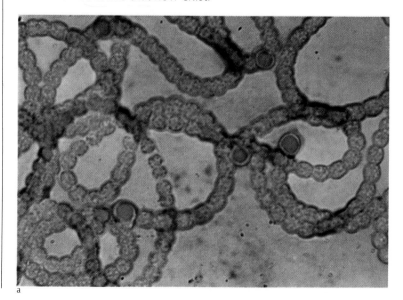

a

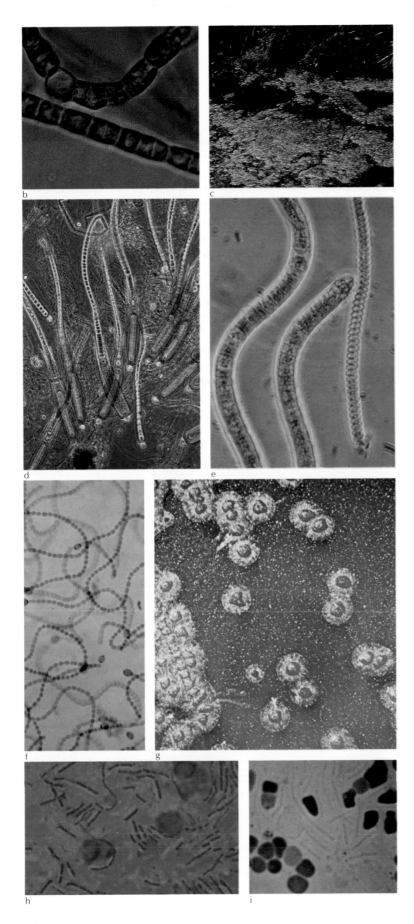

Protista

Figure 1.13 The Kingdom Protista is composed of about eleven phyla. Basically, the organisms of these phyla fall into three broad groups: algae (plantlike organisms); protozoans (animallike organisms); and funguslike protistans. Representatives of these groups are pictured here. All of the Protista are eucaryotic organisms and may be either autotrophic or heterotrophic. All have mitochondria, plastids, and nuclear membranes; cilia or flagella, when present, have the typical 9 + 2 structure. Most protistans are unicellular, but some species form multicellular colonies with different degrees of cell differentiation. The distinction between plantlike and animallike forms does not seem to be important in this kingdom.

(a) Diatoms (Phylum Chrysophyta). Note the variety in form among the types represented. These organisms are of major economic importance because they form the base of the oceanic food chain and thus support much of the marine flora and fauna.

(b) Close-up photograph of a radial diatom.

(c) Diatoms from intricate cell walls (shells) of cellulose are impregnated with silica. Thus, they literally live in a "glass house."

(d) The flagellated *Euglena* (Phylum Euglenophyta). The euglenophytes do not have cell walls and may be functional autotrophs and heterotrophs, depending on environmental factors.

(e) A slime-moldlike form of yellow-green algae, representing the Phylum Xanthophyta.

(f) Shells of various marine radiolaria (Phylum Sarcodina). Skeletons of radiolaria, which are common in the oceanic plankton, have been found in rocks formed more than 425 million years ago.

(g) Radiolarian test or shell. These marine amoebas have silica shells composed of chitin or similar materials.

(h) The Phylum Sarcodina includes both naked and shelled amoebas. Pictured here under the phase contrast microscope is an unshelled amoeba. Note the armlike extensions, called pseudopods, which enable amoebas to move about and capture food.

(i) The ciliated protozoan *Paramecium* (Phylum Ciliophora). This mobile unicellular organism is an active predator in the fresh-water pond.

(j) The stalked ciliate Stintor (Phylum Ciliophora). This organism is noted for its complex gullet and elaborate ciliation pattern. Note the conspicuous light-dark spotted pattern denoting membranelles that ring the gullet area.

The Kingdom Protista is inhabited by a large and highly varied group of organisms that over millions of years has undergone extensive adaptive radiation. In this kingdom are creatures commonly called *protozoa* (animallike organisms), *algae* (plantlike organisms), and funguslike protistans. Lying in the twilight zone between the Monera and the three other kingdoms, many of these organisms actively pursue their prey—as do animals. Although most protista are unicellular and microscopic, several form simple aggregates and some are assembled into other complex colonies that show rudimentary cell specialization.

The most primitive protozoa propel themselves through their environment by means of a thin, fiberlike extension of their cell membrane, called a "flagellum"; these protozoa are therefore called *flagellates*. The *amoeboid* members of the kingdom Protista propel themselves by sending out part of their unicellular body into footlike extensions; in effect, they simply flow about. The protozoa called *ciliates* are more complex: they move about with multiple fiberlike extensions of their cell membrane that wave back and forth like miniature oars. These appendages, called "cilia," also serve as a means of sweeping food particles to the mouthlike openings of most ciliates. The protozoa called *sporozoans* are parasites; they live inside the body of almost every animal species and have varied means of locomotion within the body of their host.

Whereas protozoa generally lack chlorophyll, their unicellular and multicellular (many-celled) relations called *algae* are characterized by the presence of this green pigment, which traps energy from the sun for photosynthesis. Algae are the base of the food chain pyramid for almost all of the world's ecosystems. Some are the "pastures" for the microscopic animals of the oceans; still another variety forms the enormous kelp that washes ashore on the west coast of North America. These organisms are considered to have diverged into three main varieties—green algae, brown algae, and red algae—based on the kind of chlorophyll they contain, the food they store, and how their cell walls are built.

The funguslike protistans include organisms that specialize in obtaining organic nutrients through absorption. Many of these organisms are parasitic and feed on algae, fungi, and plants. Funguslike protistans are relatively rare and appear to be of little importance to man, except for two species that cause destructive plant diseases in cabbages and potatoes.

Both theoretical considerations and fossil evidence suggest that the eucaryotic Protista evolved from earlier procaryotic cells. Whatever their origin, it is clear that the Protista proved remarkably successful at life in the planktonic regions of the oceans and deep lakes. The existing species of Protista include almost every imaginable variation of structure and function.

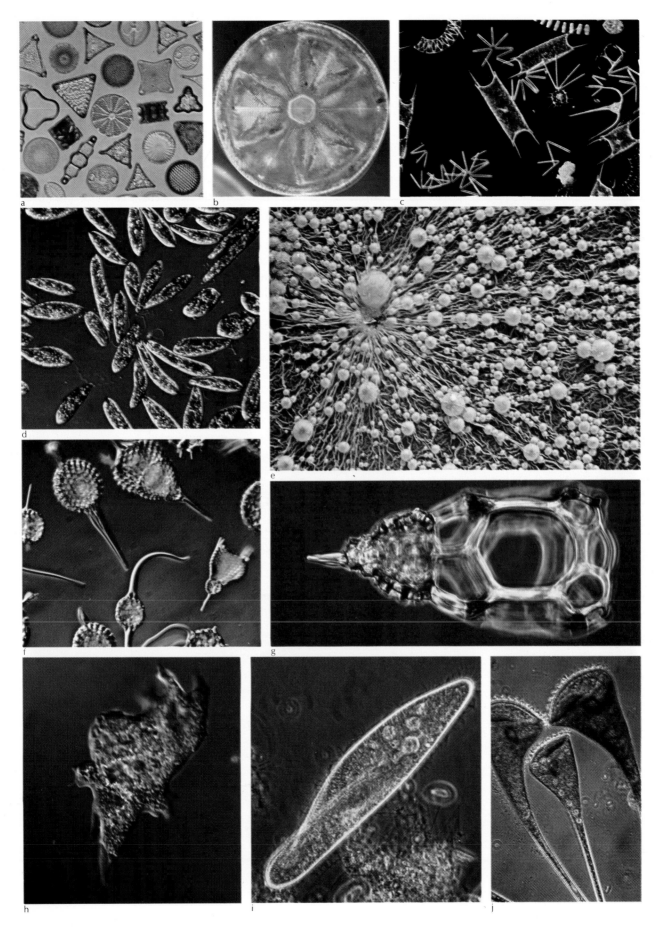

Fungi

Figure 1.14 The Kingdom Fungi is composed of two phyla: the slime molds (Phylum Myxomycophyta) and the true fungi (Phylum Eumycophyta). This kingdom includes those organisms that specialize in absorptive heterotrophy, obtaining a supply of organic nutrients by absorbing them from the environment. In one way, they might be regarded as animals that simply absorb food passively rather than ingesting it actively. However, because most fungi are immotile and structurally resemble algae, they have traditionally been classified in the plant kingdom. Some unicellular organisms are included in the fungi because of their close similarity to more complex organisms of this kingdom. The most complex fungi have a body composed of filaments (called hyphae), which in most species are made up of multinucleate cells or of long, multinucleate tubes without cross-walls to separate individual cells. The entire body is called the mycelium. The absorbing body of most species shows little differentiation, but the more complex species have reproductive bodies made up of differentiated tissues. Although some fungi have life stages that move in an amoeboid fashion, most species live embedded in a source of nutrients and are immotile. The life cycles of most species include sexual and asexual reproduction.

Phylum Myxomycophyta The true slime molds form a multinucleate body (plasmodium) without cell walls; they travel by amoeboid movement over damp soil or on the undersides of logs and rocks. The plasmodium feeds both by absorbing nutrients and by ingesting small food particles.

(a) The plasmodium, or vegetative body, of a slime mold. Most slime molds are free-living saprophytes, feeding on dead organic matter such as wood or leaf litter on the forest floor.

(b) Close-up photograph of the sporangia, or fruiting bodies, that develop from the slime mold plasmodium during the reproductive period.

The Kingdom Fungi has representatives everywhere on earth. These multicellular organisms obtain organic nutrients by absorbing them from the environment. For the most part they have not evolved ways of moving about; most species live embedded in a source of nutrients. Although they are relatively inconspicuous members of the living world, the true fungi play an important role in decomposing the dead bodies of other life forms, thereby returning vital materials to the nutrient cycles that characterize the biosphere. However, they are also the cause of many plant and animal diseases because of their specialized ability to absorb nutrients from living plants and animals as well. The great variety of fungi (more than 200,000 species are known, and about 1,000 new species are discovered yearly) rivals that of the plant kingdom. There is some evidence that this kingdom includes several independent evolutionary lines that emerged separately from the Kingdom Protista.

Included in the Kingdom Fungi are the multicellular slime molds, yeasts, molds, and the true fungi, which include mushrooms, and lichens.

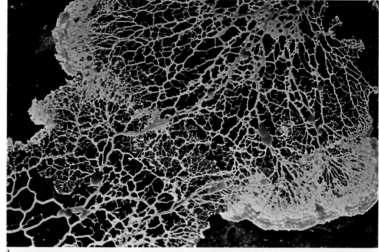
a

b

20

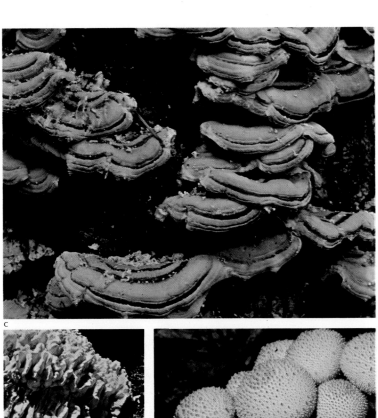

c

d

e

f

g

Phylum Eumycophyta The true fungi play an important role in the cycling of elements. Along with the heterotrophic bacteria, they are responsible for most of the breakdown of organic molecules in the bodies of dead organisms, releasing the various elements in the form of simple inorganic molecules that can be used by plants as nutrients. The true fungi lack the plasmodial stage that characterizes the slime molds. Most species secrete digestive enzymes into the environment. These enzymes catalyze the initial breakdown of complex organic molecules into substances that can be absorbed through the membrane of the fungal body.

(c) Bracket fungi growing on the bark of a tree. Certain species of bracket fungi can rot the heartwood and sapwood of living trees, causing severe damage.

(d) Bracket fungi growing on the trunk of a tree. This organism feeds on the dead bark tissue of the tree.

(e) The puffball *Lycoperdon perlatum*. This species is noted for its pear-shaped basidium.

(f) The basidium, or club-shaped fruiting body, commonly called a mushroom. These fungi act as saprophytes in breaking down dead organic matter to obtain their energy.

(g) A large, leafy type of lichen. Lichens are combinations of algae and fungi in a mutualistic relationship. The alga provides nutrients through photosynthesis; the fungus provides moisture and a slightly acidic environment, in which the alga thrives.

Plantae

Figure 1.15 The Kingdom Plantae is composed of twelve phyla or divisions. The major phyla are pictured here. Most eucaryotic, multicellular organisms that have specialized in a photosynthetic, autotrophic way of life are grouped in this kingdom. Some unicellular and heterotrophic organisms also are included in this kingdom because of close structural or biochemical resemblances to the plants. There are three phyla of algae, which incorporate the familiar seaweeds of the nearshore ocean and some species that inhabit moist areas on land. The Phylum Bryophyta includes the mosses and liverworts, organisms that have specialized somewhat for existence in moist terrestrial habitats. The remaining phyla of the kingdom are the vascular plants, organisms that are well adapted to life on land. Plants are classified by increasing organization and by the nature of their life cycles. Primitive plants have poorly-developed vascular tissue (or may lack such tissue altogether) and must live in moist environments. Their life cycle usually has a well-developed haploid stage (gametophyte). Higher plants have well-developed vascular tissue and can live in more terrestrial environments. Their life cycles are characterized by a diploid stage (sporophyte).

(a) The brown alga *Fucus* (Phylum Phaeophyta). Note the reproductive receptacles with their wartlike covering. *(Walter Dawn)*

(b) *Laminaria andersonii*, a common cosmopolitan kelp found in the state of Washington (Phylum Phaeophyta).

(c) Close-up photograph of strands of *Spirogyra*, a common fresh-water alga of the Phylum Chlorophyta.

(d) A species of red algae (Phylum Rhodophyta). Note the delicately branched body.

(e) A thallus (sheetlike) liverwort (Phylum Bryophyta). Note the globular, fingerlike reproductive units.

(f) *Lycopodium*, a club moss (Phylum Lycopodophyta). Note the terminal spore sacs.

(g) The field horsetail *Equisetum* (Phylum Arthrophyta). Note the fertile stalks with terminal sporangia and the infertile vegetative stalks.

(h) A group of temperate zone ferns in a woodland habitat (Phylum Pterophyta).

(i) Bishop pines (Phylum Coniferophyta).

(j) Seed dispersal in the dandelion *Taraxacum officinale* (Phylum Anthophyta).

(k) Passion flower (Phylum Anthophyta).

(l) Venus flytrap (Phylum Anthophyta). This insectivorous plant is a native of marsh regions, and it supplements its nitrogen intake with insect protein.

Multicellular plants apparently evolved from a primitive, fresh-water green alga. We can guess about the adaptive advantages that multicellularity would give to an organism. In nearshore waters, algae attached to a rock would be bathed in nutrients by rushing currents while fixed in a location of suitable sunlight and other conditions. For such an organism, a large, multicellular body would greatly increase the amount of photosynthesis that could be carried out. For those algae entering moist land environments, multicellularity and differentiation would provide clear advantages. Parts of the structure could be lifted into the air and specialized for photosynthesis, whereas other parts could be specialized to obtain water and mineral nutrients from the soil and to hold the organism in place.

When ancestral plants invaded the terrestrial environment, they had to contend with exposure to air, gravity, wind, and variable water supplies. In adapting to these conditions, two main lines diverged from the ancestral green algae: the *bryophytes* and the *tracheophytes*. The bryophytes include the mosses and liverworts, which grow in moist, shady places. These plants do not have leaves or stems; they do not even have roots. The tracheophytes include the ferns and seed plants. They have structures for conducting water, minerals, and foods, and they have three specialized organ systems: leaves for photosynthesis, stems for conduction of materials, and roots to anchor the plant and draw nutrients from the soil.

a

22

Animalia

Figure 1.16 There are more than twenty phyla in the Kingdom Animalia. Representatives of the major phyla are pictured here. All animals are eucaryotic, multicellular, and heterotrophic organisms.

Phylum Porifera (sponges) These tissue-level organisms have radial symmetry and possess flagellated collar cells for feeding. In different groups of sponges, the skeleton is made of protein fibers; sharp spicules of calcium carbonate or of silica; or mixtures of these materials.

   (a) A simple vase-shaped sponge. These organisms are filter-feeders, straining out planktonic organisms from the water.

Phylum Cnidaria (coelenterates) Includes jellyfishes, corals, and sea anemones. These tissue-level organisms also have radial symmetry; they possess a digestive tract with a mouth (but no anus) and specialized stinging cells, which aid in the capture of food.

   (b) The medusa, or jellyfish.

   (c) Sea anemone from the Cayman Islands, British West Indies.

   (d) Star coral on the Cayman Islands, British West Indies.

Phylum Platyhelminthes (flatworms) Includes flukes, planarians, and tapeworms. This phylum and all those that follow are composed of bilaterally symmetrical organ-level organisms. Flatworms have an incomplete digestive system (or it may be absent altogether); they have no body cavity.

   (e) The fresh-water planarian spontaneously dividing. (*Courtesy Carolina Biological Supply Company*)

Phylum Aschelminthes (pseudocoelomate organisms) Includes roundworms, rotifers, and entoprocts. Members of this phylum and all those that follow have a complete digestive tract. Another advance is the appearance of a false coelom, a liquid-filled space (not lined by special cells) between the body wall and the digestive tract.

   (f) This microscopic multicellular organism— the rotifer—shows a complex system of development. The rotifer body has a head, trunk, and tapered foot with cement glands for attachment to a substrate.

   (g) A free-living nematode roundworm.

   (h) A living entoproct. Note the ring of ciliated tentacles, the lophophore, that surrounds the mouth.

Phylum Brachiopoda (lampshells) These organisms and all those that follow have a body cavity that is a true coelom, lined with a special layer of cells.

   (i) In this brachiopod, or lampshell, all internal structures except for a portion of the lophophore are hidden by the partially closed valves.

Phylum Annelida (segmented worms) The body of an annelid is marked off into a series of disclike or ringlike segments (without jointed appendages).

   (j) Copulating earthworms. Each worm is transferring sperm to its partner that will later be used to fertilize the eggs when they are laid in the cocoon.

   (k) Two leeches attached to the fin of a fish host.

   (l) Polychaete tubeworm attached to star coral.

The evolutionary theme of the animal kingdom is *ingestion* as a way of feeding. Like fungi, animals obtain their food from other organisms. Animals, however, are multicellular organisms that *ingest* food—taking tissues of other organisms in through their mouths for digestion within their bodies, rather than absorbing organic substances through the outer body surface as fungi do.

In the function of living communities, plants are producers that create food; animals are consumers that use some of this food; and fungi and bacteria are decomposers that break down dead remains of both plants and animals into inorganic substances. There are varied ways in which animals act as consumers. *Herbivores* eat living plant tissues; *carnivores* eat other animals, obtaining food energy from plants secondhand (or thirdhand or fourthhand); *fungivores* feed on the tissues of fungi; *scavengers* feed on the dead remains of other organisms; *omnivores* (man, for example) feed on some combination of living plants or animals or other foods; and *parasites* live on or in other organisms and take their food from their hosts. Many aquatic animals are *filter feeders*, straining from the water a mixture of different kinds of small organisms and dead particles. Many filter feeders are omnivores, but some specialize in particular kinds of food drawn from the water.

Ingestion as a way of life has led to the evolution of some of the specialized structures considered to be characteristic of animals. First, most animals possess a mouth and a digestive tract that has glands or glandular cells, which secrete chemicals to assist in the digestion of food. Most animals have other organ systems that supplement the essential function of ingestion and digestion: a system for gas exchange; an excretory system to dispose of the waste products; and a circulatory system to transport food, oxygen, and wastes within the organism. In addition, the majority of animals have organ systems that make possible movement to seek out suitable food, to pursue food organisms, or to escape being eaten by carnivores. These organ systems include the muscular system, the skeletal system, the sensory system, and the nervous system.

There are more than 1 million species of animals. They vary from the organizationally simple sponges through a profusion of species of increasing structural and functional complexity to the higher vertebrates, including the birds and mammals. They range in size from microscopic rotifers and roundworms to the most gigantic of all animals in the history of life, the great blue whale.

The examples of life forms presented so far are no more than a random sampling of the immense range of life. Moreover, the somewhat static classification system of five kingdoms may be as arbitrary: further revisions may result as more is learned about the evolution of life. Life is a continuous process, and may not lend itself to discrete categorical frameworks. A classification system merely provides a means of summarizing existing knowledge of the world and provides a stimulus for new ways of thinking.

Animalia

Phylum Mollusca (molluscs) Includes snails, clams, chitons, squids, and toothshells. Although the name Mollusca means "soft-bodied," the most familiar molluscs are known for the hard shells that provide skeletal support and protection for the soft body. Molluscs usually are unsegmented and are characterized by the development of a massive muscular organ, or foot, behind the mouth.

(a) The white-lined nudibranch is a gastropod member of the Phylum Mollusca. Note the fingerlike gills on its back.

(b) A sea slug laying large gelatinous eggs. Many eggs must be produced in order for a few to survive predators.

(c) The Great Scallop, a representative molluscan bivalve.

(d) A marine nudibranch laying a whorl of eggs.

(e) A squid, representing the cephalopod class of molluscs. Squids are noted for their ability to undergo rapid color change and therefore can escape detection by their enemies.

Phylum Arthropoda (arthropods) Includes spiders, crabs, barnacles, shrimps, and insects. Arthropods are distinguished by the possession of jointed appendages; they also have segmented, chitinous, external skeletons covering their bodies.

(f) The fiddler crab (*Uca minax*). These intertidal animals exhibit rhythmic change in coloration that can be correlated with light and tidal cycles.

(g) The sowbug, closing up as an avoidance response.

(h) A garden spider, a representative of the Class Arachnida.

(i) A centipede, representing Class Chilopoda.

(j) A millipede, representing Class Diplopoda.

(k) Ladybug beetles mating. In many arthropods, sperm from the male is transferred directly to the female by specially modified appendages.

(l) The harvest ant, representing Class Insecta.

a b c d e f g h i j

k

l

m

n

o

Phylum Echinodermata (echinoderms) Includes star-fishes, sea urchins, sand dollars, brittle stars, and sea lilies. Echinoderms are unusual in their body organization. Although they develop from larvae with bilateral symmetry, the adults appear radially symmetrical or nearly so. Echinoderms lack segmentation, heads, and distinct brains. A true coelom and a circulatory system are present, but most forms have no special systems for excretion or external respiration. There is a unique water-vascular system with water tubes connected to numerous tube-feet that can be expanded and contracted by water pressure. The tube-feet are used for attachment to a substrate, for movement, and for holding food. All echinoderms live in the oceans, and there is a great variety of motile and attached forms. Most echinoderms possess hard, calcareous plates in the skin layer.

(m) Brittle, or serpent, stars. These echinoderms have a reduced number of tube-feet and move by lateral undulations of their arms.

(n) The sea urchin has an endoskeleton formed of welded calcium carbonate plates.

(o) Sea cucumbers have tube-feet restricted to two parallel rows; they move about the bottom of the sea feeding on dead organic matter.

Animalia

Phylum Chordata (chordates)   Includes some invertebrates (the tunicate and the amphioxus) and all vertebrates. Chordates are characterized by the possession of gill slits in the pharynx; a longitudinal stiffening rod—the notochord; and a dorsal, hollow nerve cord at some time in their development. Most chordates have segmented bodies, although the segments are difficult to recognize in the higher forms. Chordates and echinoderms have very similar early development patterns. Note the body plans of the fish, the male frog, and the pigeon.

  (a) The sea grape tunicate. Note the incurrent and excurrent siphons, which are used for feeding.

  (b) Blue cronies, representing the true bony fishes.

  (c) The box turtle, of the Class Reptilia.

  (d) The amphibian tree frog.

  (e) Mangrove snake, a representative of the reptilian class.

  (f) Rhinoceros viper, a reptile.

  (g) Avocet, a shore bird of the Class Aves.

  (h) White mynah bird.

  (i) The reef-dwelling lion fish, a true bony fish.

  (j) Green tree frog, a representative amphibian.

  (k) The salamander, one of the most primitive terrestrial vertebrates.

  (l) Blue-crowned pigeon.

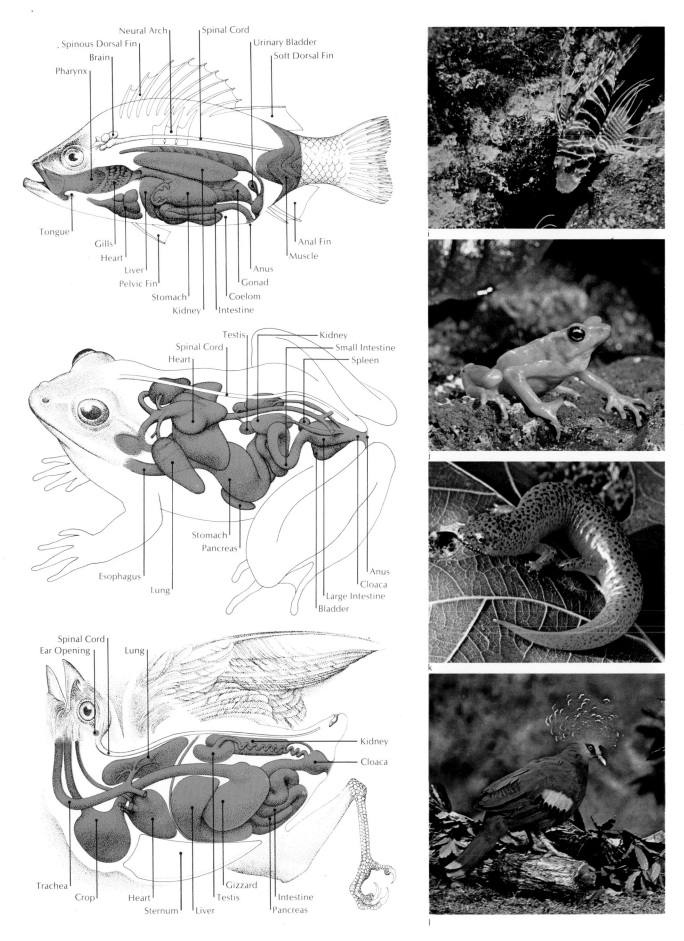

Pharynx
Spinous Dorsal Fin
Brain
Neural Arch
Spinal Cord
Urinary Bladder
Soft Dorsal Fin

Tongue
Gills
Heart
Liver
Pelvic Fin
Stomach
Kidney
Intestine
Coelom
Gonad
Anus
Anal Fin
Muscle

Heart
Spinal Cord
Testis
Kidney
Small Intestine
Spleen

Esophagus
Lung
Stomach
Pancreas
Bladder
Large Intestine
Cloaca
Anus

Spinal Cord
Ear Opening
Lung
Kidney
Cloaca

Trachea
Crop
Heart
Sternum
Liver
Testis
Gizzard
Pancreas
Intestine

i

j

k

l

29

Animalia

Phylum Chordata (continued) Note the body plan of the cat at right. All photographs represent members of the Class Mammalia. This class is thought to have evolved from a group of reptiles different from those that gave rise to birds. Like birds, the mammals developed an efficient circulatory system and a complex nervous system, as well as the ability to maintain a steady internal temperature. Hair protects mammals in the same way that feathers protect birds. In most mammals, the embryo is nourished within the mother's body rather than being enclosed in an egg. After birth, a young mammal is relatively helpless for some time and is generally fed on milk from the mother's mammary glands. Fossils of primitive mammals have been found in rocks formed about 180 million years ago. These early mammals were very small creatures that apparently remained rather rare and inconspicuous in a landscape dominated by reptiles. About 70 million years ago, within a relatively short period of time, many groups of reptiles and other animals became extinct or greatly diminished while the numbers of mammals greatly increased. The reasons for the ending of the "Age of Reptiles" and the beginning of the "Age of Mammals" are still very much a matter of debate.

(a) The Tasmanian gray kangaroo is a marsupial mammal.
(b) The white-footed deer mouse is a placental mammal.
(c) A hippopotamus, an even-toed, hooved mammal (an artiodactyl).
(d) Snow leopard, a carnivorous mammal.
(e) Giraffe and baby.
(f) Gnu and baby.
(g) Siamang and baby.
(h) Three-week-old baby cougar cubs with their mother.
(i) Mother and child.

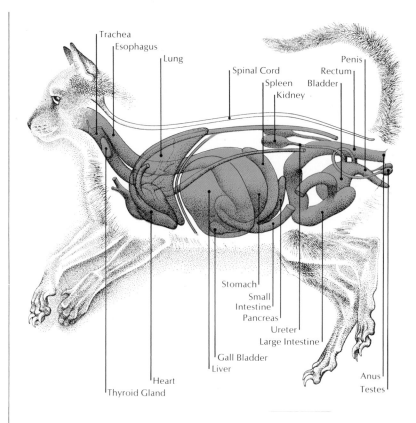

30

d

e

f

g

h

i

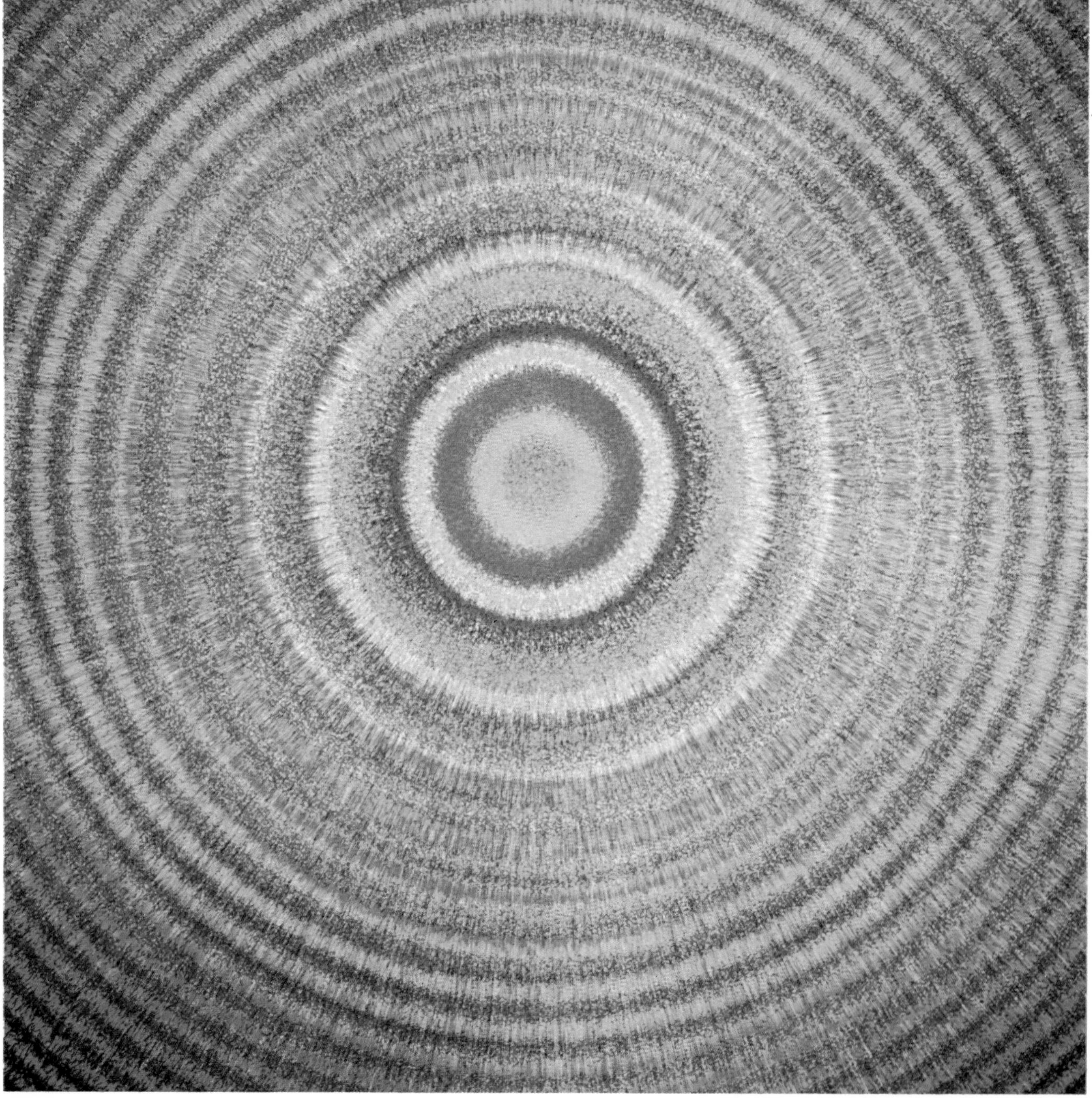

Is it really possible to conceive of something 5 billion years old? A number of such magnitude almost has no meaning for us, and yet it was almost that long ago that our earth was formed and the stage was set for a most remarkable event—the origin of life. Perhaps every organism alive today is related directly to the first living system to appear on earth about 3 billion years ago in primordial time; without question, the evolution of the first living system is related directly to the evolution of the earth itself.

When we speak of some event separated from our consciousness by a span of billions of years, a picture of that event must necessarily be based on conjecture. Even so, it is possible to reconstruct at least part of that picture by going back to the origin of the solar system and tracing, step by step, the conditions that may have led to the formation of life on earth. Our solar system is thought to have formed from a cloud of gas, dust, and solid bodies swirling about in turbulent eddies, perturbing the motions of one another by their gravitational pulls. Matter collided with matter, producing explosive bursts of light and heat. During that time magnetic and electric fields pervaded the entire solar system, accelerating and guiding charged particles through it. It is likely that energy from bolts of lightning continually agitated the swirling matter, reaching proportions that would dwarf any in our earthly experience. In this tumultuous beginning one of those eddies would become our earth; others would eventually form the rest of the planets and satellites of our solar system.

How is it possible to propose such a dramatic beginning for the story of the origin of life? Although there is controversy about the exact nature of the original cloud that gave rise to the solar system, we can make educated guesses about what happened by assuming that the physical laws governing the structure and substance of the natural world were as valid and applicable then as they are today.

BUILDING BLOCKS OF NATURE
In all the world there are only about 100 or so different kinds of *atoms*—tiny spheres of matter whose behavior is governed by relatively simple laws. Each particular kind of atom behaves in its own special way: each atom of hydrogen, for example, behaves the same way as every other atom of hydrogen throughout the world.

Nevertheless, the different species of atoms are similar to one another in certain important respects. If one atom approaches another atom, they are pulled closer together. In some cases the pull is very strong; in others, the pull is weaker. If an oxygen atom comes close to a hydrogen atom they attract each other and draw together. But once they get close enough, the attraction diminishes and a powerful repulsive force resists any effort to push them closer together.

This behavior comes about because of the internal structure of atoms, which happens to be different for each of the more than

2
Life As a Product of the Chemical Earth

Figure 2.1 The solar system may have condensed out of an enormous galactic dust cloud that was probably a remnant of a nova explosion.

100 different types of atoms. This structure determines what kinds of molecules an atom will make when combined with other atoms as well as what kind of changes will occur within the molecules.

At the center of each atom is its *nucleus*, a minute "core" that is 100,000 times smaller than the atom itself. If a baseball represented an atom, the nucleus would be a speck much smaller than the sharp point of a pin. Even so, the nucleus carries nearly all of the atom's weight. The "body" of an atom—the sphere that surrounds the pointlike nucleus—is made up of one or more particles called *electrons*, which are distributed around the nucleus in an electron "cloud" (Figure 2.2). A hydrogen atom has only one electron; a uranium atom has ninety-two. Although the electrons make up nearly all of the volume of an atom, they comprise only about 1/4,000 of the atom's weight; most of the remaining weight is contained in the nucleus.

An atom is held together by electric forces. The amount of electrification can exist only in combinations of what is called the "electric charge." Each electron in the atom carries exactly one unit of electric charge—but the nature of its charge is opposite that of the nucleus. There are one or more particles called *protons* in the nucleus. Each proton has one unit of *positive* electric charge, whereas each electron has one unit of *negative* electric charge. The nucleus may also contain *neutrons*, particles that carry *no* electric charge. One of the laws of nature is that like charges repel; unlike charges attract. It is the electric force of attraction between the positive charge of the nucleus and the negative charge of each electron that pulls the atom together.

Suppose you start out with a bare nucleus of oxygen, which has eight units of positive electric charge. This nucleus will pull toward itself eight electrons, each with one unit of negative charge. When these eight electrons have been accumulated, the negative electricity of the electrons cancels the effect of the positive electricity of the nucleus. The atom is then said to be electrically *neutral*.

Each atomic nucleus carries a certain *number* of protons, or units of positive electric charge. The nucleus of a hydrogen atom carries one; an oxygen atom carries eight; a uranium atom carries ninety-two. The *atomic number* of an atom is merely the number of protons, or positive electric charges, carried by the nucleus of the atom. Hydrogen has atomic number 1; oxygen, atomic number 8; uranium, atomic number 92.

The nature of each kind of atom is determined completely by the behavior of its electron cloud. And that behavior depends only on the *number* of electrons present in the cloud. An atom of oxygen has a nucleus with eight units of positive charge that attracts eight electrons. By the nature of the electric forces and of the electrons themselves, the eight electrons will *always* gather into a cloud of a particular size and a particular shape surrounding the

Figure 2.2   The electron cloud is a fuzzy, nearly spherical ball in which there are a number of electrons—from one to several hundred, depending on the kind of atom. Each electron of the atom is represented as part of the broad "smear," something like a cloud of smoke. The cloud has substance and rigidity so two atoms cannot be pushed together. The nucleus is a tiny space with a diameter only 1/100,000 as large as the diameter of the atom. However, the nucleus contains over 99 percent of the weight of the atom.

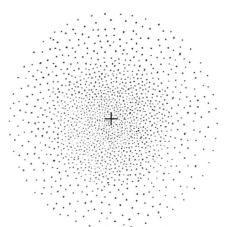

Figure 2.3   Electron shell configuration of the elements lithium, boron, and oxygen. Each of the atoms shown contains a central area, which represents the innermost electron shell. Within this central area (but not shown in this figure) is the atomic nucleus with its corresponding protons and neutrons. Atoms of heavier and heavier substances contain increasing numbers of electrons and therefore contain more electron shells. Hydrogen, with its single electron, contains only one shell. All of the elements in this figure contain two shells, but none are complete—that is, none contain the maximum number of eight electrons. Lithium has seven "empty spaces"; boron has five; and oxygen has two. All are therefore capable of reacting with the electrons of other atoms to fill out their electron shells.

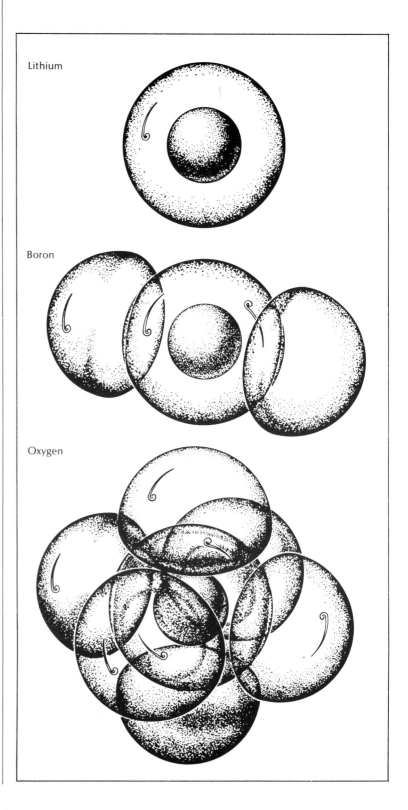

Lithium

Boron

Oxygen

nucleus to make an atom of oxygen. Because all electrons behave according to the same laws of nature, all oxygen atoms will be the same everywhere.

Electrons of similar energies are grouped together in *shells* (Figure 2.3). There is a limit to the number of electrons that can be grouped into a given shell. The maximum number for the first shell is 2; the second 8; the third 18; the fourth 32; and so on. It is the *outermost* shell that influences the behavior of an atom most strongly because it is the shell that interacts most strongly with other atoms. If this shell does not contain the maximum number of electrons, then the atom is said to be "reactive." For example, an atom of hydrogen, which has only one shell, contains a single electron. Even though this atom is electrically neutral, it still has room in its electron shell for another electron and it can react with another atom to acquire it. An atom of neon, on the other hand, has the maximum number of eight electrons completing its second (and outer) shell. For that reason atomic neon is said to be "stable." It does not react with other atoms.

Atoms of heavier and heavier substances contain increasing numbers of electron shells. But it is the *second* shell that is important when we consider living systems. The reason is that the most common atoms found in living substances have fewer than ten electrons. Carbon has six electrons, nitrogen has seven, and oxygen has eight. In other words, although the atoms making up these substances have complete first shells of two electrons, they have incomplete second shells: carbon has four "empty" spaces, nitrogen has three, and oxygen has two. All are capable of reacting with the electrons of other kinds of atoms to fill out their electron shells. It is a vitally important characteristic, as you will soon see.

It is on the basis of such laws that we can make reasonable assumptions about events that occurred long ago. Whether or not a particular atom in the primordial earth joined together with another atom to make a molecule depended at that time, as it does today, on the behavior of electrons circling about the nucleus. Atom joined with atom to form new molecules. The sequence in which atoms joined together as well as the kinds of molecules that were produced was determined then, as it is today, by the specific number and the arrangement of electrons in each atom.

BUILDING BLOCKS COME TOGETHER: CHEMICAL BONDS

A long time ago, before the atomic picture of matter had been worked out, people had identified many unique substances that could not be broken down by any chemical means into new and different substances. These they named "chemical elements." Today we know that an *element* is simply a substance made of one particular kind of atom. The atoms of some elements may join together to form *molecules;* and two or more *different kinds* of atoms bound together constitute a *compound.* The forces binding

Figure 2.4 John Dalton's table of the elements, derived in 1803. The idea that all things are made up of tiny, indivisible particles called atoms was suggested as early as the fifth century B.C., but it was not until the early nineteenth century that Dalton, an English chemist, formulated an "atomic theory." He postulated that all elements are composed of atoms and that each element contains only a single kind of atom. He also stated that in every reaction in which a chemical compound is formed, the atoms can be rearranged but their individual qualities will not change. When Dalton outlined a new system of symbols for the elements, he assigned a number to each element based on comparative (rather than absolute) weight. Thus the lightest hydrogen atom was assigned the base number 1; nitrogen, which was more than four times as heavy, was given the number 4.2; and so on. Although his relative atomic weights later proved to be erroneous, his ideas survived.

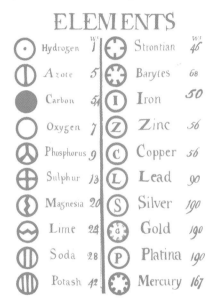

Figure 2.5   Periodic table of the elements, first developed by the Russian chemist Dmitri Mendeléev in 1869 on the basis of the sixty-three elements then known. Over the past century, many new elements have been discovered and added to the table. The elements are listed in order of increasing atomic number, and each vertical column of the table includes elements of similar chemical properties.

Group	I	II											III	IV	V	VI	VII	O
Period																		
1	H 1																	He 2
2	Li 3	Be 4											B 5	C 6	N 7	O 8	F 9	Ne 10
3	Na 11	Mg 12			Transition Elements								Al 13	Si 14	P 15	S 16	Cl 17	Ar 18
4	K 19	Ca 20	Sc 21	Ti 22	V 23	Cr 24	Mn 25	Fe 26	Co 27	Ni 28	Cu 29	Zn 30	Ga 31	Ge 32	As 33	Se 34	Br 35	Kr 36
5	Rb 37	Sr 38	Y 39	Zr 40	Nb 41	Mo 42	Tc 43	Ru 44	Rh 45	Pd 46	Ag 47	Cd 48	In 49	Sn 50	Sb 51	Te 52	I 53	Xe 54
6	Cs 55	Ba 56	* 57-71	Hf 72	Ta 73	W 74	Re 75	Os 76	Ir 77	Pt 78	Au 79	Hg 80	Tl 81	Pb 82	Bi 83	Po 84	At 85	Rn 86
7	Fr 87	Ra 88	‡ 89-															

*	La 57	Ce 58	Pr 59	Nd 60	Pm 61	Sm 62	Eu 63	Gd 64	Tb 65	Dy 66	Ho 67	Er 68	Tm 69	Yb 70	Lu 71
‡	Ac 89	Th 90	Pa 91	U 92	Np 93	Pu 94	Am 95	Cm 96	Bk 97	Cf 98	Es 99	Fm 100	Md 101	(?) 102	

substances together are electric in nature, however, even though we traditionally speak of "chemical" bonds.

Chemical bonding is an *energy relationship*. When atoms (or molecules) are heated to higher temperatures, more energy is added to them. And with more energy there is more motion. If the temperature is great enough, atoms will move about rapidly enough to collide with one another. And it is during their collisions that atoms can join together. The molecule that is created consists of two or more atoms that exhibit new and unique properties. The molecule is stable because the "empty spaces" in the outer shells of the constituent atoms have been filled. For example, the helium atom, with two electrons filling its first shell, is very stable. But the hydrogen atom has only one electron in this shell and is much more reactive chemically. It readily becomes bonded to other atoms to form molecules. Hydrogen gas at normal temperatures is made up not of individual hydrogen atoms but of diatomic (two-atom) molecules composed of two hydrogen atoms bound together.

The formation of this bond can be understood by visualizing what happens as two hydrogen atoms approach each other. Each hydrogen atom has a "vacancy" in its outer shell. When the two atoms are close together, the unfilled shells overlap. In the region of overlap, each electron can be shared by both nuclei. Because each electron is attracted to both nuclei and is able to move within a stable pathway, there are strong attractive forces holding the two

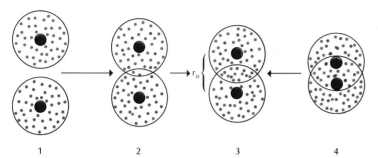

Figure 2.6   Formation of a covalent bond. From left to right, (1) two free atoms are moving in space; (2) the atoms are starting to interact, and the energy of the pair is decreasing as they share their electrons; (3) the bond is fully formed, and the energy of the pair of atoms is minimized at its most stable internuclear distance, r_0; (4) any attempt to push the atoms closer together causes the energy of the system to rise rapidly because of internuclear repulsion.

1   2   3   4

atoms together. In fact, the two electrons apparently correlate their movements in such a way that they stay far apart from each other, minimizing the repulsive forces between them and helping to hold the molecule together.

Consider the bond between two hydrogen atoms, which may be represented as H—H. Because electrons are shared equally by the two nuclei, each atom has approximated the stable electron configuration of a full outermost shell. When electrons are shared between atoms in this way, we say the bond is *covalent*.

Sometimes the bonding electrons are not shared at all but are held entirely by the nucleus of one atom. The only thing holding the two atoms together is the electric attraction between the atom that has lost an electron (the "positive ion") and the atom that has gained an electron (the "negative ion"). Such *ionic bonding* does not really produce single molecules.

A simple notation has been worked out for depicting the chemical bonds involving the electrons of the outermost shell. Consider the hydrogen atom, with its single electron. It may be written as H·. And the hydrogen molecule, with its *shared* pair of electrons, may be written as H:H. A helium atom has two electrons filling its outer shell, so it is written as He: . If two helium atoms approach each other, little overlap can occur because each outer shell is fully occupied. That is why helium atoms do not form bonds with other atoms but instead remain independent.

An oxygen atom has six electrons in its outer shell, with room for two more. You would expect oxygen to form covalent bonds, then, because it could share two electrons with other atoms. For example, each of the unpaired electrons could be shared with a hydrogen atom. The resulting molecule would be H_2O, the water molecule, which may be written as:

$$:\overset{..}{\underset{.}{O}}· + H· + H· \rightarrow :\overset{..}{\underset{..}{O}}:H$$
$$H$$

In a water molecule, the oxygen nucleus exerts a much stronger attractive force than does the hydrogen nucleus. The shared electrons (with their negative charge) are more likely to be closer to the oxygen nucleus. In other words, the part of the molecule near the oxygen nucleus has an excess of negative charge,

Figure 2.7   Certain elements have been "chosen" by evolution to form the building blocks of all living systems; others have been completely ignored. In this diagram, the chemical composition of the human body is compared with the approximate compositions of seawater, the earth's crust, and the total universe. The percentages are based on the total number of atoms in each category (numbers have been rounded off so totals do not exactly equal 100). Elements that are boxed in the human body column appear in at least one other column.

PERCENT OF TOTAL NUMBER OF ATOMS			
HUMAN BODY		**SEAWATER**	
H	63	H	66
O	25.5	O	33
C	9.5	Cl	.33
N	1.4	Na	.28
Ca	.31	Mg	.033
P	.22	S	.017
Cl	.03	Ca	.006
K	.06	K	.006
S	.05	C	.0014
Na	.03	Br	.0005
Mg	.01		
ALL OTHERS < .01		ALL OTHERS < .1	
EARTH'S CRUST		**UNIVERSE**	
O	47	H	91
Si	28	He	9.1
Al	7.9	O	.057
Fe	4.5	N	.042
Ca	3.5	C	.021
Na	2.5	Si	.003
K	2.5	Ne	.003
Mg	2.2	Mg	.002
Ti	.46	Fe	.002
H	.22	S	.001
C	.19		
ALL OTHERS < .1		ALL OTHERS < .01	

Al	Aluminum	He	Helium	O	Oxygen
B	Boron	H	Hydrogen	K	Potassium
Br	Bromine	Fe	Iron	Si	Silicon
Ca	Calcium	Mg	Magnesium	Na	Sodium
C	Carbon	Ne	Neon	S	Sulfur
Cl	Chlorine	N	Nitrogen	Ti	Titanium

and the parts near the hydrogen nucleus have an excess of positive charge. Whenever two nuclei do not share the electrons equally, we say a *polar covalent bond* exists.

Oxygen, like hydrogen, forms a diatomic molecule (O_2). In this case, two pairs of electrons must be shared between two oxygen atoms:

$$:\ddot{O}: + :\ddot{O}: \rightarrow :\ddot{O}::\ddot{O}:$$

Such a double bond is stronger than a single bond (where only one pair of electrons is shared) and draws the two nuclei closer together. This double bond may be represented in the following way: $O{=}O$.

Nitrogen, with only five electrons in its second shell, can share three of its electrons with a corresponding number of electrons in other kinds of atoms. With hydrogen atoms it forms the molecule ammonia (NH_3):

$$:\dot{N}\cdot + 3H\cdot \rightarrow \begin{matrix} & H \\ :\ddot{N}:H \\ & H \end{matrix}$$

Like hydrogen and oxygen, nitrogen exists as a diatomic molecule (N_2). In this case, the two atoms share three pairs of electrons, forming a triple bond that may be represented as

$$\ddot{N}:::\ddot{N} \qquad \text{or} \qquad N{\equiv}N$$

Because four electrons of carbon are found in its outer shell, they can pair up with four hydrogen atoms to form methane (CH_4):

$$\cdot\dot{C}\cdot + 4H\cdot \rightarrow \begin{matrix} & H \\ H:\ddot{C}:H \\ & H \end{matrix}$$

The fact that the carbon atom forms strong, stable, covalent bonds with other carbon atoms is essential to the chemistry of living systems. In fact, long *chains* of carbon atoms may be formed, with various other atoms bonded to the remaining unpaired electrons of the carbon atoms. The hydrocarbon pentane is such a chain:

$$\begin{matrix} & H & H & H & H & H \\ H:\ddot{C}:\ddot{C}:\ddot{C}:\ddot{C}:\ddot{C}:H \\ & H & H & H & H & H \end{matrix}$$

Other atoms such as silicon are capable of forming strongly bonded chains, but only the carbon atom forms strong chains even when various other atoms are bonded to its remaining unpaired electrons. There are seemingly endless possibilities for joining carbon atoms to one another and to other atoms. In fact, the study of carbon-containing (or organic) compounds is an entire field in itself, a field known as *organic chemistry*.

BUILDING BLOCKS IN THE EARLY EARTH

Given such knowledge of the behavior of matter, we can make reasonable assumptions about the origin of life on earth. As the pre-earth whirled through space, its gravitational pull became increasingly stronger as the gas, dust, and solids comprising it began to contract. Heavier atoms such as iron and nickel were drawn toward the center. Lighter atoms such as silicon, phosphorus, and aluminum "floated" on the heavier matter. And still lighter atoms of hydrogen, oxygen, nitrogen, and carbon gravitated to the outside of the whirling mass. A slow process of solidification began that is still continuing today. As the mass drew together, the increasing pressure generated enough heat to bring matter to the molten state. Some of the solid compounds decomposed, which freed lighter atoms as gases. As the early crust of the earth began to form, the gases were driven out from the interior to form the early atmosphere.

There is little doubt that the early atmosphere differed from the one surrounding us today. In all probability hydrogen was the most abundant (as well as the most reactive) atom present. Hydrogen is believed to have combined with itself and with other atoms in various ways to produce molecular hydrogen (H_2), water (H_2O) in the form of steam, methane (CH_4), and ammonia (NH_3). Perhaps formaldehyde (CH_2O) and hydrogen cyanide (HCN) were formed as well. But the exact composition of the early atmosphere is still a puzzle.

Nor is the temperature of the early earth known. It seems likely that the amount of energy reaching the earth from the sun was considerably less than it is today. About 4 billion years ago the sun probably gave off about 40 percent less energy than it does now. But large amounts of methane and ammonia, if present in the early atmosphere, would be able to capture a large amount of the sun's energy, which would keep the temperatures above subfreezing levels. The reactions leading to the synthesis of organic compounds would proceed slowly at the lower temperatures. But even though thousands of years might have been required to form the organic compounds, they nevertheless could form and they could exist for significant periods of time.

Almost all biological compounds are energetically unstable. Under any conditions they decompose more-or-less rapidly. How, then, can they exist in significant quantities? The reactions that build up, or *synthesize*, molecules are possible only in the presence of a source of energy. Today the primary source of energy at

Figure 2.8   Geologic time scale. Before the advent of radiometric (or absolute) dating, geologists developed a relative time scale based on major units of sedimentary rocks and their distinct fossils. These stratigraphic units became incorporated into geologic systems, which were often named after the area in which the fossils were found. These systems are the basic time-stratigraphic units of historical geology. Geologists have since correlated these units to absolute time using various methods of radioactive isotope decay. It can be seen that the composition of the early or pre-Cambrian atmosphere was drastically different than it is today. The evolution of life and more complex organisms is related to these atmospheric conditions. Prior to 600 million years ago, the atmosphere was lower in oxygen—only in the evolution of algae (which produce oxygen) 2 to 3 billion years ago did the atmospheric composition begin to change. Eventually this evolution of oxygen led to conditions that allowed organisms to diversify and spread first through the oceans and later to the land. In this figure, the time scale is distorted to show the post-Cambrian (the last 600 million years) as three-fourths of the table; in reality, the pre-Cambrian period occupied 4 billion years.

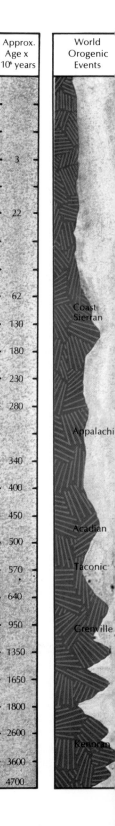

Geologic Succession				Approx. Age x 10⁶ years	World Orogenic Events
Eon	Era	System-Period			
Phanerozoic	Cenozoic	Quarternary	Recent		
			Pleistocene		
		Tertiary — Neogene	Pliocene	3	
			Miocene		
			Oligocene	22	
		Tertiary — Paleogene	Eocene		
			Paleocene		
	Mesozoic	Cretaceous		62	Coast Sierran
		Jurassic		130	
		Triassic		180	
	Paleozoic	Permian		230	
		Carboniferous — Pennsylvanian		280	
		Carboniferous — Mississippian			Appalachi
		Devonian		340	
		Silurian		400	
		Ordovician		450	Acadian
		Cambrian		500	
		Mediocarian		570	Taconic
Precambrian	Proterozoic	Upper		640	
				950	Grenville
				1350	
		Middle		1650	
		Lower		1800	
		Archean		2600	
		No Record		3600	Kenoran
				4700	

Biological Evolution

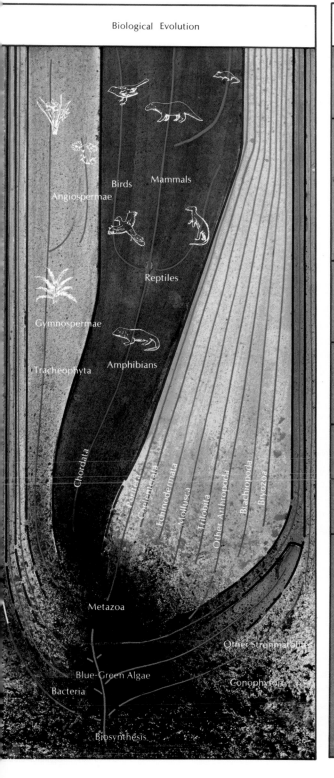

Angiospermae

Birds Mammals

Gymnospermae

Reptiles

Tracheophyta Amphibians

Chordata

Porifera
Coelenterata
Echinodermata
Mollusca
Trilobita
Other Arthropoda
Brachiopoda
Bryozoa

Metazoa

Other Stromatolites

Blue-Green Algae
Conophyton
Bacteria

Biosynthesis

Age

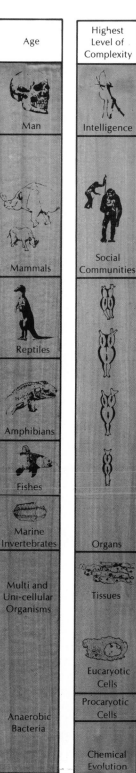

Man

Mammals

Reptiles

Amphibians

Fishes

Marine
Invertebrates

Multi and
Uni-cellular
Organisms

Anaerobic
Bacteria

Highest Level of Complexity

Intelligence

Social
Communities

Organs

Tissues

Eucaryotic
Cells

Procaryotic
Cells

Chemical
Evolution

Atmospheric Evolution

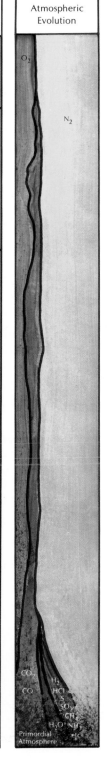

O_2

N_2

CO_2 H_2
CO HCl
S
SO_2
CH_4
H_2O NH_3
Primordial etc.
Atmosphere

the earth's surface is light from the sun. But only about 3/100 percent of solar energy is in the form of *ultraviolet* light, which is particularly advantageous in the synthesis of the kinds of molecules that supposedly were found in the early atmosphere. Perhaps electric discharges such as lightning were sources of the energy needed in the synthesis of the simpler organic compounds. Or perhaps temperature changes brought about by thunderclaps or by meteors passing through the atmosphere gave rise to organic synthesis; both cause brief and sudden temperature rises, followed by rapid cooling. In any event, once simple organic compounds had formed, they would be able to absorb wavelengths longer and more abundant than ultraviolet. And given that capability, such compounds would have at their disposal greater quantities of energy to carry out further syntheses.

These reactions are believed to have occurred in the early ocean, a watery environment often called an "organic soup." This early ocean began to form as the earth cooled and the vast layers of steam condensed into huge clouds. Further cooling and condensation led to the formation of raindrops, and the deluge began. For thousands of years it rained. Every depression, every crevice filled with water which, as it fell, carried ammonia, methane, hydrogen cyanide, and the other components of the early earth into the forming seas. Minerals and salts were leeched out of the solid rocks, creating the wet, warm, salty environment that was to nourish life.

The chemical building blocks of life exist at different levels of complexity. The first is the atomic level, which includes the atoms of elements found in all living creatures: primarily carbon, hydrogen, oxygen, and nitrogen; smaller amounts of sulfur and phosphorus; and at least trace amounts of many other elements. As you have seen, these atoms presumably were constituents of the early earth and were capable of reacting to form simple combinations of atoms. The next level is the molecular level, where atoms are bound together by means of chemical bonds to form the smaller biochemical building blocks: *sugars, organic acids* and *alcohols, amino acids*, and nitrogen-containing ring compounds called purine and pyrimidine *bases*. All these molecules could be formed from the early earth's constituents.

Why do we believe these molecular building blocks probably formed on the early earth? The chemical conditions that have been proposed for the early earth and its atmosphere have been examined in the laboratory. Mixtures of sources of carbon (in the form of methane or hydrogen cyanide), hydrogen and oxygen (as water), nitrogen (as ammonia or hydrogen cyanide), and other presumed trace substances were sealed in flasks (Figure 2.9). The required chemical bonds were induced by adding energy in the form of electric sparks, shock waves, or heat, and the "soup" was left to react for a week or so. After various periods, the contents of the flasks were analyzed chemically. In many cases, the simple constituents had indeed reacted to form many of the molecular building blocks of life.

Each of the building blocks can play a role in reactions leading to still larger molecules. Presumably that capacity developed at a later point in time. Nevertheless, many of the building blocks were important in their own right: once broken down, they could act as efficient sources of chemical energy. For example, energy is needed to create the chemical bonds between carbon, hydrogen, and oxygen when those atoms come together in the form of sugar. When such bonds are produced, they trap much of that required energy, and breakage of the bonds will release it in the form of chemical energy. Still other kinds of building blocks are capable of transferring the liberated energy to a synthesis that requires energy. This conservation and flow of energy in biological molecules was begun even before life, as such, existed.

In order for more complex reactions to have occurred, the proper molecules must have been in the right place at the right time with an available source of energy. In the organic soup, the concentrations of simple organic molecules formed by electric discharges, ultraviolet light, and shock waves were probably far too low to permit the occurrence of more complex syntheses at significant reaction rates. What mechanism might have served to concentrate the organic chemicals, thereby permitting the formation of larger molecules?

The simplest concentration mechanism would be the evaporation that would occur in isolated tide pools and shallow lakes. But salts, too, would become concentrated in such areas and might interfere with some key reactions; moreover, some important organic compounds such as hydrogen cyanide would evaporate with the water. When ocean water becomes frozen, salts and organic compounds are concentrated in the remaining liquid. And here again the presence of concentrated salts may interfere with reactions, and low temperatures would lead to slow reaction rates.

A more likely means of concentration would be the surface binding, or *adsorption*, of organic molecules on minerals and clays or on the water-air surface. Some experimental syntheses have been carried out on mineral surfaces. In some cases the surface not only concentrates organic chemicals, it also serves to speed up the reaction. In a similar way, large droplets of dense organic materials could serve as a means of concentrating organic molecules.

Several concentration mechanisms may have been involved in the origin of life. Certainly two conditions existing on the early earth helped the process along. First of all, the early atmosphere could not have contained "free" oxygen; in other words, most of the oxygen atoms must have been bound into molecules with other atoms. In fact, because an oxygen atmosphere tends to decompose large molecules, life probably could not originate under

Figure 2.9 Stanley Miller performed the first experiment in which molecular building blocks of life were created in the laboratory. As a graduate student in 1952, Miller mixed together certain gases that were probably present in the atmosphere of the early earth. These gases—methane, water vapor, ammonia, and hydrogen—were circulated through a glass bowl containing an electric discharge. After one week, Miller discovered that the water contained a number of different amino acids and other organic compounds. In later experiments, different combinations of organic compounds have been created in the laboratory from a number of different gas mixtures. Various sources of energy—such as ultraviolet irradiation, heat, and bombardment by alpha particles—have been employed to manufacture these building blocks. The experimental results show that the molecular building blocks could have been created in a number of different ways when the early earth was forming.

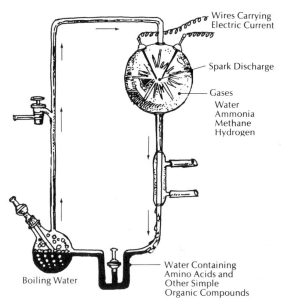

Wires Carrying Electric Current

Spark Discharge

Gases
Water
Ammonia
Methane
Hydrogen

Boiling Water

Water Containing Amino Acids and Other Simple Organic Compounds

the conditions that exist today. Secondly, there was no *decay* as we know it today on the early earth: decay is a product of living organisms, which had not yet evolved. As a result, more and more molecules on the early earth accumulated and became increasingly complex.

The evidence that building blocks of *every* important kind of large biological molecule could have been formed and concentrated under primordial conditions is as yet incomplete. However, the evidence does suggest that the appropriate mechanisms were available for the spontaneous formation of all the building blocks necessary for biological *macromolecules*. Previously nonexistent molecules could have emerged from combinations of the simpler molecules found in the organic soup. In their turn, these molecules could have formed the basis for the next level of biochemical complexity.

The history of the third level of biochemical complexity, after the atomic and simple molecular levels, is less well understood. The molecular building blocks in systems alive today are linked together to form large molecules of repeating subunits. The simple molecules making up those individual subunits are called *monomers*; the large molecules in which they become linked in a repetitive way are called *polymers*. Different kinds of polymers require different subunits, but the basic mechanism of polymerization is similar for all biological macromolecules. The same mechanism could have operated in the ancient seas.

Two molecular building blocks join together by means of a chemical bond between them. As you read earlier, a covalent bond involves the availability of a pair of electrons, one from each reactant. But each molecular building block is reasonably stable because all of its available bonds have been filled. How can "vacancies" be made to exist so that polymers can form? Each reactant must lose one or more atoms. That is how the electrons involved in the initial bond become available for the formation of a new bond. In biological molecules, the "lost" atoms are often hydrogen and oxygen atoms, which are given off in the form of water molecules.

Consider, for example, a molecule of the sugar *glucose*. Two molecules of glucose can be joined together, thus producing one water molecule and one molecule of a *disaccharide* sugar. This process is called *dehydration synthesis* (Figure 2.10). Each of the sugars in the disaccharide still has reactive groups present and can be joined to additional glucose units. In this way, large *polysaccharides*, containing hundreds or thousands of sugar subunits, can be formed.

In addition to polysaccharides, the other major classes of biological macromolecules are the *lipids* (fats), the *proteins*, and the *nucleic acids*, which are also assembled by means of dehydration synthesis. Each kind of macromolecule carries out functions that

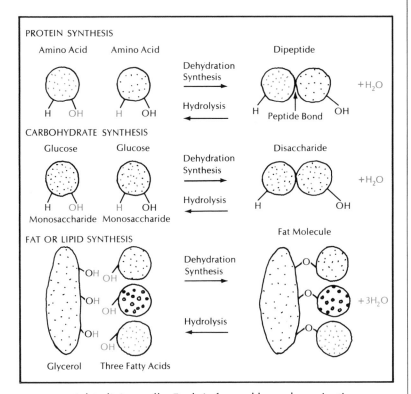

PROTEIN SYNTHESIS

Amino Acid   Amino Acid   Dipeptide

Dehydration
Synthesis

Hydrolysis

H   OH   H   OH   H   OH
                               Peptide Bond   +H$_2$O

CARBOHYDRATE SYNTHESIS

Glucose   Glucose   Disaccharide

Dehydration
Synthesis

Hydrolysis

H   OH   H   OH   H   OH   +H$_2$O
Monosaccharide Monosaccharide

FAT OR LIPID SYNTHESIS   Fat Molecule

OH OH
                  Dehydration
                  Synthesis

OH
OH
                  Hydrolysis   +3H$_2$O

OH
OH

Glycerol   Three Fatty Acids

Figure 2.10 Diagrams showing dehydration synthesis and hydrolysis. The main categories of organic compounds in living systems are proteins, fats, carbohydrates, and nucleic acids. All of these compounds may act as building materials and energy sources. In addition, proteins may act as enzymes and nucleic acids may act as information carriers. *Dehydration synthesis* involves the build-up of organic compounds by the removal of water molecules. Energy and specific enzymes are required to carry out the reactions in dehydration synthesis. The opposite reaction—*hydrolysis*—involves the breakdown of organic compounds by the addition of water. In hydrolysis, some energy is released and specific enzymes are required. Examples of dehydration synthesis and hydrolysis are shown in the three diagrams at left. In *protein synthesis* (above), proteins (which are composed of up to twenty amino acids) are arranged in a set sequence determined by nucleotide sequence within nucleic acid molecules. In *carbohydrate synthesis* (middle), molecules with six carbon atoms (such as glucose) are joined together. Cellulose, starch, and glycogen are examples of long, complex chains of glucose molecules. In *fat synthesis* (below), a single glycerol molecule is hooked to three fatty acid molecules to form a complete fat molecule (triglyceride).

are essential to living cells. Each is formed by polymerization reactions that split off water. And presumably all of those reactions occurred in the organic soup.

Now rates of chemical reactions vary enormously. They are sensitive to conditions of temperature, pressure, concentrations of the reactants, the physical state of the reactants (solid, liquid, gas, solution, and so forth), and the presence or absence of an agent called a *catalyst*. A catalyst is a material that increases the rate of a chemical reaction without being consumed by the reaction. It does interact with the reactants at some stage in the reaction, but the interaction is temporary. The catalyst facilitates the reaction by altering the configuration of the molecules or by bringing them into the proper spatial relationship to change reactants into products. In the case of most living organisms, the chemical reactions take place at atmospheric pressure and moderate temperature and each reaction is speeded up by its own catalyst, a complex macromolecule called an *enzyme* (Figure 2.11).

Certain proteins act as these specific organic catalysts, or enzymes, that are necessary for practically every chemical reaction in living systems. Without them, the reactions cannot occur within the narrow limits of temperature, pressure, and acidity that are compatible with life. Proteins, then, are necessary for the synthesis of macromolecules in living things. But this requirement is puzzling. Although proteins are necessary for living syntheses, the

Figure 2.11   Characteristics of enzymes. All enzymes are proteins. Each enzyme is specific for a given reaction and is not used up in the reaction. Various other enzyme characteristics are shown in this series of graphs. (A) An enzyme has its maximum activity at a specific pH (relative acid-base concentration). (B) An enzyme has its maximum activity at a specific temperature. At a higher temperature, enzyme activity is higher, but some enzyme molecules will become deactivated. (C) Reactions will proceed very slowly without enzymes but will speed up dramatically with the aid of specific enzymes. (D) The activity of most enzymes is aided by the presence of a coenzyme (an organic or inorganic substance that makes an enzyme more effective). (E) Enzyme activity also slows in the presence of inhibitors, or poisons. (F) For a given enzyme concentration, the reaction rate will increase by raising the substrate concentration up to a maximum velocity. (G) An enzyme lowers the energy barrier over which reactants (A + B) must pass in order to form products, AB. The amount of energy released in the overall reaction is not altered by the enzyme, nor is the final equilibrium, which is governed solely by the difference in energy between reactants and products.

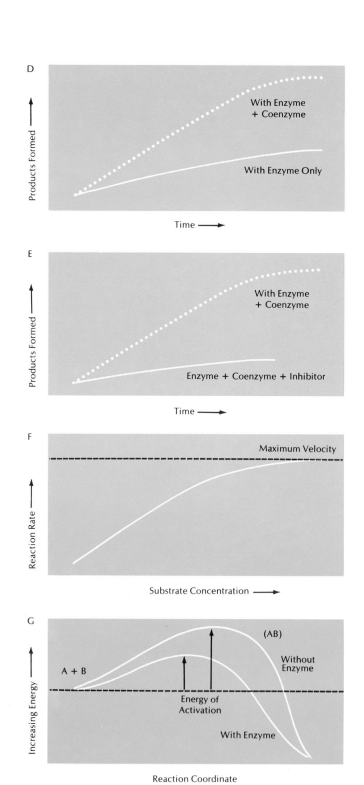

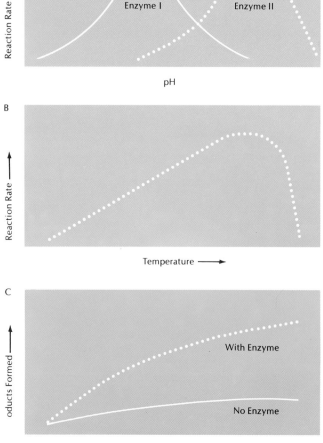

Figure 2.12   Complementary hydrogen bonding of the nitrogen-containing base pairs in DNA. Two of the bases in DNA—adenine and guanine—are *purines* (two-ring bases). The other two—cytosine and thymine—are *pyrimidines* (one-ring bases). Because of the shapes and hydrogen bonding of the base groups, only two pairings are possible: adenine with thymine and guanine with cytosine.

details of their structure are the *product* of another class of macromolecule, the nucleic acids. Like the chicken-or-egg problem, this paradox of which came first—the protein or the nucleic acid—has not been resolved.

THE ORIGIN OF LIVING SYSTEMS

To what degree of size, organization, and complexity must chemical syntheses proceed before a living system is formed? Could the process of biochemical evolution have consisted merely of the gradual synthesis of more and more complex organic molecules? At what point should the line be drawn to say that a living creature has been formed? Probably the central feature of living systems is their ability to reproduce themselves. All living systems utilize the flow of energy through the universe to reproduce and to maintain life. A living system uses this energy to produce complex molecules and to maintain highly integrated and ordered structures. In most nonliving systems, such structures steadily break down into more probable, random aggregations of matter, with a concurrent release of energy.

Reproduction and energy flow in living systems are interlocked concepts. The flow of energy in living systems is reasonably well understood. It is apparent that living systems are not energetically unique—they, too, must obey all chemical and physical laws. The most pressing and difficult questions concerning the origin of life center on the formation of the simplest system capable of reproducing itself.

So far, two quite promising theoretical approaches have been advanced, although the experimental evidence in either case is sketchy. The first proposes that the transition from nonliving to living involved an original macromolecule, capable of self-reproduction. The alternative theory favors aggregates of macromolecules as the starting point.

The Theory of the Original Macromolecule

Biochemists have isolated *deoxyribonucleic acid* (DNA) as the chemical constituent of living systems that provides the information for self-reproduction. In addition to directing its own duplication, or *replication*, DNA indirectly is responsible for the synthesis of proteins in all living systems (Chapter 5). These proteins include the catalysts that make available the energy and the precursor molecules that are needed in the replication of DNA. How does the theory of the original macromolecule account for the origin of such a DNA-protein system?

It is possible to demonstrate experimentally that DNA can duplicate itself in a solution containing the appropriate precursor molecules and a few enzymes. Under appropriate circumstances a single DNA-enzyme system might be able to function as a living system—a relatively simple system in comparison with a whole

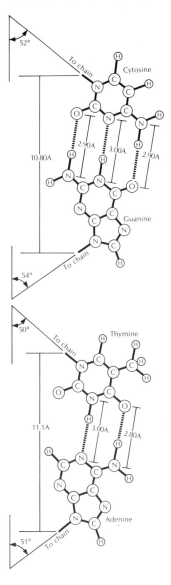

cell, to be sure—but even then it would be too complex and too efficient to have originated spontaneously. Surely this system is the product of a long evolution.

Before the DNA-enzyme system could have arisen, nucleotides—the building blocks that could be put together to form the DNA—must have been formed in high concentrations. The nucleotides, in turn, would require precursor molecules of sugar, phosphoric acid, and purine and pyrimidine bases, all of which must have been created by other mechanisms. The mechanisms producing these precursor molecules probably did not show the properties of self-replication; accordingly, those molecules might not be considered living systems.

In one sense, then, the beginning of life has been "defined" to be an arbitrary point in time. Continuous chemical evolution must have led to relatively complex macromolecular systems before the first self-duplicating system could be built. This evolution, however, would have differed from the evolution that has existed since the creation of the DNA-enzyme system, for the ability to incorporate and transmit changes would not have existed at that time. The important point is that *life appeared as a major step in a continuous process of chemical evolution.*

Now it could be argued that the present chemical system for transmitting genetic information has evolved so far from the primitive genetic system that it bears little resemblance to the original. The first genetic system might have used nucleic acids with different sugars and bases; it might not have been built upon a sugar-phosphate backbone. Although it is possible that other self-duplicating polymers preceded nucleic acids, no experimental models have yet been proposed and tested to substantiate this theory. The simplest assumption is that the first genetic material was similar to DNA.

There is another tempting idea: Suppose the first living organism consisted of a nucleic acid associated with an enzymatic protein that could catalyze the process of duplicating the strands of the nucleic acid. Such a system would be capable of replicating itself through the use of preformed organic molecules (nucleotides) available in its environment. Such a system could also evolve, through the mechanism of natural selection, into a more complex organism. Indeed, the very existence of this sort of primitive "life-system" conceivably could have caused enough changes in the environment to enhance its further evolution.

The original primitive organism need not have been too efficient. As long as replication was accurate enough to prevent the system from incorporating too many changes—or *mutations*—it would not become extinct. Natural selection would select for the most favorable mutations. The replication of this early organism might have been slow, perhaps with a small "polypeptide" (a polymer of two amino acids is called a "peptide") acting as a relatively inefficient catalyst. Here, too, selection would favor mutations that produced more efficient catalysts.

This idea is attractive but it is not as simple as it might appear. How could even such a simple entity as a self-duplicating polynucleotide be produced by random chemical combinations without the aid of enzymes? On the primitive earth, an appropriate polynucleotide could be produced only if random polymerization of nucleotides occurred often enough. Catalysts that promote polymerization of nucleotides would be required before significant amounts of nucleic acid could form. Catalysts of this sort might have been formed from the amino acids that more than likely were abundant on the primitive earth. Such primitive enzymes would not have to be as specific or as efficient as modern enzymes; relatively small polypeptide molecules might have served the function. It is also quite possible that some other small molecule served as a catalyst.

Perhaps macromolecules continued to form at random until one happened to acquire the structure necessary for organizing amino acids into a simple protein that could act as a catalyst for DNA replication. Perhaps a part of the structure of the first successful self-replicating macromolecule was capable of catalyzing the process of replication. It is even conceivable that the first living molecule was a protein capable of using some as-yet undiscovered mechanism to duplicate its own structure — or that the first nucleic acid somehow possessed its own enzymatic activity, which it lost after catalyzing the formation of more efficient protein catalysts.

The Theory of Molecular Aggregates

What if the first organisms were not individual molecules at all but *aggregates* of molecules, separated from their environment by a definite boundary? Imagine, as the Russian biochemist A. Oparin did in 1936, that colloidal droplets of proteins and other complex molecules might have formed in the primitive ocean. (A *colloid* is a system of finely divided matter suspended in some medium, much like the particles of butterfat that are dispersed throughout homogenized milk.) In these droplets a series of linked chemical reactions might have been carried out. Complex molecules would have been accumulated from the environment. This living matter would grow in size and finally split into smaller droplets that would repeat the process, and would thus represent a kind of primitive "metabolism." But Oparin's model does not provide for a detailed mechanism by which such a droplet could replicate, mutate, and therefore evolve into a more complicated system.

On the other hand, Sidney Fox of the University of Miami later outlined the mechanism by which an "aggregate" complex could have progressed to a living cell. By subjecting collections of amino acids to mild heat treatment, he obtained small spherical aggregates of protein that display three properties (Figure 2.14). First

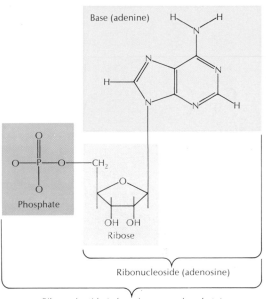

Figure 2.13 A ribonucleotide is a compound composed of one of four purine or pyrimidine bases, the sugar ribose, and a phosphate group. Each ribonucleotide forms a monomer segment of the larger polymer ribonucleic acid (RNA).

Figure 2.14   Proteinoid microspheres of uniform size under low magnification. Sidney Fox and his co-workers have proposed that primitive cells could have developed from aggregates of these tiny microscopic droplets.

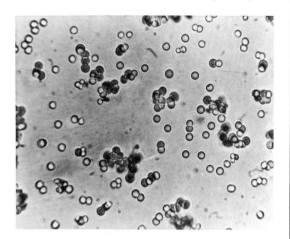

of all, the protein spheres can *change in size*, depending on the medium in which they are placed. This behavior is suggestive of a property of living cells, whereby substances can selectively enter or leave cells according to the concentration in which the substances exist within and outside of the cells. Secondly, the spheres are capable of *fusing* with one another. Because different sorts of spheres are produced from different collections of amino acids, fusion could act to bring together different chemical properties. And some of those spheres might be better equipped to exist in the environment than were the original spheres. Both of these characteristics promote a chemical tendency toward self-organization. In addition, as the protein spheres form, they exhibit a primitive but generalized sort of *metabolism*—including the production of a molecule that is a precursor for nucleic acids. Many such aggregates could have been formed, many could have fused, and some could have evolved toward the greater complexity that characterizes cells.

Whether the first form of life was a molecule of nucleic acid and protein (a *nucleoprotein*) or an aggregate of such molecules, a number of further steps must have occurred. At some point in time (either earlier for aggregates or later for the single nucleoprotein), a number of these units must have fused together. In so doing, a surface boundary (or *membrane*) must have formed, permitting the selective formation of an *internal* environment that differed from the *external* environment. Today this property is characteristic of all cell membranes, and "selective permeability" is seen in all living systems (Chapter 4). In ancient times, differences in permeability would have resulted in the absorption and maintenance of different molecules inside the unit than outside of it. This differentiation in turn might have increased the likelihood that all the large molecules characteristic of living things would be built. As these simple units grew in size by selectively incorporating external materials and perhaps by fusion as well, the probability of a primitive sort of reproduction by fragmentation, or "budding," would increase. The newly budded units having enough raw materials would repeat the growth and budding cycle, even though many small buds would perish by fragmentation below the level of molecular complexity needed for survival.

Eventually an aggregate might have been formed that had a far greater survival potential than most. Perhaps it resembled a primitive bacterium; perhaps not. But over immense spans of time one of its offspring may have carried that resemblance and thus formed the crucial link in the chain leading to the diversity of life forms that now inhabit the earth.

CHARACTERISTICS OF LIFE

The differences between "living" and "nonliving" systems may seem obvious, until you attempt to describe them. The same kinds of atoms are present in both. The iron in human hemoglobin is composed of the same kind of atom as the iron in a kitchen knife or in the hull of a submarine. The carbon atoms that make up so much of our flesh and bones are no different from the carbon atoms in a diamond or in a "lead" pencil. The same must be said for *all atoms* in living systems—they are identical to their counterparts in lifeless things. By means of radioactive tracers you could follow a group of sodium atoms and find that at one time they occur in lifeless salt deposits in the ground, at another they are vital components of a living cell, and at still another they might be in tears of frustration that pour down the cheek and splash on the ground, where once again they are lifeless salt, as at the start.

Nor are the differences to be found at the molecular level. Although it is true that today some molecules are produced only by living things and are essential to life, they are not alive in themselves. An ounce of hemoglobin will remain inert and lifeless in a test tube for as long as you care to observe it. The same is true of molecules of nucleic acid, or glycogen, or a pound of lard. You may even make a mixture of all the different kinds of molecules that occur in living matter, and still you would not have "life."

What, then, are the differences between life and nonlife? Perhaps the question has no straightforward answer. The difference between "point of departure" and "destination," for example, would be difficult to describe for someone planning a trip around the world, starting from, say, New York and ending up in New York. To find the answer to that sort of question you have to search for what is known as a "working definition." Consider another example: "How does a painter differ from a musician?" Here you intuitively approach the question in terms of what each might look like, or wear, or eat for dinner. Similarly, the differences between living and nonliving matter are to be found by defining what living and nonliving matter *do*. The working definition avoids an inevitable dilemma in searching for a simple distinction between life and nonlife—namely, the difficulty in defining the differences between a living person and a corpse. In terms of chemical composition there are no well-defined differences.

Living and nonliving matter both constitute chemical systems. As chemical systems they may be as energetically "inert" as a rock lying in a field, or they may be undergoing energy transformations. According to the *second law of thermodynamics*, all energy transformations result in a reduction in the amount of free energy available. Burning wood is a chemical reaction, for example, in which chemical bond energy is transformed into heat energy. The chemical bond energy in wood exists primarily in the covalent bonds within the individual molecules of the polysaccharide cellulose. In the presence of free oxygen at the so-called kindling temperature, these bonds are broken and replaced by lower energy bonds with oxygen. The resulting products are heat, carbon dioxide, and wa-

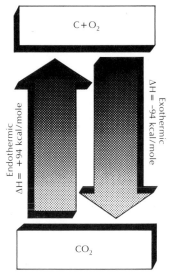

Figure 2.15 Energy relationships taking place during the combustion of coal. In this chemical reaction, which involves the breakdown of a substance, bonds within the original molecules are broken and new bonds are formed to create molecules of the products. A result of these changes is the release of energy. When coal is burned, carbon combines with oxygen to form the gas carbon dioxide. For every molecule of carbon burned, a molecule of CO_2 is formed and energy is released. This energy may be given off as heat or light. In living systems, burning of organic compounds will release energy that will be available for some cellular activity.

In a reverse reaction, energy from some outside source is required.

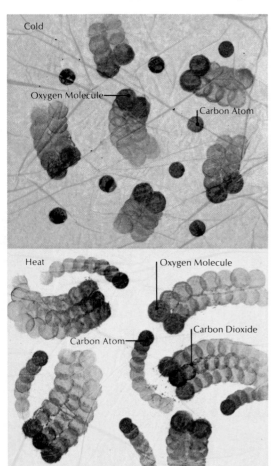

Figure 2.16   In the collision theory of chemical reactions, molecules, atoms, or ions of the reactants collide with one another. The force of the collision depends on the speed of the molecules involved, and speed is a function of temperature. In these diagrams, individual oxygen molecules and carbon atoms are interacting. At a low temperature (upper diagram), the oxygen molecules move slowly and merely bounce off the carbon atoms. The energy of the collision is not sufficient to break the double bond in the oxygen molecules so no chemical change occurs. At a high temperature, however, the molecules and atoms move rapidly, and a collision occurs with enough energy to break the double oxygen bond (lower diagram).

ter. The amount of energy available to do work in carbon dioxide and water is considerably less than in cellulose (wood). Heat is a useful form of energy as long as there is a difference in temperature between two objects. But when heat is produced continuously within a closed system, the difference diminishes in time to zero, at which point free energy is no longer available.

Living matter is a natural chemical system that obeys both the first and the second laws of thermodynamics. In obedience with the *first law of thermodynamics*, which states that energy can be neither created nor destroyed, life cannot *create* new energy; it can only transform one form of energy into another. Chemical energy in food, for instance, can be transformed into body heat. To *sustain* the complex organization of life, cells must obey the second law and continuously obtain a flow of energy from the environment. A green plant uses energy from the sun to build high-energy sugar or cellulose from carbon dioxide and water, which is the direct opposite of burning wood. Viruses use the energy of the cells that they invade to build more viruses. Animals use the chemical energy from other erstwhile living things to build the molecules that are characteristic of themselves or to perform their individual chemical and mechanical functions.

Metabolism

The sum total of the chemical transformations by which living systems regulate themselves in terms of the second law of thermodynamics is called *metabolism*. For the most part, metabolic reactions involve two steps. First, they derive energy from outside sources either directly (as in photosynthesis) or indirectly by slowly burning, or "oxidizing," organic molecules. Second, they use this energy to build up new, different, and often complex organic molecules. Metabolic reactions whereby complex organic molecules are broken down to simpler ones, usually with the release of energy, are called "catabolic" processes. Metabolic processes that entail the building up or synthesis of organic molecules are called "anabolic" and generally use energy. In all cases that biologists know of, both kinds of processes are regulated by the biological catalysts called enzymes. In operational terms, the most fundamental characteristic of living substance is metabolism, *the complex of chemical reactions by which living substance derives and utilizes matter and energy for its own use.*

One must not lose sight of the totality of energy transformations in the universe, however. Although living matter such as a green plant appears to run contrary to the second law of thermodynamics, it does so only as long as it does not die and only as long as there is any energy differential between itself and its environment. But plants do die and are consumed, and their stored energy is dissipated. Moreover, the primary energy differential on which green plants and, for that matter, most living things on earth depend is that which exists between the sun and the earth. But with time the differential diminishes, and when the day arrives that the sun has burned itself out, the differential will be zero and life will cease to exist. Even if a substitute source of energy were found, such as terrestrial thermonuclear energy, it, too, would be finite in time and the course of life on earth would eventually be played out.

Reproduction

All living systems must maintain an active metabolism as well as be able to reproduce and to mutate. *Reproduction* is a direct expression of metabolism. But is it in itself a characteristic of life? Consider viruses: they are able to reproduce and mutate. A single virus particle may invade a living cell and, using the energy and raw materials of that cell to replicate its nucleic acid and to synthesize a new protein coat, it may reproduce itself a hundredfold within a matter of minutes. The host cell is often destroyed as a consequence, and a large number of virus progeny emerge from it, ready to invade other cells. Nevertheless, viruses themselves do not maintain an active metabolism but rather rely on the metabolism of the host cell.

Unlike viruses, all the diverse forms of life are *cellular*, which means they consist either of single cells or of groups or organized masses of cells. A *cell*, in turn, is a functional unit of life that at the very least must include DNA and another nucleic acid, RNA; proteins; lipids; carbohydrates; and a limiting cell membrane. The structure and arrangement of these and other components of cells are considered in more depth in Chapter 4; of immediate concern at this point of the discussion is the concept of reproduction as a distinguishing characteristic of life.

Reproduction is a cellular phenomenon even though its most essential feature—the replication of nucleic acid—is also charac-

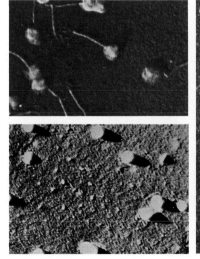
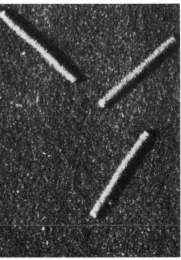

Figure 2.17 Viruses exist in a wide variety of sizes, structures, and types. Most viruses are spherical, but some are bricklike, rodlike, cubic, or irregular in shape, as shown in this figure. The influenza virus is shown at the lower left; the T-5 bacteriophage can be seen at the upper left; and the tobacco mosaic virus is shown at right.

Figure 2.18   The electron micrograph at left shows T-4 bacteriophages adsorbed on an *E. coli* B cell wall. A typical bacteriophage (bacterial virus) has a large hexagonal head surrounded by a protein coat. At the base of the head is a protein collar with a tail assembly attached, consisting of a sheath and tail fibers. In this micrograph, the phages are connected to the bacterial cell wall by these tail fibers. When a phage attaches to the bacterial cell wall, an enzyme produced by that phage will eventually digest a hole in the wall. Then the sheath contracts, the head collapses, and the DNA contents of the bacteriophage head are extruded into the cytoplasm of the bacterial cell.

This viral DNA takes over the direction of the cellular mechanism and directs the production of 100 or more new viruses in less than 30 minutes.

The electron micrograph at right shows a T-4 infected *E. coli*. Note how the bacteriophage protein remains as an empty "ghost" attached to the outer surface of the cell wall.

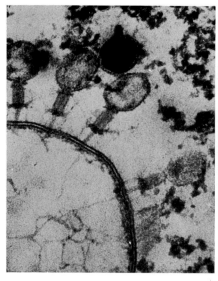
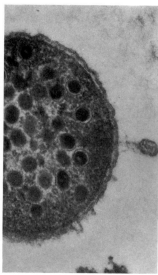

teristic of viruses. In viruses the nucleic acid may be either DNA or RNA, but in cellular forms the essential nucleic acid involved in initiating reproduction is invariably DNA. Only DNA carries the code for the synthesis of enzymes and other proteins that make metabolism and therefore growth possible. It encodes the "information" a cell will need for any and all of its functions throughout its lifetime. After DNA replication occurs and the other cell components are duplicated, the cell divides into two identical "daughter" cells (Chapter 6). The daughter cells then typically grow to approximate the volume of the cell from which they arose. It is primarily through the process of cell division, followed each time by the restoration of cell volume, that growth occurs in cells. The two phenomena, reproduction and growth, are therefore virtually inseparable.

The nonliving world may show signs of growth as well, but never in a manner like the reproduction or growth in living systems. Stalactites "grow" in a cave when calcium carbonate precipitates out of the water that seeps down through limestone and drips from the cave's ceiling. The water contains calcium hydroxide, dissolved from the limestone. When the calcium hydroxide comes in contact with the air in the cave, it reacts with carbon dioxide in the air to form insoluble calcium carbonate and thus a stalactite grows. Inorganic crystals can also evidence growth, but again it is in a manner quite different from biological growth: it involves merely the rearrangement or packing of molecules already present into geometric patterns. Crystalline growth may be dramatic, as it is when the element sulfur is placed in a shallow dish containing the solvent carbon disulfide. When the solvent is allowed to evaporate slowly, the crystals of sulfur grow to form branches that terminate in flowerlike extensions. A more rapidly

Figure 2.19   A T-2 *Escherichia coli* bacteriophage that has been osmotically shocked. In the early twentieth century, the work of several bacteriologists revealed that certain viruses attack bacteria. These bacterial viruses were subsequently named ''bacteriophage.'' In this micrograph of the T-2 *E. coli* bacteriophage, the protein coat is evident in the center, and the DNA originally contained within it appears to stream out in a single continuous thread.

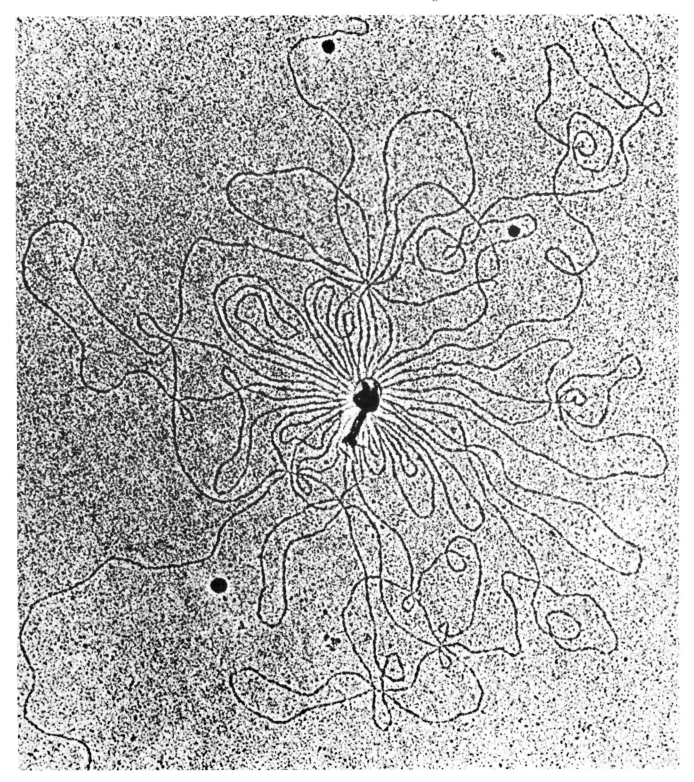

occurring example of crystalline growth occurs in an alcoholic solution of benzoic acid. When a drop of it is placed on a microscope slide, covered with a cover glass, and then viewed through the low power of the microscope in dark field, needlelike crystals shoot out in all directions in a fascinating display that lasts about 5 minutes or less. But this is growth by layering and is more analogous to a chimney that "grows" in height as the bricklayer lays tier upon tier. In biological growth, each essential component is duplicated by complex chemical reactions from chemically different precursor substances.

Irritability

Living systems also exhibit *irritability*. The meaning of this word has changed somewhat since it was first used in the biological context. It refers to the ability of a system to become sufficiently altered, when confronted by a change in its environment, so that it responds in some dynamic way to the change. Of course this characteristic is displayed by nonliving matter as well, but the *quality* of the response is vastly different and the intensity of the alteration of the environment needed to evoke the response is in general of a totally different magnitude. A few examples will illustrate these points. An amateur potter may place a sample of his handicraft in

Figure 2.20 A *tropism* may be defined as an involuntary response to a physical stimulus. In a normally growing green plant, light and gravity are two primary stimuli that act to produce a tropism. *Phototropism* reflects the influence of light upon growth. *Geotropism* — in which gravity has an effect on growth — is exhibited in this figure. When the plant is turned on its side, it grows upward, away from the pull of gravity.

a kiln in order to fire it to permanent hardness. But much to his dismay, his work of art explodes when it reaches 1500°F. That single experience is ordinarily enough to tell him his method must be altered; perhaps he will thoroughly dry his samples before firing the next time. Or consider what happens when a living organism is exposed to a rise in temperature of only 5°F. Generally it will respond in some very obvious way. For example, if the temperature of a classroom climbs from a tolerable 75°F to a sweltering 80°F, students become fidgety, inattentive, and irritable. If tropical fish are transferred suddenly from a container at 80°F to a tank at 77°F, or vice versa, a number of them may die from temperature shock. Consider, finally, the sensitivity to light. Most living systems are sensitive to and respond to slight changes in light intensity (Figure 2.20). Few nonliving systems are; in fact, it has required considerable ingenuity, research, and inventiveness for man to develop inanimate photosensitive compounds such as photographic emulsions. In the world of living systems, photosensitivity is the rule rather than the exception.

Even the simplest unicellular organisms exhibit different forms of irritability. An amoeba is a tiny microorganism with relatively few intracellular structures that are located in what appears to be a formless mass of cellular material. Although it lacks apparent specialized structures for receiving or sending sensory messages, the amoeba does show a regular behavior pattern. It responds to small food particles or to certain chemicals with feeding behavior and responds to almost any other stimulus by withdrawing. Unicellular organisms with specialized structures called cilia exhibit far more intricate responses when confronted by certain kinds of stimuli. For example, a paramecium that encounters an object or a noxious chemical as it swims along will back up by reversing its ciliary beat and will then move forward in a different direction. If microelectrodes are attached to this organism, it is possible to depolarize the anterior end so that the paramecium causes its cilia to beat in reverse. Evidently some tactual and chemical stimuli cause such depolarizations, thus controlling the cell's behavior.

Many years ago the physiologist F. Lillie devised a nonliving model that simulated a living nerve. It was called "Lillie's iron wire nerve" and it consisted of a wire of soft iron, such as that used in a coat hanger, treated with concentrated nitric acid until it acquired a thin coating of oxide. The coated wire was carefully transferred to a cylinder of dilute nitric acid. When an immersed place on the wire was scratched lightly with a file, the oxide coating flew off dramatically, starting at the point of the scratch and progressing along the entire length of the wire in both directions. Examples of this kind of sensitivity in the nonliving world are extremely rare and most often must be contrived by a member of the living world, such as Lillie. Among all living systems this degree of

Figure 2.21 Evolution of the horse during the Cenozoic Era, beginning over 60 million years ago. The important stages of development between the earliest known ancestor (*Eohippus*) and the present-day horse are shown in this series of restored paintings. Over time there was a loss in the number of side toes; there were changes in tooth structure; the brain became more developed; and horses increased in overall size. (a) *Eohippus* lived about 58 million years ago during the Eocene epoch. This species was distinguished by the presence of four toes on each front foot and three toes on each rear foot.

(b) *Mesohippus*, which lived during the Oligocene epoch about 35 million years ago, had three toes on each of the front and the rear feet. (c) *Hypohippus* lived during the top of the Miocene epoch about 20 million years ago and had one main toe and two side toes on each foot. (d) *Hipparion* lived about 13 million years ago in the Pliocene eopch. (e) *Equus scotti* was a species of the modern horse that lived in the Pleistocene epoch about 2 million years ago. By this time, horses had single toes on each foot. (f) This photograph shows a representative of the present-day horse.

sensitivity to a stimulus is not only commonplace but universal.

The real point of irritability as a characteristic of life is that the response, which is the expression of irritation, is *purposeful*. The response of a living cell or organism to a given environmental change is normally one that enables it to adjust in a beneficial way to the change. When a classroom gets too hot, a student sweats and this response cools him; if it gets still hotter, he breathes more rapidly and may fan himself. A seedling's stem will grow toward the light, the source of its energy; its root will grow toward the gravitational center of the earth to reach water and minerals. That is what "irritability" is all about. It is one of the phenomena that separates living chemical systems from the nonliving.

Adaptation

The kinds of adaptive responses categorized as irritability are, for the most part, short-lived and reversible. All living systems evidence another category of response, one that is long-lived and often irreversible. It is the response of genetic and evolutionary *adaptation* to a changing environment. The basic mechanisms that are believed to underlie evolutionary adaptation were described in Chapter 1. In this context, however, adaptation can be viewed as a *group*, rather than *individual*, response. Selection, in the face of environmental change, will elevate certain members of the group to the important role of producing much of the next generation. If reproduction is considered a device for perpetuating the individual, then evolutionary adaptation works to perpetuate a group that exhibits a changing stability in relation to a changing world. Adaptive change works by way of differential reproduction which, in its turn, has its roots in slightly different patterns of metabolism that fit in less well or better than the average pattern of the group.

Whatever the exact sequence in which living things were initially formed, the connections between organization, metabolism, reproduction, irritability, and adaptation were established. These characteristics are visible today and, if the theory of the unity of life is valid, they were also present in much the same way in the earliest cellular forms of life that evolved on earth.

Living systems, like the nonliving, function in a universe that obeys certain fundamental laws. These are laws governing the behavior of matter and energy. Taken as a whole, matter and energy are finite in amount although the forms they take may change from time to time. This concept is embodied in the first law of thermodynamics and has been alluded to as the cycling of matter and the flow of energy. It is within this much, much broader context that living systems play their part.

THE FLOW OF ENERGY THROUGH THE BIOSPHERE

Every atom and molecule incorporated into living systems has a part in some cycle in nature. We often interrupt some of these cycles when, for example, we use steel caskets and concrete vaults to prevent the soil from reaching the atoms and molecules of dead men. And in a larger context, some ecologists are beginning to worry that technological alteration of the earth's surface may disrupt one or more of the cycles and cut off the supply of some atoms necessary for all living systems. But so far we have succeeded only in delaying the passage of certain atoms through the natural cycles. Like it or not, we are part of a scheme of recycling atoms that includes all of the living and nonliving systems that have ever existed. Each atom in our bodies has had a place in hundreds or thousands of other systems in the past.

Energy, on the other hand, does not cycle. It flows one-way from the usable, useful state to the unusable state, and it is impossible to recycle energy that has reached that point. Nevertheless, even though energy cannot be created, it can be changed from one form to another.

When the substance called gasoline burns in an automobile engine, the chemical energy of gasoline is converted into heat, and part of that "thermal energy" is then converted into "mechanical energy," or energy of motion, which is used to move the car. In other words, chemical energy can be converted into mechanical energy, but some thermal energy is always lost in the conversion. Living and nonliving systems are capable of converting chemical energy *directly* into some other form of energy—such as electric energy in nerves, mechanical energy in muscle, or even light energy in a firefly—but even in these processes some energy is lost as heat during the conversion.

Living systems are said to exist in an "open steady state." They are "open" to the environment in the sense that they exchange matter and energy with their surroundings: Food comes into and waste products leave an organism. Living systems are in a "steady state" in the sense that they achieve a dynamic equilibrium or balance: Even the bones you had a minute ago are a little different in specific atomic composition than the bones you have now. Perhaps an atom of calcium left an arm bone and entered the blood. Of course it will soon be replaced by another atom of calcium so

3

Energy Flow and Early Living Systems

Figure 3.1   J. Willard Gibbs (1839–1903), the American physicist who derived the second law of thermodynamics.

that the bone can continue to do the same things—in fact, it will look the same at both points in time.

The interaction of an open, dynamic steady state with the universe becomes more complicated when it is thought of in terms of the second law of thermodynamics. This law may be interpreted to mean that whenever energy is converted from one form to another, some useful energy takes the form of thermal energy (heat) and cannot be retrieved. Said another way, the entropy of the universe constantly increases. "Entropy" is not an easy word to define, for its precise definition involves complex concepts of physics and mathematics. In simple English, it is often described with such words as "useless energy," "randomness," "nonpredictability," and "disorder." A state of high entropy is sometimes referred to as one that can be maintained with little or no input of energy or matter. In a living system, death is the state of highest entropy because dead systems require neither energy nor matter to remain dead. Living systems are highly organized and ordered, so they are in a state of relatively low entropy. And they require a continuous input of matter and energy to stay alive.

Entropy may be considered broadly as a measure of the amount of disorder or randomness in a system. Any system not supplied with energy will tend toward a state of maximum entropy; in other words, its order or structure will tend to become increasingly random, and less and less useful energy will be available within the system to do work. Even in a closed system, which neither gains nor loses energy to the surroundings, the amount of useful energy will steadily decline even as the total amount of energy remains constant.

All systems isolated from energetic inputs will proceed to a state of maximum entropy. The energy of order will be lost, giving way to disorder. The ability of the system to do useful work will progressively decline. A rock is a thermodynamic system; so is an oak tree, and so is a man. The state of highest entropy is attained when the man or the tree dies; a state of lower entropy exists during the organized chemical life of either one.

If "system" is now defined to include all living things collectively, then the biosphere itself may be regarded as a thermodynamic system. Because it is made up of highly organized macromolecules that are organized, in turn, into larger and more complex cells and organisms, *the biosphere is a system of low entropy and high free energy.* Its maintenance requires the continuous input of energy from the surroundings. The only major source of useful energy for the biosphere is radiant energy from the sun. The green plants, through a process called "photosynthesis," convert radiant energy into chemical energy, which can be stored and transmitted from one organism to another. In the cells of plants and animals, the chemical energy is converted to useful forms of chemical energy in a process called "cell respiration." Some of this energy is used to synthesize more macromolecules and to organize them into the complex structures of living systems as organisms grow and reproduce.

The energy in the universe, then, may be seen as a whole in its flow from useful energy to entropy, and living systems can be seen as unique, temporary places of delay. Only in living systems does order not revert quickly to disorder; only in living systems is an attempt made at the conservation of energy in its useful state.

THE USE OF ENERGY BY EARLY LIVING SYSTEMS

Even though the first systems of life may have been small and relatively simple, they, too, were part of the universal cycling of matter, and they must have been able to efficiently convert energy from one useful form to another. In short, they had to have food in order to survive.

Finding food, or metabolic fuel, was probably not a difficult task for the earliest organisms. They existed in a primitive ocean (or "organic soup") rich in macromolecules and other complex chemicals that had appeared as a direct consequence of the prevailing physical and chemical conditions. Because food was abundant and because natural selection is a process that works with the existing environment, rather than designing for the future, the most likely adaptation a living creature could make would be to absorb and use that food as a source of matter and energy.

The Early Heterotrophs

Extracting matter and energy from more-or-less complex organic molecules is called *heterotrophic nutrition*. The primary difference between primitive "heterotrophs" and modern ones, such as ourselves, is merely the level of complexity of the source of these organic molecules. We basically depend on a *biotic* supply of nourishment—food that was made by and formed a part of another living system such as a cow or a green pea. The primitive heterotrophs, however, being the only living systems then in existence, presumably used the one food source available: the organic molecules of the primitive soup.

When life began there were few creatures and there could not have been much of a competitive scramble for food. Although there might have been great variation in metabolic pathways, the biochemistry of the early inhabitants of the earth need not have been complex. They might have required only the building blocks that were available, the mechanisms necessary to transfer external food to their internal environment, and the enzymatic reactions necessary to use the building blocks for reproduction. The operation of simple metabolism, however, implies a mechanism for harvesting and capturing energy from chemical reactions that *liberate* energy (*exergonic* reactions) and using it for reactions of biosynthesis that require the *intake* of energy (*endergonic* reactions).

Figure 3.2 Diagram illustrating the principle of entropy. Two metal blocks are placed in an insulated enclosure that prevents heat from entering or leaving the system of the two blocks. Initially, the block on the left is hot, and the one on the right is cool. After a time, however, the heat flows into the cool block, and the temperature becomes uniform throughout the two blocks, an example of the increasing entropy of the system. At this point, it is impossible for the system to return to its original state without the expenditure of some energy from an outside source.

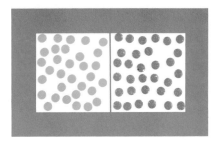

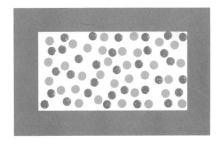

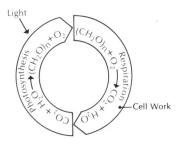

Figure 3.3 Energy cycle of the biosphere. Photosynthetic organisms utilize light energy from the sun to synthesize large carbohydrate molecules from simpler compounds. Oxygen is produced as an end product. Respiring organisms degrade the large carbohydrate molecules synthesized by plants, utilizing the energy obtained to sustain life functions and produce water and carbon dioxide as end products.

Figure 3.4   Heterotrophic organisms can only partially satisfy their nutritional needs by making use of certain raw materials in the physical environment in which they live (air, soil, or water). Additional needed organic molecules must be secured from the biological environment (other organisms). Two ways in which heterotrophs take in high-energy molecules from the biological environment are *holotrophism* and *saprotrophism*. Holotrophs swallow other organisms; this means of nutrition is characteristic of animals. Saprotrophs find high-energy molecules in dead or decaying organisms and absorb them. In this diagram, the arrows indicate the directions in which various nutrients are flowing.

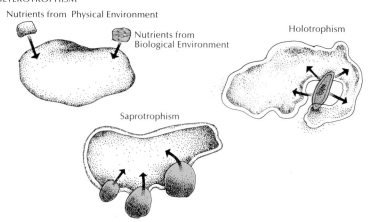

HETEROTROPHISM

Nutrients from Physical Environment

Nutrients from Biological Environment

Holotrophism

Saprotrophism

Without this mechanism they could not have reproduced or even lived long without consuming themselves.

One of the most versatile and universal units of metabolic chemical energy in all living systems is a molecule called *adenosine triphosphate*, which is often written simply as "ATP." This molecule is composed of a unit of adenosine with three repeating phosphate "groups" attached by energy-rich chemical bonds. When ATP reacts with water, the bond between the terminal phosphate group and the phosphate group to which it is attached is broken and the terminal phosphate is released along with a quantity of energy. What remains is a molecule called *adenosine diphosphate* ("ADP") and inorganic phosphate. This reaction may be depicted in the following way:

$$ATP + H_2O \xrightarrow[\substack{\text{with specific} \\ \text{enzyme}}]{} ADP + \text{inorganic phosphate} + \text{energy}$$

This reaction proceeds readily in the presence of the proper enzyme or acids, and heat energy is liberated. In living systems, the reaction is coupled with an energy-requiring process so that work can be done. The energy "currency" of life is ATP. Its value is measured in kilocalorie "coins" of chemical energy per mole of ATP. The amount of energy released is appropriate for most metabolic, muscular, and other biological processes. In fact, the use of larger packages of energy would not only be wasteful, it would cause cellular damage because the cells would not be able to withstand the associated temperature changes.

Molecules of ATP may have been present in the organic soup and could have been used by the primitive organisms. But the supply of randomly created ATP could not have supported life for long. Early in the history of life, organisms must have acquired some mechanism that would convert the chemical bond energy of

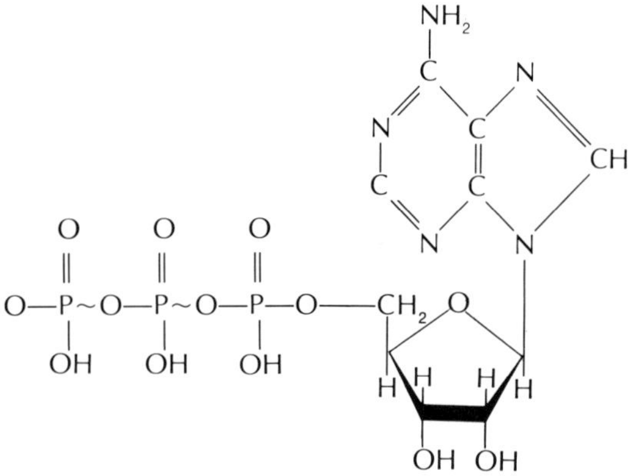

Figure 3.5   The structural formula of a molecule of adenosine triphosphate (ATP). Each molecule is composed of a unit of adenosine connected to three repeating phosphate groups by energy-rich chemical bonds.

other organic molecules into the common coinage of ATP, which could then be used throughout a living system.

In present-day living organisms, ATP is formed as the result of three major metabolic processes. First of all, in organisms that live in oxygen-free environments, glucose is partially broken down in a process called *substrate-level phosphorylation*, or fermentation. Small amounts of ATP are formed, which are sufficient to support life. Secondly, in creatures able to carry out photosynthesis, light energy is used to convert ADP to ATP by the addition of a phosphate group. This process is called *photosynthetic phosphorylation*. Thirdly, in plant and animal organisms that live and thrive in an oxygen-containing environment, substrate-level phosphorylation is linked to *oxidative phosphorylation*. In this process, oxygen is used and large amounts of ATP are produced.

In the earliest heterotrophs, as well as in almost all present-day organisms, substrate-level phosphorylation, or fermentation, was used to transfer energy. Fermentation in yeast cells and bacteria yields two molecules of ethyl alcohol and two molecules of carbon dioxide from a molecule of the sugar glucose. In other cells, particularly muscle cells, fermentation yields two lactic acid molecules from a glucose molecule. In both cases, the net energy released is used to convert two molecules of ADP to two molecules of ATP, which can then be used to donate energy to other reactions in living systems that require energy. A more general term for fermentation is "glycolysis," which will be discussed later in the chapter. The overall reactions and energy relationships described above may be written as follows:

Overall Reactions:

$$C_6H_{12}O_6 \xrightarrow[\text{with specific enzymes}]{\text{in yeast cells}} 2C_2H_5OH + 2CO_2 + \text{energy}$$
glucose   ethyl   carbon
                 alcohol   dioxide

or

$$C_6H_{12}O_6 \xrightarrow[\text{with specific enzymes}]{\text{in muscle cells}} 2C_3H_6O_3 + \text{energy}$$
glucose   lactic
                 acid

Energy Relationships:

$$2ADP + 2P_i + \text{energy} \xrightarrow{\text{with specific enzyme}} 2ATP + 2H_2O$$
adenosine   inorganic   adenosine   water
diphosphate phosphate   triphosphate

Far less energy is obtained from these reactions than could be obtained through the complete breakdown of the glucose to carbon dioxide. Most of the chemical energy of the glucose remains stored in its end products, alcohol or lactic acid. In most present-day organisms, these fermentation reactions are used only when

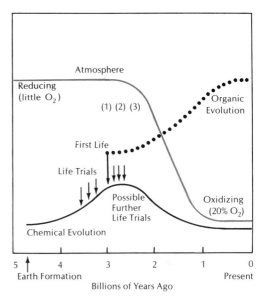

Figure 3.6   Graph depicting the probable origin of life and course of evolutionary events in relation to the history of the earth. 1 = substrate-level phosphorylation, 2 = photosynthetic phosphorylation, and 3 = oxidative phosphorylation.

there is a shortage of oxygen, which they need for respiration. A few bacteria and yeasts make exclusive use of such fermentations to survive in oxygen-free environments.

Glycolysis may have been important to early organisms, which developed in and had to survive in the absence of oxygen. The reactions used by present-day organisms involve complex systems with enzymatically controlled intermediate steps and are probably not the same in detail as those used by the earliest organisms. For example, a number of separate steps, each catalyzed by a specific enzyme, are now required to carry out the process of glycolysis. Primitive organisms must have used simpler and less efficient methods of trapping and converting some of the chemical energy of the organic soup into ATP.

Eventually the supply of chemical energy from randomly produced molecules was depleted by heterotrophic nutrition and its availability for reactions was reduced.

At first most of the ultraviolet light in solar radiation reached the earth's surface, where it could provide energy for organic synthe-

Figure 3.7   During intense muscular activity, animals often supplement the energy from aerobic (oxygen-requiring) respiration by another pathway, anaerobic (oxygen-free) glycolysis. This process results in the accumulation of toxic lactic acid in the muscles. By using glycolysis during exertion, a muscle acquires an oxygen debt; eventually all the lactic acid formed under anaerobic conditions must be reoxidized. During the resting period, faster breathing occurs, which "repays" the oxygen debt and causes the complete oxidation of the lactic acid. In mammals, the lactic acid is carried to the liver, returned to the muscle, and transformed into glycogen.

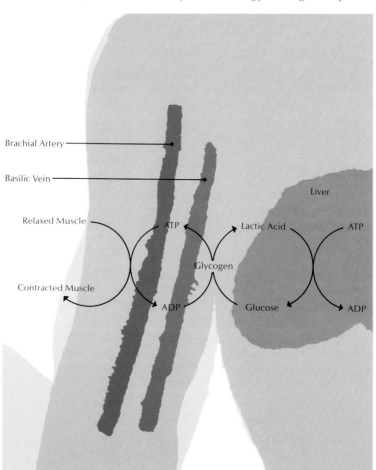

ses. However, the light molecules of hydrogen gas must have gradually escaped from the earth's gravitational pull. Small amounts of oxygen could have been formed in the atmosphere as the ultraviolet light broke down water vapor molecules. (Only as long as enough free hydrogen was available would the water molecules re-form by the combination of hydrogen and oxygen.) The bulk of oxygen, however, was produced by photosynthesis much later in time. The resultant layer of oxygen and ozone would absorb some of the ultraviolet light high in the atmosphere, and the rate at which organic molecules were synthesized in the organic soup would decrease accordingly.

This barrier to ultraviolet radiation would have some advantages for living systems. But ultraviolet radiation can cause damage to nucleic acids and thus produce mutations; in fact, the existence of modern land organisms is made possible only by the shielding effects of the oxygen and ozone in the atmosphere. As long as unshielded ultraviolet radiation reached the surface, even the simplest organisms would have been forced to remain several meters below the surface of the primitive ocean.

As traces of free oxygen began to appear in the atmosphere, drastic changes must have occurred on the earth. The rate of spontaneous synthesis of organic molecules must have decreased still further, and a strong selective advantage would have accrued to those organisms capable of using the available food with greater efficiency. Furthermore, at that time oxygen must have acted as a poison to many organisms, even in minute doses. Even today cells exist that can be killed quickly by a few parts per million of oxygen dissolved in their fluid medium. At the time when free oxygen began to accumulate, the existing cells either had to exist in deeper levels of the primitive ocean or they had to develop new modes of metabolism that could cope with oxygen. Only those organisms having sufficient genetic variability to permit either of these courses of action would survive; the others would become extinct.

The Appearance of Autotrophs

Because of the dwindling food reserves in the organic soup, a radically new and different form of nutrition became advantageous. Little by little, organisms called "autotrophs" appeared. These organisms had somehow acquired the means to *make* their own food out of simpler precursors and they had acquired the ability to cope with oxygen, qualities that enabled them to utilize a process called *autotrophic nutrition*. The early heterotrophs remained in subsurface and other oxygen-free regions, but the autotrophs were free to proliferate in the changing environment.

The crux of autotrophic nutrition lies in the *source* of energy. Heterotrophs (including ourselves) can only use energy that is already in the form of chemical bonds; in other words, they require

Figure 3.8 In contrast to heterotrophic organisms, which derive nutrients from both the physical environment and the biological environment, autotrophs obtain all raw materials from the physical environment alone. These low-energy materials may be transformed into high-energy organic molecules by one of two processes, each employing a different source of energy. In *photosynthesis*, light energy from the sun is used; in *chemosynthesis*, which is carried out only in certain bacteria, chemical bond energy is derived from certain environmental chemicals.

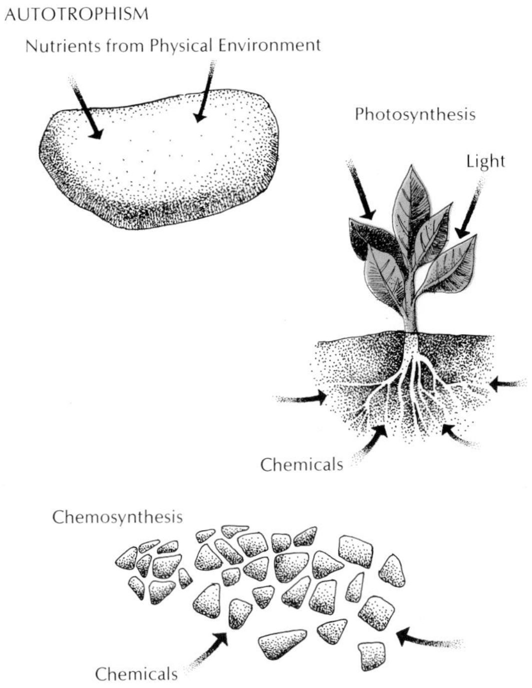

AUTOTROPHISM

Nutrients from Physical Environment

Photosynthesis

Light

Chemicals

Chemosynthesis

Chemicals

chemical energy in bonds of organic molecules at the beginning of their energy cycle. Most autotrophs, however, can successfully use light energy and convert it into chemical bond energy. For example, we as heterotrophs must eat energy-rich foods such as polysaccharides (starches) and lipids (fats). We derive energy by breaking the bonds holding such molecules together, and we incorporate this energy into ATP, which we then use for growth, movement, reproduction, and other vital functions. As part of the breakdown process, we excrete end products of low energy, such as carbon dioxide and water. Autotrophs convert these low-energy molecules in the environment to higher energy molecules by the direct use of solar energy. For example, an autotrophic green plant can convert the heterotrophic end products—carbon dioxide and water —into sugars and other high-energy molecules, which can provide the ATP needed to support life processes.

The development of autotrophic nutrition led to increasing numbers of autotrophs and their molecular by-products (glucose and oxygen), which had two major effects on the living community. First, the original heterotrophs now had a new source of organic food molecules to replace the abiotic molecules of the organic soup. In effect, their continued life and future evolution was assured as long as they could avoid the toxic effects of oxygen. Second, there arose a selective advantage in being able to live with oxygen without having to expend needed energy to burrow into mud or rocks, or to remain beneath the seas.

In the competition for survival, organisms arose that not only were unharmed by oxygen but actually required it. A threefold interlocking community developed. The autotrophs used the end products of the heterotrophs—principally carbon dioxide. All the heterotrophs used the products of autotrophic nutrition—principally organic molecules such as glucose. Those heterotrophs requiring oxygen-free (*anaerobic*) conditions were not crowded out; and oxygen-requiring (*aerobic*) heterotrophs stabilized the levels of oxygen in the atmosphere by using it as rapidly as the autotrophs produced it. Each part of this whole depended on and contributed toward the survival of the other parts although perhaps, in their role as solar energy transformers, the autotrophs can be considered today to be the most essential component of the community.

PHOTOSYNTHESIS AND LIFE ON EARTH

Before the advent of photosynthesis, the primitive heterotrophs were subjected to the combined pressures of a dwindling food supply and an increasing level of poisonous, atmospheric oxygen. The evolution of photosynthesis as a complex yet efficient pathway of energy transformation brought with it a new era for the earth and all its living inhabitants.

Photosynthetic organisms produce molecular oxygen (as an end product) and consume carbon dioxide. As the number of auto-

Figure 3.9 All green plants except algae and bryophytes possess specialized conducting, or vascular, tissues. Vascular plants have two distinct types of vascular tissue. *Xylem* carries water and mineral nutrients from roots to photosynthetic cells; and *phloem* carries organic products of photosynthesis from photosynthetic cells to the rest of the plant body. The development of xylem and phloem has permitted vascular plants to live in many habitats on land, with roots deep into the soil to trap moisture and leaves high into the air to capture sunlight. These ferns, which are growing in a temperate rain forest habitat, are a type of vascular plant limited to a relatively moist and shady land environment.

trophic organisms increased, the levels of carbon dioxide and oxygen shifted dramatically. Carbon dioxide, once a major constituent of the atmosphere, became a minor part of it; and free oxygen, which had previously been nonexistent, became increasingly important as a selective force in changing the rate and kind of future evolution.

The early earth had been bombarded with cosmic and ultraviolet radiation. Although these forms of energy presumably had been the nurturing forces of life, they also had the destructive effect of causing molecules to break apart. In the days before life, the positive effects of great amounts of radiation outweighed the negative. But once life arose and living creatures began to multiply, their annihilation by radiation probably would have established an equilibrium that would have maintained life at a primitive and precarious level.

On earth today we are partially protected from immense showers of radioactivity by a layer of ozone in the upper atmosphere, as described earlier. The rates of mutation and evolution are thereby lowered but so, too, is the loss of organisms that would ensue from massive energy input to the earth's surface. The oxygen that is converted to ozone is derived from the process of photosynthesis and, in a broader sense, so is the greater stability of the terrestrial environment.

As a metabolic process, photosynthetic phosphorylation (photosynthesis) involves two series of reactions—the *light reactions* (which require light to proceed) and the *dark reactions* (which do not require light). First, in the light reactions, specific wavelengths of light energy are captured and converted into chemical energy, which is stored in chemical bonds in ATP molecules and in a molecular reducing agent that will be described below. Second, in the dark reactions, the energy in ATP and in the reducing agent is used to convert carbon dioxide and water into complex organic molecules. One of these molecules, the sugar glucose, can be polymerized into complex starch and cellulose molecules or converted into other kinds of organic molecular building blocks. Both series of reactions require a specific array of enzymes, pigments, and other vital elements and compounds for the conversions.

The simple sugar glucose is typical of the organic molecules that photosynthesis produces. The overall generalized reaction necessary to synthesize glucose may be written as follows:

$$6CO_2 + 12H_2O \longrightarrow C_6H_{12}O_6 + 6O_2 + 6H_2O$$

carbon dioxide   water   1) specific wavelengths of light energy in photon packets   glucose   oxygen (from water on left)   water

2) specific enzymes

3) photosynthetic pigments

Before this reaction can proceed as shown, energy must be supplied. In photosynthesis, the initial energy is supplied by *photons*,

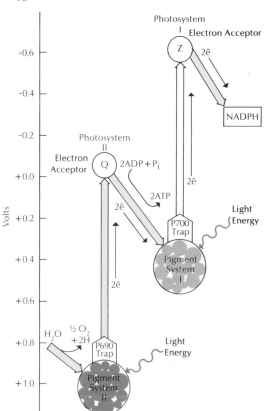

Figure 3.10 The two-step relationships of the light reactions in photosynthesis. These reactions occur in the grana of chloroplasts, which are located in the cells of eucaryotic plantlike protistans and multicellular plants. Specific wavelengths of light energy in photon packets activate electrons in green pigments such as chlorophylls *a* and *b*. This light energy is transferred to chemical bond energy in the compounds ATP and NADPH. These compounds can be used as high-energy sources in the dark reactions to synthesize complex energy-containing organic molecules such as glucose. As a by-product of the light reactions, molecular oxygen is liberated from water molecules.

Figure 3.11 The structural formula of the simple sugar glucose. Most of the organic molecules created through photosynthesis are sugars, of which glucose is an example. Plants can polymerize these sugars to form polysaccharides such as starch and cellulose.

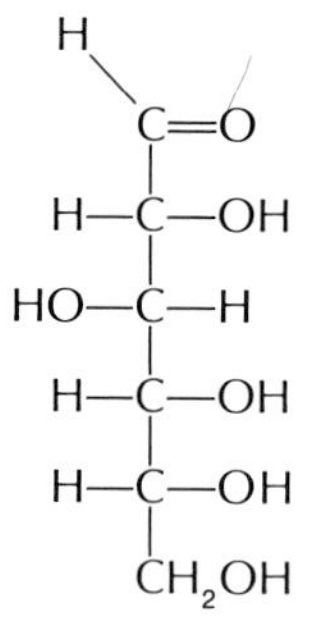

```
        H
         \
          C=O
          |
     H—C—OH
          |
   HO—C—H
          |
     H—C—OH
          |
     H—C—OH
          |
        CH₂OH
```

which are incredibly small packets of light energy of specific wavelengths. Part of this energy is used to reduce the carbon that enters the reaction in the form of carbon dioxide. *Reduction* may be considered as a gain of electrons, whereas *oxidation*, its opposite, signifies a loss of electrons. Both reactions are processes of electron transfer in which a certain amount of energy is absorbed or released as electrons are transferred from one molecule to another. Because hydrogen usually is the source or destination of electrons in living systems, the term "hydrogenation" is often used interchangeably with "reduction."

The fact that energy resides in electrons means that the molecule acquiring the electrons (and being reduced) is acquiring energy. The act of reduction is always accompanied by its opposite: oxidation of an electron donor molecule. A molecule being oxidized loses electrons and thereby loses energy. During photosynthesis, carbon in the form of carbon dioxide (CO_2) is reduced to the more energetic sugar glucose ($C_6H_{12}O_6$). In contrast, water (H_2O) loses electrons and hydrogen and is oxidized to oxygen (O_2). In other words, the oxygen in the sugar originates in CO_2, whereas molecular oxygen (O_2) is derived from water.

The Light Reactions

Four hydrogen atoms and eight photons are needed to process each molecule of carbon dioxide. Apparently each hydrogen atom is pushed "uphill" energetically in two steps, each representing a push by one photon. This energetic pushing, together with the trapping of the energized hydrogen, is the heart of the light reaction. Many details of the two-step process in this reaction remain obscure. For example, its precise physical nature is not understood. And how the two steps become coordinated so effectively that the overall photosynthetic process can proceed as efficiently as it does is still a puzzle. Furthermore, the details of the enzymatic splitting of water to release the hydrogen are not clear.

What *is* known is that each of the two energy pushes involves a separate set of reactions. These sets are known as Photosystem II and Photosystem I. (The Roman numerals I and II were assigned in order of the detailed study of the steps, not in order of their occurrence in the photosynthetic process.) Hydrogen atoms are removed from the water (leaving the oxygen in some unstable intermediate form) and pushed up to the higher energy level of Photosystem II (Figure 3.10). The hydrogen atoms then travel down a series of reactions in which part of the acquired energy is used to convert ADP and inorganic phosphate to ATP. Now the hydrogen atoms are given another push to an even higher energy level in Photosystem I. Finally, the hydrogen atoms pass into another reaction sequence, which results in the conversion of a molecule called "nicotinamide adenine dinucleotide phosphate" (mercifully abbreviated "NADP") to its reduced state "NADPH."

NADPH is the reducing agent mentioned earlier that is required to reduce the carbon dioxide.

The net result of the light reaction, then, is the transfer of hydrogen atoms from water molecules to the powerful biological reducing compound NADPH, with the production of another energy-rich substance, ATP. Oxygen is released as an end product of this activity. In the dark reaction, NADPH acts as a reducing agent donor for the conversion of carbon dioxide to carbohydrate, and ATP provides chemical energy for the conversion to glucose. The nature of the intermediate molecule from which the hydrogen atoms are removed at the beginning of the light-limited stage is not yet completely known, but the net result of the process is a flow of energy from sunlight to sugar by means of ATP and the reducing agent NADPH.

The reactions of the light phase are made possible by *pigments*, which are molecules that are able to absorb the energy of certain kinds of photons. When a pigment molecule absorbs a photon, one or more electrons in the molecule becomes excited and is promoted to a higher energy level. The electron can leave the pigment molecule and be accepted by another molecule. This action causes the pigment molecule to become oxidized, and it can now react with another reducing agent and return to its original state. In photosynthesis, water is the reducing agent and oxygen is liberated.

The evolution of the light reaction occurred with the evolution of pigment molecules that are able photochemically to trap sunlight. Primitive photosynthetic bacteria existing today have only

Figure 3.12   The absorption spectrum of chlorophyll compared with the light energy emitted from the sun. Note that the maximum energy comes through at about 4,500 angstroms in the blue (UV) region. Chlorophyll usually absorbs about 6,500 angstroms in the red (infrared) region.

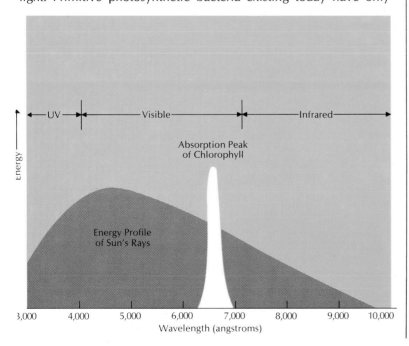

Figure 3.13 The structural formula of chlorophyll. In chlorophyll *a*, the pigment that is found in all plants, X = −CH$_3$; in chlorophyll *b*, another pigment found in green land plants and green algae, X = −CHO.

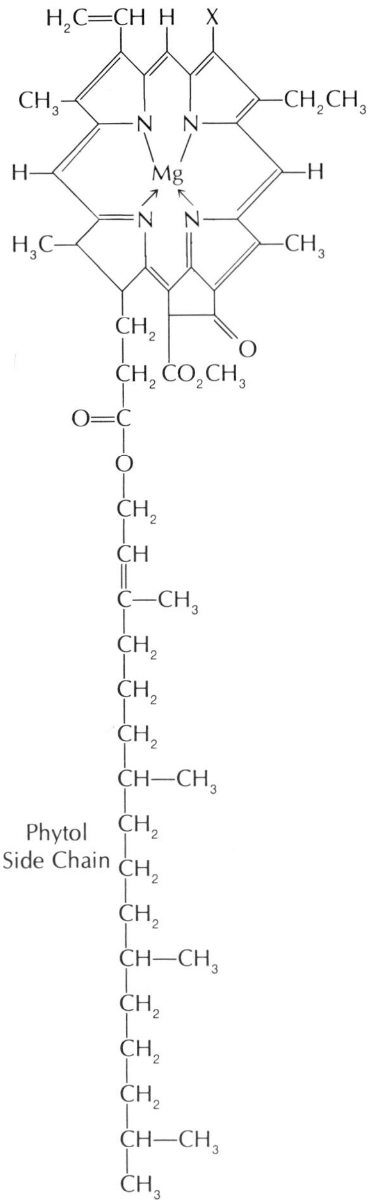

H$_2$C=CH H X

CH$_3$ CH$_2$CH$_3$

N   N

H   Mg   H

N   N

H$_3$C   CH$_3$

CH$_2$

CH$_2$ CO$_2$CH$_3$ O

O=C

O

CH$_2$

CH

C—CH$_3$

CH$_2$

CH$_2$

CH$_2$

CH—CH$_3$

CH$_2$

Phytol
Side Chain   CH$_2$

CH$_2$

CH—CH$_3$

CH$_2$

CH$_2$

CH$_2$

CH—CH$_3$

CH$_3$

that single pigment system and only exhibit the reactions of Photosystem I. In such a situation, light energy is trapped by the pigments of Photosystem I, and hydrogen obtained from something other than water is passed uphill to a receiver molecule. This molecule releases the energetic hydrogen, which then falls in a stepwise series of molecular reactions back to a low-energy level. In the process ATP is formed, but water is not split, NADPH is not formed, and oxygen is not released.

This mechanism served to generate ATP from sunlight, but it had one serious drawback. In order to proceed, the organism would have required surface sunlight; but it is likely that the energy on the surface of the earth was great enough to cause the spontaneous breakdown of ATP to ADP. For the process to have been effective, any ATP generated would have had to be used immediately. Because an organism's requirements for energy do not necessarily coincide with the time of maximum sunlight, mutations that permitted a more stable fixation of light energy would have had a selective advantage.

One mechanism for the more stable and efficient use of sunlight resulted from the evolution of *organized* systems of pigments. A pigment is a molecule that is able to reflect or transmit color or blackness because of its ability to capture photons of specific wavelengths. A black pigment can absorb photons of all wavelengths of visible light; it neither reflects nor transmits any of that light so it appears black. A colored pigment such as chlorophyll absorbs photons corresponding to the violet, blue, and red wavelengths of white light, but it reflects rather than absorbs the green. (One pigment, chlorophyll *a*, occurs in all plants and is present in slightly modified form in photosynthesizing bacteria.) Reflected light is not available to the pigment.

A few biological pigments have another important energy characteristic: Having absorbed light energy, they can transform it into chemical bond energy, which is then available to the organism. The evolution of a variety of pigments such as the yellowish carotenes or the red or blue pigments found in the less primitive photosynthesizers permitted the capture of more light because each pigment absorbs (and reflects) different wavelengths of sunlight. However, only chlorophyll *a* acts directly to transfer energy from sunlight to energy-rich compounds used in enzymatic reactions. The functions of the other pigments are believed to be "light traps" that pass on the energy to chlorophyll *a*.

Greater efficiency was eventually achieved through evolution, as photosynthetic pigments and enzymes became organized spatially within subcellular structures called "chloroplasts" (Chapter 4).

The Dark Reactions

Autotrophic nutrition probably evolved slowly as a result of a gradual decrease in high-energy food molecules in the primitive soup. Photosynthesis was the end product of this evolution, but several intermediate steps may have occurred in the past that resemble certain processes taking place on earth today.

For example, some bacteria use carbon dioxide as the molecular raw material from which they synthesize complex carbon-containing molecules. Such "fixation" of atmospheric carbon dioxide in organic molecules was probably followed by a similar ability to "fix" atmospheric nitrogen. Both abilities were selected for because of the depletion of available food substances. These metabolic pathways were probably followed by the ability to make certain molecular derivations of two kinds of molecules—isoprene and porphyrin—both of which could act as protective agents against oxygen. For instance, the process of transforming carbon dioxide into organic molecules is incorporated as the second half of photosynthesis—the dark reaction. It was the further evolution of isoprene and porphyrin derivatives that led to chlorophyll and other molecules necessary for the first half of photosynthesis—the light reactions.

The association of carbon dioxide fixation with the light reactions occurred by means of modifications in the light reactions in that NADPH and ATP could be fed into the dark series. As a result, an energy-rich molecule more stable than ATP could be made and stored for future use. Moreover, such a molecule could be *transported within* the body of the organism so that the parts not directly exposed to sunlight would also have an available source of energy. This molecule is glucose.

Several sequences of enzymatic or dark reactions are involved in the photosynthetic process. The sequence occurring between Photosystems II and I and that occurring after Photosystem I have been discussed as part of the light reactions. Other enzymatic reactions are involved in the formation of an intermediate molecule from the water molecule and the conversion of the intermediate to oxygen after the removal of the hydrogen atoms.

The dark reactions occur in three major stages. First, carbon dioxide is incorporated into a carbon dioxide acceptor. Second, this complex is reduced by means of the ATP and the NADPH produced during the light reactions. Finally, the reduced complex is transformed into the six-carbon sugar glucose.

Each of these stages consists of many enzymatically regulated reactions which, in total, constitute a metabolic cycle. To keep the cycle going, all that is needed is carbon dioxide, energy in the form of ATP, reducing power in the form of NADPH, and specific enzymes. By-products of the cycle are ADP, inorganic phosphate, and NADP (all of which return to the light reactions), and sugar. The cycle has been named the *Calvin cycle*, in honor of Melvin Calvin whose recent work was instrumental in clarifying the details of the dark reactions. Figure 3.16 shows how each trip around the cycle requires one molecule of carbon dioxide and produces 1/6 of a

Figure 3.14 Experimental evidence proving that liberated oxygen in photosynthesis originated from from the *photolysis*, or splitting, of water molecules. Originally it was assumed that the photosynthetic process involves the removal of oxygen from carbon dioxide, followed by hydration of the remaining carbon. According to this mechanism, the oxygen molecules are formed from oxygen atoms in the CO_2 molecules. It is possible to test this hypothesis by means of isotope tracer experiments involving isotopes of oxygen. Water can be prepared with radioactive oxygen-18 rather than normal oxygen-16 (red type represents radioactive oxygen). When this heavy water is used in photosynthesis, the liberated oxygen gas is entirely made up of oxygen-18. It appears that the oxygen gas is formed from oxygen atoms in the water. The mechanism of photosynthesis therefore must involve the photolysis of water, followed by the reduction, or hydrogenation, of carbon dioxide.

$$CO_2 + H_2O \longrightarrow O_2 + [CH_2O]_n$$

$$CO_2 + H_2O \longrightarrow O_2 + [CH_2O]_n$$

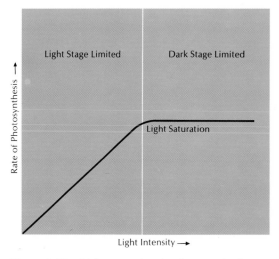

Figure 3.15 Light saturation in photosynthesis. In the presence of adequate amounts of CO_2 and H_2O, it would be expected that the *rate* of photosynthesis—measured by the rate of production of oxygen—would be proportional to the *intensity* of light available (the rate at which photons are being made available to carry out the reaction). Experiments show that this proportionality is valid only up to a certain point, beyond which the rate of photosynthesis remains constant regardless of further increases in light intensity.

Figure 3.16   The photosynthetic formation of glucose from carbon dioxide via the Calvin cycle. Inputs are shaded in brown; products are in yellow. Clockwise from the left, the abbreviations designate the following terms:

PGA = 3-phosphoglyceric acid
TP = glyceraldehyde 3-phosphate
DHAP = dihydroxyacetone phosphate
FDP = fructose 1,6-diphosphate
FMP = fructose 6-phosphate
GMP = glucose 6-phosphate
X5P = xylulose 5-phosphate
E4P = erythrose 4-phosphate
SDP = sedoheptulose 1,7-diphosphate
S7P = sedoheptulose 7-phosphate
R5P = ribose 5-phosphate
Ru5P = ribulose 5-phosphate
RuDP = ribulose 1,5-diphosphate

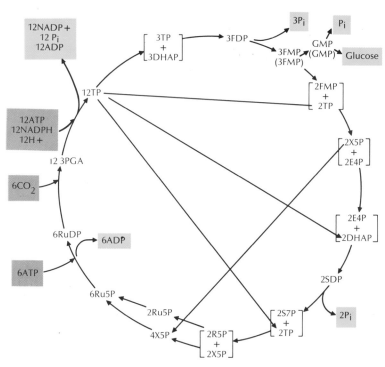

molecule of glucose. Six turns of the cycle produce one molecule of glucose. Many glucose units can be converted not only to starch; the products of the Calvin cycle can be converted also to oils, other sugars, amino acids, and all the other special molecules needed for life.

AEROBIC RESPIRATION

The pathways of energy extraction used by the earliest oxygen-independent, or anaerobic, organisms were probably like the "modern" biochemical sequences of glycolysis, or fermentation. Glucose is degraded and some of its energy is utilized by the production of ATP. However, the amount of ATP so produced is small; most of the energy of the sugar molecule remains trapped in the end products lactic acid or alcohol. More efficient extraction of energy must come through a further breakdown of these substances and the liberation of the energy they contain. The ultimate end products of the complete breakdown of glucose are carbon dioxide, water, and large amounts of energy.

Autotrophs construct sugar from carbon dioxide and water by photosynthesis, and from this substance they derive practically everything else they need for life. When a heterotrophic animal eats a photosynthetic plant, sugar and its polymers, and plantmade proteins and lipids enter the metabolic machinery of the animal's cells. However, animal cells cannot use these macromolecules directly to build more animal. Some of the plant substances must be degraded, and the energy that the photosynthetic process incorporated into these substances must be extracted. Many usable small molecules (monomers) must be salvaged, and complex molecules must be rebuilt that are characteristically "animal." Similarly, plant cells must also *use* the energy in some of the glucose that they make or store in order to do the work of growth and reproduction. Therefore, the breakdown of glucose—the stable energy store—to carbon dioxide and water is characteristic of autotrophs as well as of heterotrophs.

Aerobic respiration, then, is a series of processes in which one molecule of glucose is broken down to carbon dioxide and water. It is accompanied by the synthesis of thirty-eight molecules of ATP and the consumption of oxygen. Why go through the bother of converting the energy of ATP into the energy of glucose during photosynthesis if the process now goes "backward" from glucose to ATP? Part of the answer may be found by comparing the storage potential of these molecules. ATP is an extremely reactive substance. In short order it will lose its terminal phosphate and revert to ADP. Glucose, on the other hand, is an extremely stable molecule. It can be dissolved in blood or cytoplasm without breaking down. It is therefore an ideal substance for transferring energy from one organism to another, or, in its polymerized forms of starch or glycogen, for storing energy within an organism for long periods of time.

As a result of the evolution of cellular oxidative respiration, chemical energy in the form of glucose was converted to ATP in

Figure 3.17 The Krebs citric acid cycle and the electron transport chain. Major intermediate compounds, each requiring a specific enzyme to continue the reaction series, are shown in the Krebs cycle. Note the function of NAD in transporting high-energy electrons (e^-) to the respiratory chain, where they are passed along to various cofactors, ultimately reducing oxygen and forming metabolic water.

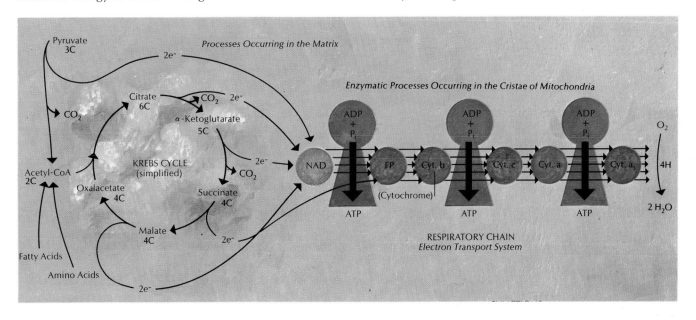

an amount much greater than that produced by fermentation alone. A second result was the liberation of carbon dioxide and water as end products of the process. These end products were cycled back to photosynthetic autotrophs, to be used again as vehicles of energy flow. Finally, in terms of the evolution of organisms, cellular respiration provided a use for oxygen. After the evolution of this process, oxygen was no longer a poison for most heterotrophs. It became a vital substance—the final oxidizing agent—that was required for life.

The process evolved as a pathway of enzymatically controlled chemical reactions that liberated the energy of glucose in small packages of ATP rather than in one explosive burst, as would occur if a teaspoonful of glucose were ignited. This process was more efficient because the loss of energy as heat was minimized. Actually, a certain amount of heat produced is a necessary end product because enzymes work more rapidly at higher temperatures, as long as those temperatures are not high enough to damage cells. The evolution of oxidative respiration not only provided an efficient means of bundling energy, it also produced a small amount of necessary heat. In all probability the evolution of aerobic respiration provided the impetus for the flourishing and diversification of living things. With a far greater supply of potential energy and a stabilized solution to the oxygen problem, the two great challenges of the early earth were dealt with.

The simple sugar glucose is typical of the six-carbon organic sugars broken down in aerobic respiration, which occurs in most advanced heterotrophs and green plants. The overall reaction depicting aerobic respiration has two parts—substrate-level phosphorylation (glycolysis) and oxidative phosphorylation. The overall reaction including both parts necessary to completely break down glucose may be written as follows:

Overall Reaction:

$$C_6H_{12}O_6 + 6O_2 \xrightarrow{\text{with specific enzymes}} 6CO_2 + 6H_2O + \text{energy}$$

glucose   oxygen   carbon   water
                                dioxide

Energy Relationships:

$$38ADP + 38P_i + \text{energy} \xrightarrow{\text{with specific enzymes in cells}} 38ATP + 38H_2O$$

adenosine   inorganic   adenosine   water
diphosphate   phosphate   triphosphate

Glycolysis

The glycolytic pathway begins with the input of simple six-carbon sugars such as glucose. It ends with the net production of two molecules of a three-carbon acid (pyruvic acid) for each six-carbon sugar and two molecules of reduced nicotinamide adenine dinucleotide (NADH). What has happened? The six-carbon unit has been partially dismantled in a series of reactions, linked together by enzymes that use the product of one reaction as the beginning molecule (substrate) for the next. In energy terms, however, the dismantling has resulted in the liberation of a small part of the energy originally found in glucose and in its conversion to ATP and the reducing agent NADH.

If glucose is stable enough to be the means of energy transfer between organisms, what must be done during glycolysis to cause the splitting and partial release of energy? In order to make glucose unstable enough to split, more energy must be added initially to the molecule.

Glucose is twice energized by acquiring one high-energy phosphate group from ATP at each end. The molecule is then ready to be cleaved into two three-carbon fragments and the energy utilized. However, the biological advantages of glycolysis are that glucose is cleaved under mild conditions and the energy is conserved by neatly controlled enzymes. The energy is trapped and used to convert two molecules of ADP into two molecules of ATP. One further "energy event" follows the splitting of glucose. Two hydrogen atoms are taken from each three-carbon fragment. Because the addition of hydrogen during photosynthesis is a process of energy addition, its removal (oxidation) represents a release of energy from the glucose fragments. As these fragments are oxidized, the hydrogen atoms are transferred to the nucleotide molecule NAD (similar to the NADP encountered in photosynthesis) and NAD is reduced to a more energetic form, $NADH_2$.

The glucose fragments are then further converted into two more stable small molecules of pyruvic acid, and once again two molecules of ATP are generated by substrate-level phosphorylation; in other words, by the controlled release of energy present in the molecules undergoing transformation.

The energy balance for each molecule of glucose entering the pathway and transformed into two three-carbon molecules of pyruvic acid is:

2 molecules ATP used to energize glucose
4 molecules ATP gained by substrate-level phosphorylation
Net: 2 molecules ATP + energy of 2 molecules of $NADH_2$

Oxidative Phosphorylation

The evolution of oxidative phosphorylation was the key to aerobic respiration. The three-carbon pyruvic acid molecule that is the end product of glycolysis undergoes further enzymatic oxidation. In this process, hydrogen atoms, electrons, and one carbon atom in the form of carbon dioxide are removed. Acetic acid, the two-carbon fragment that is formed, combines with a complex nucleotide called coenzyme A (CoA) and is ready to enter the first of two metabolic pathways as the complex acetyl-coenzyme A (acetyl-

Figure 3.18 Schematic diagram illustrating the sequential action of the glycolytic chain of enzymes. The product of one reaction becomes the substrate of the next enzyme. Thus, the chain of enzymes can be considered a type of disassembly line that takes in glucose at one end and turns out lactic acid in muscle cells when molecular oxygen is not present. (Refer to the key for specific enzymes and intermediate compounds in the series.)

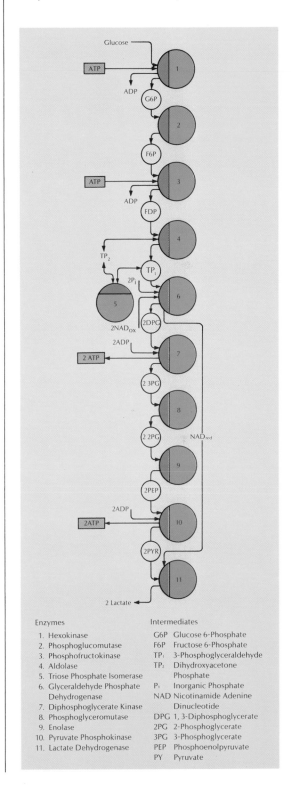

Enzymes
1. Hexokinase
2. Phosphoglucomutase
3. Phosphofructokinase
4. Aldolase
5. Triose Phosphate Isomerase
6. Glyceraldehyde Phosphate Dehydrogenase
7. Diphosphoglycerate Kinase
8. Phosphoglyceromutase
9. Enolase
10. Pyruvate Phosphokinase
11. Lactate Dehydrogenase

Intermediates
G6P Glucose 6-Phosphate
F6P Fructose 6-Phosphate
TP₁ 3-Phosphoglyceraldehyde
TP₂ Dihydroxyacetone Phosphate
Pᵢ Inorganic Phosphate
NAD Nicotinamide Adenine Dinucleotide
DPG 1, 3-Diphosphoglycerate
2PG 2-Phosphoglycerate
3PG 3-Phosphoglycerate
PEP Phosphoenolpyruvate
PY Pyruvate

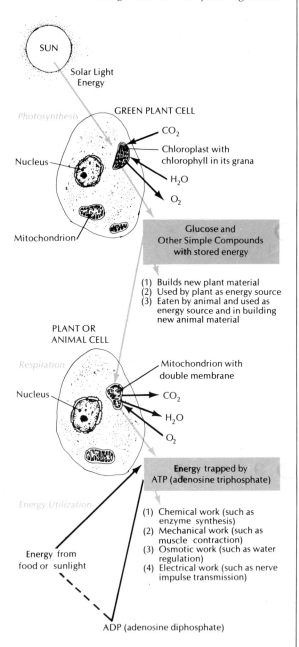

Figure 3.19   Diagram showing energy flow through cells of eucaryotic organisms.

CoA). The first pathway is a cycle of enzymatic reactions that receives the two-carbon acetyl fragment from the coenzyme A nucleotide. This cycle removes hydrogen atoms and electrons from the acetyl fragment and causes its oxidation to two molecules of carbon dioxide.

The cycle consists of ten enzymatically controlled reactions, as shown in Figure 3.17. It is called the *Krebs citric acid cycle* after Hans Krebs, who did much work leading to its discovery. In this cycle, pyruvic acid is systematically dismantled. The carbon and oxygen remnants are released as the end product carbon dioxide. One molecule of ATP is produced directly, and the hydrogen atoms are used to reduce NAD to $NADH_2$ or to reduce the related nucleotide molecule flavin adenine dinucleotide (FAD) to $FADH_2$.

The Krebs citric acid cycle does more than dismantle pyruvic acid. In effect, it acts as a final common pathway of much of metabolism because breakdown products of proteins and lipids (as well as sugar) can enter at appropriate stages and small building blocks for proteins can be formed. Although the reactions of the cycle do not *use* free oxygen, they will not function unless oxygen is available because the next stage of respiration *is* oxygen-dependent and it controls the citric acid cycle.

During the Krebs citric acid cycle, energy has traveled from the chemical bonds of pyruvic acid to the chemical bond holding hydrogen to $NADH_2$ (or $FADH_2$). The molecules of $NADH_2$ produced during glycolysis are not added to those produced in the Krebs citric acid cycle, and the total supply of energy-rich molecules moves into a complex series of oxidation-reduction reactions known as the electron transport chain.

Electron Transport Chain

In this reaction, the electrons used to reduce molecules in the preceding two stages of respiration are now passed along a series of molecules called "oxidation-reduction carriers." In many ways the series resembles the cascade of water in descending levels of waterfalls: as the water descends it loses energy at each level. The high-energy electrons are passed down the series of electron carriers, and during the passage some of the energy is captured by incorporating it into ATP; the rest is lost as heat.

There are five way-stations along the chain. At the first station the electrons are picked up by specific molecules, which thereby become reduced. The electrons then continue their journey as each reduced molecule returns to its oxidized state. The same process is repeated four more times. Between three of the stations the energy liberated by the electrons in their downhill journey is used to join ADP and inorganic phosphate to make ATP. Finally, after the electrons have lost most of their potential energy, they are picked up by a final electron acceptor—oxygen. This substance—the oxygen that we breathe and that was so injurious to the early heterotrophs—accepts two hydrogen atoms and two electrons to produce water, a final end product of respiration.

For every molecule of glucose that is completely oxidized during respiration, thirty-eight molecules of ATP are produced. Two of these are derived from glycolysis, and the rest, from the Krebs citric acid cycle and the electron transport chain. The difference in energy-extracting ability before and after the evolution of aerobic respiration was therefore enormous, and only with this more complete use of photosynthetic products was the more complete evolution of life assured.

The highly organized intricacy of the process can perhaps be appreciated by comparing the living system to a gasoline engine. The complete burning, or oxidation, of a mole of glucose (180 grams) releases 686 kilocalories as heat energy. The 38 moles of ATP produced from 1 mole of glucose store about 266 kilocalories. The minimum efficiency of aerobic respiration is 266/686, or about 42 percent; the remaining energy is lost as heat. The engineer designing a gasoline engine is extremely pleased if his creation works at an efficiency of 25 percent. Fully 75 percent of the chemical energy of gasoline is lost as heat and is *not* transformed into the useful work of the engine.

The rate at which the Krebs cycle reactions occur appears to be limited chiefly by the ratio of ATP to ADP in the organism. When a great deal of work is being done, ATP is converted to ADP and the ADP concentration rises. With an abundance of ADP available, the reactions of the Krebs cycle proceed rapidly, turning the ADP back into ATP. If little work is being done, the ATP concentration rises and the ADP concentration decreases, so the reactions cannot proceed very rapidly. This feedback mechanism regulates the rate of glucose oxidation to match the amount of work being done.

The precise stages in the evolution of aerobic respiration are not known. It appears that the first reactions involved iron bound to a variety of organic compounds called chelates. One of the most important of these iron-binding compounds is the porphyrin ring.

The porphyrin ring forms the basis of some of the oxidation-reduction molecules of the electron transport chain. It also binds oxygen in hemoglobin (the oxygen transporter in red blood cells) and myoglobin (the pigment in muscles that serves as an oxygen store); and is the active group in chlorophyll, where magnesium rather than iron is present. Because chlorophyll and the oxidation-reduction molecules both are crucial mediators in the capture and use of living energy, the evolution of the porphyrin ring to handle energy was a crucial event of biochemical evolution.

THE ROAD TO CELLULAR DEVELOPMENT

If any "purpose" exists in the evolutionary development of photosynthesis and aerobic respiration, it is the retrieval of energy and small molecules in order to make larger ones. If an organism is to

Figure 3.20 The photograph above depicts the common blue-green alga *Anabaena*, noted for its "chain of beads" aggregation. Shown below are large masses of blue-green algae that often form scumlike layers on the surfaces of polluted lakes and ponds. All species of blue-green algae are photosynthetic, and all contain the photosynthetic pigments chlorophyll *a* and phycocyanin (a blue pigment). The pigments are arranged on infoldings of the cytoplasm (the living matter inside the cell).

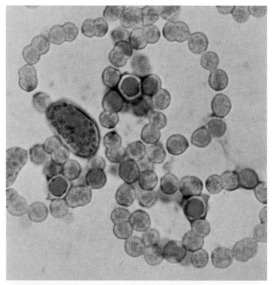

remain alive, the flow of useful energy must be maintained. However, organisms do more than sit about, occupying living space. The essential characteristic associated with life is multifaceted dynamism. To move, to communicate, to react, and, most of all, to grow and reproduce—all of this is to be alive.

The dynamism underlying life is complex and encompasses the two stages of photosynthesis, the three stages of respiration, and the dependency of each stage on the others. The light and dark reactions are coupled molecularly, for each reaction needs the products of the other. Glycolysis, the Krebs citric acid cycle, and the electron transport chain are similarly coupled. The overall processes of photosynthesis and respiration are coupled by oxygen, carbon dioxide, water, and sugar. Only by such intricate recyclings of molecules within an open steady state is the progress toward total entropy in the biosphere held off for a while.

In order to cycle matter simultaneously and to delay the conversion of useful energy to its useless state, structure and function within living creatures had to be closely intertwined. Many of the reactions of photosynthesis and respiration are linear, stepwise pathways. Others are cyclic, but they still are pathways because each reaction requires the product or products of its predecessor.

The way in which living creatures impose structural order on biochemical order has been through the evolution of membranes and granules of greater and greater complexity in which enzymes can be organized in definite arrays, as in assembly lines of production in a factory. The evolution of the first living cells undoubtedly included the development of a surface membrane to form a boundary between the internal world of the cell and the outside environment. As long as abiotic supplies of energetic molecules were available, there may have been no strong selective pressures toward the evolution of complex internal cellular membranes.

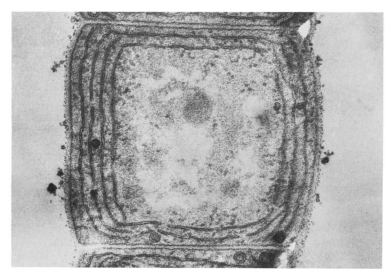

Figure 3.21 In this electron micrograph of the blue-green alga *Anabaena*, the outer cell wall and the layers of photosynthetic membranes are clearly visible. (x 6,800)

Figure 3.22   Diagram of the general structure of a blue-green alga. Note that the nuclear material, composed of DNA fibrils, is localized in a nuclear region or "nucleoid" but remains unbounded by a nuclear membrane. The cell wall is of cellulose; in many species, the outer portion of the cell wall becomes covered with a slimy substance, which apparently protects the cell from dehydration and facilitates intercellular interaction.

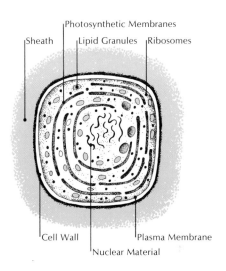

Most reactions could have proceeded in the absence of a great deal of internal structure because much of the work of biosynthesis had been done abiotically. Bacteria, present-day representatives of those early forms of life, possess only a small variety of internal membranes and granules, nucleic acids, proteins, lipids, and small organic molecules. On these structures and in the parts of a bacterial cell, the reactions of life occur.

The development of so complex a process as photosynthesis required a more elaborate structure as its framework. The cells of blue-green algae can perhaps provide a clue as to structural evolution because these organisms are believed to be similar to forms that were ancestral to higher organisms. The cells of the blue-green alga *Anabaena* (Figure 3.21) illustrate the principle of increasing organization through membranes. Membranes to house and organize the metabolic machinery of photosynthesis are present in an orderly, albeit diffuse, array.

Cells such as those of bacteria and blue-green algae lack much further membrane development. For instance, the genetic material is not delimited by a membrane, and the enzymes for respiration either are loose in the cell or they are bound to the multipurpose cell membrane.

However, as simple as those cells are or might have been, the important biochemical steps in the flow of energy are exhibited by these organisms. With their appearance life was firmly established on earth. The next stages of cellular development were to be concerned with the greater efficiency and adaptability of cells by an increase in their structural complexity and compartmentalization. This is the story of the development of the cells of all organisms except the bacteria and blue-green algae, and it constitutes the second unit of this book.

Unit II
The Continuity of Life

At the base of the entire process whereby the envelope of the biosphere spreads its web over the face of the earth stands the mechanism of reproduction which is typical of life. Sooner or later each cell divides (by mitotic or amitotic division) and gives birth to another cell similar to itself. First, a single centre; then two. Everything in the subsequent development of life stems from this potent primordial phenomenon.

In itself, cell division seems to be due to the simple need of the living particle to find a remedy for its molecular fragility and for the structural difficulties involved in continued growth. The process is one of rejuvenation and shedding. The more limited groups of atoms, the micromolecules, have an almost indefinite longevity, and with it an equivalent rigidity. The cell, continually in the toils of assimilation, must split in two to continue to exist. At first sight reproduction appears as a simple process thought up by nature to ensure the permanence of the unstable in the case of these vast molecular edifices.

But, as always happens in the world, what was at first a happy accident or means of survival, is promptly transformed and used as an instrument of progress and conquest. Life at first seems to have reproduced itself in self-defence; but this was a mere prelude to its vast conquests.

—Pierre Teilhard de Chardin (1961)

In 1838–1839 the botanist Matthias Schleiden and the zoologist Theodor Schwann brought together into a unified theory a series of ideas that they had been developing on the cellular nature of living systems. Schleiden, after considerable microscopic analysis of plant structures, wrote that all higher plants "are aggregates of fully individualized, independent, separate beings, namely the cells themselves." He discussed his findings with his friend, Theodor Schwann, who greatly extended Schleiden's conclusions after doing his own extensive work, primarily with animal tissues. He postulated that all organisms—from oak trees to tigers to people—are made up of individual cells. The fertilized egg culminating in an organism—whether the large egg of a bird, the smaller egg of a frog or a fish, or the microscopic egg of a mammal—is a single cell with a nucleus, cytoplasm, and a surrounding membrane. That cell develops, said Schwann, through the creation of new cells. He concluded that every animal and plant is composed entirely of cells and of substances produced by cells, and he also perceived that to some extent cells are independent living units even though they are subordinate to the whole organism.

Over the century that followed, the significance of these postulates became increasingly clear. Whereas the theory of evolution provided rational explanations for the multitudinous forms of life on earth, the cell theory provided an elegant framework for those explanations. With the theory of cellular organization, many of the basic reasons why life takes the form that it does, and functions as it does, could be understood without sorting through an infinite number of its representatives.

Today the cell is recognized as the basic functional unit of all living systems. An individual cell is clearly a distinct unit because it is bounded by a membrane separating it from other cells or from the outside environment. But a definition of the cell needs more than the mere recognition of a physical boundary. It requires an understanding that there are a minimum number of universal subcellular systems. All cells contain *genetic material* in the form of DNA molecules, which hold all of the information needed for everything the cell does and becomes, and for coordination of all that potential implies. All cells contain *messenger molecules*, which convey the genetic information to the ribosomes where it is expressed as protein synthesis. All cells contain water and other kinds of *molecules*, large and small. And all cells are equipped with mechanisms for energy transfer that do the vital work of energy extraction and conversion, growth, reproduction, and irritability. In short, *a cell is the simplest unit that can exist as an independent living system.*

Refinements in the cell theory as originally set forth by Schleiden and Schwann have come about largely as a result of our ability to visualize microscopic and submicroscopic matter with increasingly sophisticated magnifying devices. The physical proper-

4

The Fundamental Unit of Life

Figure 4.1 (above)   A size comparison chart of different types of cells. The ostrich egg and the other bird eggs are reduced in size by one-half.

Figure 4.2 (below)   Table showing common units of the metric system.

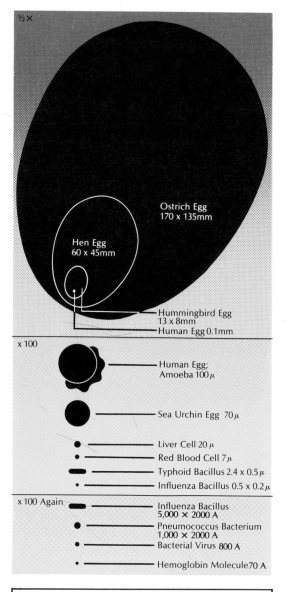

½ ×

Ostrich Egg
170 x 135mm

Hen Egg
60 x 45mm

Hummingbird Egg
13 x 8mm
Human Egg 0.1mm

x 100

Human Egg;
Amoeba 100 μ

Sea Urchin Egg 70 μ

Liver Cell 20 μ
Red Blood Cell 7 μ
Typhoid Bacillus 2.4 x 0.5 μ
Influenza Bacillus 0.5 x 0.2 μ

x 100 Again

Influenza Bacillus
5,000 × 2000 A
Pneumococcus Bacterium
1,000 × 2000 A
Bacterial Virus 800 A
Hemoglobin Molecule 70 A

1 centimeter (cm) = 0.4 inch = 10 millimeters
1 millimeter (mm) = 1/10 cm = 1,000 microns
1 micron (μ) = 1/10,000 cm = 1/1,000 mm
1 angstrom (A) = 1/10,000 μ = .000000004 inch

ties of visible light prevent us from using even the best light microscope to resolve structures less than 1/5 micron in width. For a hundred years biologists strained their eyes, squinting at cellular structures that could barely be seen as anything more than colored dots. About 1955, however, the newly developed electron microscope was applied to biological materials. Objects 1,000 times smaller than the smallest objects visible in the light microscope can be resolved with the electron microscope, and further improvements may be possible (Figure 4.3). Structures that formerly appeared only as specks were found to have intricately detailed, regular structures. The electron microscope has made it clear that the cell—once regarded as a simple body—is an intricately organized and complex assemblage of macromolecules.

PROCARYOTIC AND EUCARYOTIC CELLS

The bacterium *Escherichia coli*, which is found in the intestines of man and other animals, is a unicellular organism weighing about two hundred-billionths of a gram. Seventy percent of it is water. Within one *E. coli* cell are between 3,000 and 6,000 different kinds of molecules (Table 4.1). The *E. coli* is one example of a kind of cell called *procaryotic*, a word derived from the Greek word meaning "before a nucleus." A "typical" procaryotic cell is

Table 4.1

Approximate Chemical Composition of a Rapidly Dividing *E. coli* Cell

Component	Number of Different Kinds	Average Molecular Weight	Approximate Number of Molecules Per Cell	Percentage of Total Cell Weight
Water (H₂O)	1	18	40,000,000,000	70
Inorganic ions	20	40	250,000,000	1
Carbohydrates*	200	150	200,000,000	3
Amino acids*	100	120	30,000,000	0.4
Nucleotides*	200	300	12,000,000	0.4
Lipids*	50	750	25,000,000	2
Other small molecules	200	150	15,000,000	0.2
Proteins	2,000–3,000	40,000	1,000,000	15
Nucleic acids				
DNA	1	2,500,000,000	4	1
RNA				6
16s rRNA	1	500,000	30,000	
23s rRNA	1	1,000,000	30,000	
tRNA	40	25,000	400,000	
mRNA	1,000	1,000,000	1,000	

*Including precursors.

Source: James D. Watson, *Molecular Biology of the Gene*, 2nd ed. (New York: Benjamin, 1970), p. 85.

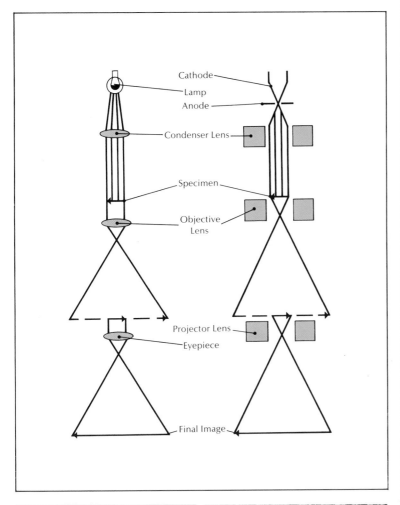

Figure 4.3 In order to emphasize the similarities between the light microscope (left) and the electron microscope (right), the light microscope is inverted, and the dimensions and details of both have been somewhat distorted. In the light microscope, light from a hot filament (or from the sun) is passed through a condenser lens to produce a parallel beam of light. This beam passes through the specimen and is then focused by the objective lens. An eyepiece lens is used to magnify the image. Light microscopes reached their theoretical limits of resolution late in the 1800s, when new instruments with oil immersion lenses and condenser lenses became available. The first experimental electron microscopes were built in the early 1930s, and commercial models became available in 1939. This type of microscope is based on the fact that a beam of electrons has wave properties with very short wavelengths. The electrons are drawn from a hot filament by an electric field. The beam of electrons is focused by magnetic fields, which are produced by electromagnets. A visible image is produced when the electrons strike a coated screen, whose molecules emit visible light when struck by electrons. In most electron microscopes this screen swings out of the way so that the electrons can fall directly onto photographic film and produce a micrograph. Because electrons are scattered by gas molecules, clear images are formed only if a vacuum is maintained within the electron microscope. Because electrons are scattered so easily, the specimen must be very thin—a few hundred angstroms or less—or most of the electrons will be scattered and a uniformly dark image will result. With a very thin specimen, most of the electrons pass through the specimen except where they are scattered by the heavy atoms of a metal stain. In the light microscope, the image is focused by moving the glass lenses. In the electron microscope, the focal length of the magnetic lenses is changed by altering the current flowing through the electromagnets.

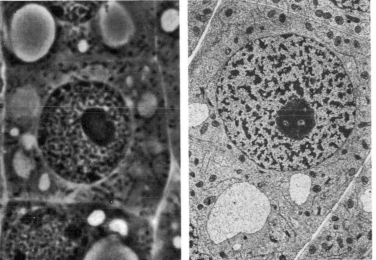

Figure 4.4 Comparison between a light micrograph of an onion root cell (left) and an electron micrograph of the same type of cell (right). Both micrographs have a linear magnification of × 1,000—that is, the distance between two points in the image is 1,000 times as great as the distance between the corresponding points in the actual specimen. The image obtained with the light microscope is blurry and indistinct, whereas much finer detail can be seen in the electron micrograph.

Figure 4.5   Generalized diagram of a procaryotic cell. Note the absence of a membrane-enclosed nucleus.

Figure 4.6   Diagrams of bacteria denoting both shape and growth pattern—chains (streptococcus); small groupings; or irregular clusters (staphylococcus). Most bacterial cells are 2 to 5 microns in length, although a few kinds are as long as 100 microns or as short as 1/2 micron.

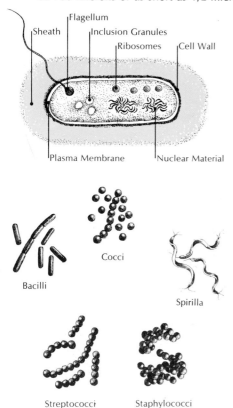

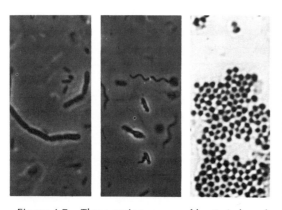

Figure 4.7   Three major groups of bacteria based upon their overall shapes. At left are rodlike *bacilli.* Corkscrew-shaped *spirilla* are shown in the middle photograph; note the whiplike extensions called flagella on these bacteria. At right are predominately spherical *cocci.*

diagrammed in Figure 4.5, but even this figure cannot convey the true complexity of one living cell. For example, the small granules labeled "ribosomes" act as the site of protein manufacture; an *E. coli* cell contains between 20,000 and 30,000 of them. About a million protein molecules are continuously being made on these ribosomes, then are used elsewhere within the cell, and eventually are broken down by the cell.

Beyond the fact that so many things are going on in so small an entity is the staggering fact that *each and every part of the cell is coordinated with every other part.*

The two major groups of procaryotic organisms are *bacteria* and *blue-green algae.* All other organisms are composed of *eucaryotic* cells, which means they have an organized, membrane-enclosed nucleus in which much of the cellular DNA is found.

Procaryotic and eucaryotic cells both contain the basic subcellular systems discussed earlier. Unlike the eucaryotic cell, however, the procaryotic cell has no nuclear and intracellular membranes, and its DNA is not kept apart from the rest of the cellular substance by a nuclear membrane. But even though they are simpler organisms, procaryotic cells, too, are capable of maintaining themselves and reproducing. In fact, much of what biologists have learned about genetic and cellular control mechanisms has been accomplished by studying viruses and bacteria such as *E. coli,* as described in Chapter 5. However, it is believed that most of these mechanisms also are found in eucaryotic cells.

Why, then, do we make such a clear distinction between procaryotic and eucaryotic cells? Apparently more than a difference in cellular anatomy may be involved. Recently, questions about the evolutionary history of cells have come about as an extension of the questions concerning the origin of life. According to the "classical theory" of cell evolution, primitive bacteria gave rise to cells capable of photosynthesis; such cells were, therefore, autotrophs. These cells then gave rise to all other organisms by the stepwise addition of processes for oxidative phosphorylation, movement by means of a whiplike flagellum, a more intricate system of internal membranes to handle the increasing complexity of advantageous chemical reactions, and a process of nuclear cell division to assure equal distribution of cellular information when cells divide. Once these processes evolved, the stage was set for cell-to-cell associations. And these associations culminated in multicellular and therefore larger organisms. In this view, eucaryotic organisms (including people) were derived by stepwise evolution from one kind of bacterium-like ancestor. The problem is, are there cells that are intermediate in structure between procaryotic and eucaryotic forms of life? None have been found. And some people speculate they never will be found because they never existed.

In contrast to the classical theory, the "symbiotic theory" would have us trace our cellular ancestry to *more than one* ancient or-

Figure 4.8   Three-dimensional generalized drawing of a eucaryotic animal cell. All eucaryotic cells are surrounded by a plasma membrane and contain a nucleus surrounded by a nuclear membrane. Within the cytoplasm (all the material outside the nuclear membrane, including the outer plasma membrane) are a number of organelles—specialized structures that perform particular functions and contain specialized membranes.

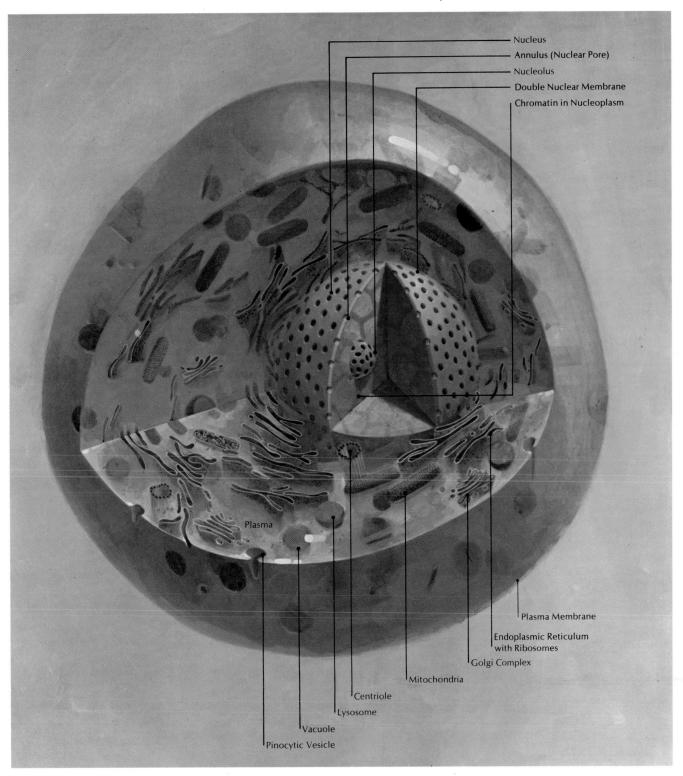

Nucleus

Annulus (Nuclear Pore)

Nucleolus

Double Nuclear Membrane

Chromatin in Nucleoplasm

Plasma

Plasma Membrane

Endoplasmic Reticulum with Ribosomes

Golgi Complex

Mitochondria

Centriole

Lysosome

Vacuole

Pinocytic Vesicle

Figure 4.9   A theoretical model for the origin of a eucaryotic cell. In such a model, several kinds of primitive procaryotic cells came together permanently; each of these primitive cells contributed to the specialized structures and functions of organelles in eucaryotic cells. The theory underlying this model has been recently criticized by Rudolf Raff and Henry Mahler of Indiana University. They contend that proto-eucaryotes derived from procaryotic symbionts would not have been able to compete with more efficient procaryotes. They also maintain that proto-mitochondria and other proto-organelles derived from primitive procaryotes would have had to have had a wholesale transfer of genes to unrelated nuclear areas. Furthermore, they claim that the fossil record and comparative biochemistry of eucaryotic and procaryotic cells do not support this model. However, many other scientists still consider this model as a possibility.

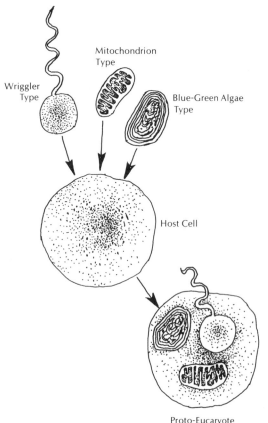

Wriggler Type

Mitochondrion Type

Blue-Green Algae Type

Host Cell

Proto-Eucaryote

ganism. In evolutionary terms, protozoans, fungi, and animals had three primordial parents; green plants had four.

The progression might have gone something like this: Ancestral, procaryotic heterotrophs gave rise to photosynthesizers. In addition, other unicellular procaryotic organisms arose. Some could at first tolerate free oxygen, and in time they would become truly dependent on oxygen for cellular respiration. Other procaryotes developed the ability to wriggle. In ancient times, most creatures could only float and be carried passively by currents. When active movement appeared, it certainly must have been selected for because the organism that could move would find more food to eat and more space in which to dwell.

The history of the modern eucaryotic line may have begun when a bacterium capable only of anaerobic fermentation of glucose to pyruvic acid engulfed—but did not digest—a smaller procaryote capable of using oxygen for respiration. In such a situation, the host cell would harbor the aerobe, and in time each would become dependent on the other. The host would be the proto-eucaryote; and the aerobic symbiont would be the proto-mitochondrion ancestor of mitochondria, one of the most important constituents of the eucaryotic cell.

Perhaps these primitive amoeba-like cells were replaced by a triple symbiont composed of a proto-eucaryote and a proto-mitochondrion that engulfed—but did not digest—a "wriggler." If the genetic material of the host became surrounded by a membrane, the first true eucaryotic cell would be formed. Such a triad would be called an "amoebo-flagellate"; it may have been the initial ancestor of all protozoans, fungi, and animals. Similarly, a quadruple symbiont arising by the association of an amoebo-flagellate and a blue-green algae cell (proto-chloroplast) may have been the ancestor of green plants.

If this theory of the symbiotic origin of eucaryotic cells is valid, then we, as animals, are colonies. The genetic information in our cells would be the derivative of DNA from three different sorts of cells, and our biochemical abilities would owe their origin to diverse ancestral sources. It is intriguing that specific DNA and ribosomes have recently been associated with the mitochondria and chloroplasts.

CELLULAR ORGANIZATION

A single eucaryotic cell may exist independently of other cells as a complete organism. Or it may be integrated with a few dozen other cells into a colonial organism or a simple animal or plant. Or it may be one of millions upon millions of interdependent cells making up a pine tree or a human being. You might imagine that a cell having to survive on its own, such as a paramecium in a pond, would be a different sort of "unit of life" than a cell that is intimately dependent on countless millions of other cells for its survival. And the many interdependent cells of a large organism do come in an enormous variety of shapes and sizes that reflect their numerous functions.

There are, however, certain constraints on the degree of cellular variation that can be achieved. For one thing, there is a minimum size for cells: each must be large enough to include at least the basic subcellular components. Secondly, for each kind of cell there is a maximum size attainable: none could ever grow to the size of, say, an elephant. The reason is that the nucleus, the information center of the cell, must be able to send its genetic commands to other areas of the cell, which contain the working parts. Finally, as a cell grows, its volume increases faster than its surface area does. Because all materials must enter and leave a cell through the cell membrane, an optimum surface-to-volume ratio must be achieved.

Broadly speaking, eucaryotic cells have two areas: the *nucleus*, usually single; and the surrounding *cytoplasm*, which includes all of the material outside the double nuclear membrane, including the cell's outer membrane. The living "stuff" inside the cytoplasm was once believed to be homogeneous, watery, and relatively formless. As the tools of the cell biologist improved, however, it became apparent that cells are highly organized systems of membranes. Within this organizational framework are cytoplasmic *organelles*, specialized particles of living substance that are present in nearly all cells. Organelles are active in the sense that they perform particular metabolic functions in the cell. In contrast, cytoplasmic *inclusions* are temporary storage structures that are relatively lifeless with respect to the metabolic activities of the cell. They include such substances as crystals, pigments, and secretory products.

All such constituents are suspended in the cell's watery matrix. Dissolved or suspended in the water itself are the ions and molecules needed to provide the conditions necessary for life—whatever those conditions are. Try as we may, no laboratory combination of chemicals yet devised has duplicated a living system.

The Nucleus

Electron micrographs show that the contents of the nucleus are separated from the surrounding cytoplasm by a double *nuclear membrane* (Figure 4.10). Each of the membranes is about 70 angstroms thick and is separated from the other by a space of about 150 to 200 angstroms. Small, round structures appear where the inner and outer membranes come together at regularly spaced intervals. These *annuli* are sometimes called "nuclear pores," but actually they may not be permanently open channels between the nucleoplasm and the cytoplasm. An annulus can be as large as 400 angstroms in diameter. Channels of that size would permit the free flow of ions, and the ionic concentrations of nucleoplasm and cy-

Figure 4.10   Electron micrograph and diagram of a cell nucleus, the most prominent feature of a eucaryotic cell. Several dense nucleoli and scattered chromatin granules can also be seen. (× 4,000)

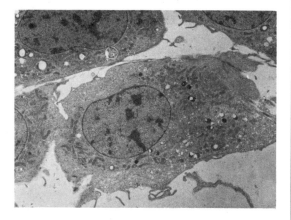

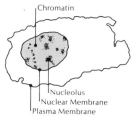

Chromatin

Nucleolus
Nuclear Membrane
Plasma Membrane

toplasm are known to be quite different. On the other hand, the annuli may be selective barriers, allowing the passage of certain large molecules and concurrently preventing the free exchange of ions between the nucleoplasm and cytoplasm.

Within the nucleus is a material called *chromatin*, which is composed of DNA in close association with RNA and protein. The chromatin gathers into the distinct, threadlike bodies called "chromosomes" only during the process of cell division. When the cell is not dividing, chromatin distribution appears to be fairly homogeneous throughout the nucleus, although it may be massed here and there in clumps or concentrated around the periphery.

Also located in the nucleus are one or more bodies called *nucleoli*, which contain large amounts of RNA and protein (Figure 4.12). The nucleolus is now known to be the site for synthesis of ribosomal RNA (Chapter 5). A small portion of chromosomal material, called the "nucleolar organizer," is situated within the nucleolus and apparently carries information that directs the formation of the nucleolus itself and of the ribosomal RNA. The nucleolus is made up of some granules about 150 angstroms in diameter and other granules about 75 angstroms across; these granules are thought to be precursors of ribosomes. No membrane is visible around the nucleolus, and the nuclear material appears to extend in light zones within the nucleolus.

The importance of the nucleus as a control center of the eucaryotic cell was suggested by Robert Brown in 1831, when he observed that every plant cell contains a nucleus. Experimental evidence to support this idea was soon provided by Edouard-Gérard Balbiani. When he removed the nuclei from protozoan cells he found that even though the enucleated cells can carry on most cel-

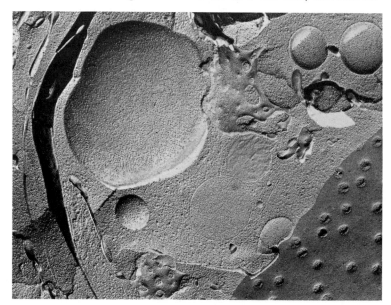

Figure 4.11   Electron micrograph view of the nuclear region of an onion root tip cell made by the freeze-etch preparation technique. This relatively new technique involves a splitting of membranes, which allows extremely detailed examination of membrane faces. (× 27,000)

Figure 4.12   Electron micrograph of part of a pancreatic cell from a bat. The nucleus is bounded by a double membrane complex with pores (arrows), and contains a dark staining nucleolus; a site for ribosome assembly. The cytoplasm contains many organelles, among which is the membrane endoplasmic reticulum with associated ribosomes used for protein synthesis. (× 22,000)

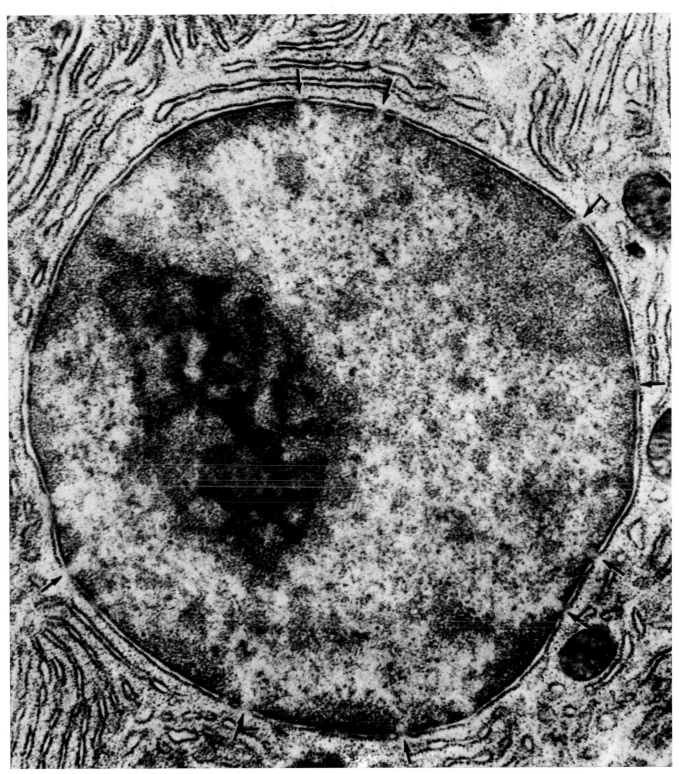

lular activities, such cells are unable to grow or divide and they are relatively shortlived.

The unicellular protozoan called *amoeba* is useful in such experiments because its nucleus is large enough to be removed easily. An enucleated amoeba moves slowly and contracts to a spherical shape; it can ingest food and react to stimuli but cannot digest the food, grow, or divide. It will live for about 20 days in the enucleated state but eventually it dies. If the nucleus from another amoeba of the same species is implanted within 3 days after enucleation, however, the amoeba recovers all normal functions and is capable of growth and division. In some cases, an enucleated cell will survive if it is implanted with the nucleus from a cell of a closely related species.

The Cytoplasm

The term "cytoplasm" encompasses *everything* in a cell other than the nucleus and the nuclear membrane. It is the main region of metabolic activity, and it is the functional expression of the genetic information contained within the nuclear chromosomes as well as in some DNA-containing cytoplasmic organelles. In addition, certain molecules that enter the cell or are synthesized within the cell are suspended in the cytoplasm. Of greatest importance in this regard are certain enzymes and molecular building blocks. Enzymes and other proteins are manufactured on the small granules called "ribosomes." Many of these proteins are transported to the cellular membrane system or to the organelles. Enzymes, such as those involved in glycolysis, remain suspended in the cytoplasm.

The cytoplasm is the part of the cell that is capable of specialization. That is why the structure and function of cytoplasmic components will differ, at least in part, from cell to cell and even from time to time. The reason is that, even though the basic structure of membranes is the same in different kinds of cells, the particular protein and carbohydrate components of a membrane can vary.

CELLULAR STRUCTURE AND FUNCTION

The nucleus and the cytoplasm are easy to classify into distinct, static categories. But the living cell is a dynamic entity; it is constantly changing and carrying out biochemical reactions. And a listing of cell parts no more completely describes a living cell than does a listing of organs describe a living human. What are the various parts of a cell? What functions do they serve? And how do the dynamic and continuous processes in cells intermesh to maintain that cell's existence? The remainder of this chapter is a beginning step toward understanding how multiple cell parts are interrelated.

Cell Membranes

An external *plasma membrane* surrounds every kind of cell, whether it is procaryotic or eucaryotic (Figure 4.13). Within eucar-

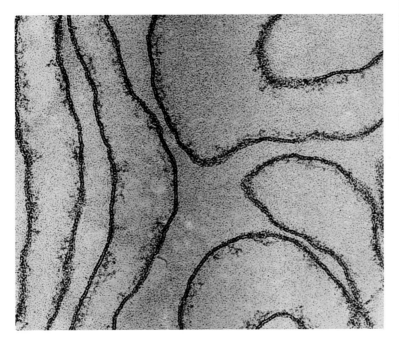

Figure 4.13   Electron micrograph showing a thin section of several red blood cell plasma membranes. Note the triple-layered configuration of these membranes—two dark layers separated by a light space. The entire membrane structure is about 75 to 80 angstroms in thickness.

yotic cells, internal membranes of similar structure enclose the nucleus, delimit most of the living internal structures, and also form sacs, stacks, and channels linking or separating one part of the cell to another.

For many years, the structural features of membranes were not visible, even using the most powerful light microscope. In the 1950s, however, techniques were developed for the study of cell membranes with the electron microscope. Using a special stain and a very high magnification, electron microscopists observed the membrane as a triple-layered structure: two electron-dense lines separated by a lightly stained space (Figure 4.14). Because this triple-layered structure was found to be characteristic of virtually *all* cellular membranes, it came to be known as the *unit membrane* configuration.

All membranes in cells are composed of lipids and proteins and, in some instances, carbohydrates. The lipids make the membrane relatively impenetrable to ions and polar molecules. Some proteins have enzymatic functions, including active transport of molecules across the membrane and metabolic reactions carried out on the membrane surface. In general, however, relatively little is known about the types and properties of proteins in membranes. The carbohydrates play an important role in chemical interactions between the cell and its surroundings.

The total membrane system of a cell is never static. The plasma membrane marking the external boundaries of the cell is more than a passive barrier that holds together the contents of the cell and protects them from the conditions of the external environment. A living cell must continually interact with its environment, obtaining materials, producing substances, and discarding waste products. For this reason, the plasma membrane serves as an "active envelope" to regulate this vital flow of materials into and out of the cell interior. Within eucaryotic cells, membranes play a similar role in maintaining the integrity and specialized conditions of each compartment while simultaneously regulating and helping along the necessary interchanges among the compartments.

Transport of Molecules Across Membranes

Membranes, then, serve as barriers that separate different compartments within the cell and also separate the cell from its external environment. They are selective barriers, transporting needed substances into cells and unwanted substances or secretions out of them. The concentrations of ions and molecules within a cell are maintained at levels suitable for the processes of life.

Molecules always are in more-or-less constant motion. They bounce off surfaces and off one another in a random manner. But

Figure 4.14 Diagram of the unit membrane, showing the triple-layered configuration. During the 1950s, techniques were developed for the study of cellular membranes with the electron microscope. Using a special stain and very high magnification, electron microscopists observed the thin membrane as a triple-layered structure, formed by two electron-dense lines separated by a lightly stained space. Each of the three layers in the structure is about 20 to 30 angstroms in thickness. Because this structure was found to be characteristic of all cell membranes, J. Robertson suggested that it be called the unit membrane configuration. Variations visible in electron micrographs suggest that multiple lipid bilayers may exist in some parts of the membrane. A possible arrangement of these lipid bilayers is shown.

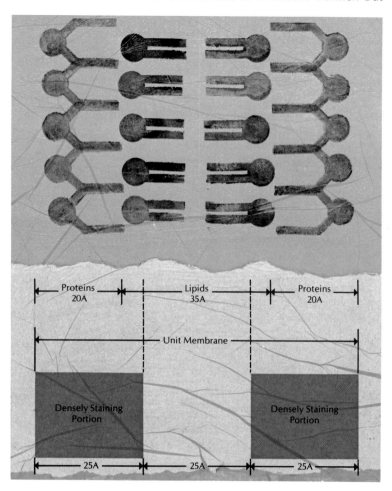

sometimes they may pass through a barrier. Molecules passing through a membrane may move by one of three basic mechanisms: diffusion (passive transport), facilitated transport, or active transport.

Diffusion occurs when there is a net movement of molecules from an area of higher concentration to an area of lower concentration. This movement may or may not occur through a membrane. Eventually, equilibrium will be reached and the two concentrations will be equal. Transport by diffusion is always passive. It requires no expenditure of energy by the cell.

Most biological membranes are semipermeable—that is, permeable to some molecules but not others. All, however, are somewhat permeable to water. Because the cytoplasm enclosed by the membrane is a highly concentrated solution of various molecules (the membrane is impermeable to most of them), the concentration of water molecules inside the cell is lower than the concentration of water molecules in pure water. For this reason, most animal cells placed in water burst because water diffuses into the cell in response to the difference in concentration. Diffusion of water or other solvent molecules across a semipermeable membrane is known as *osmosis* and is illustrated in Figure 4.15.

Nonpolar inorganic molecules (such as oxygen and carbon dioxide gases) and lipidlike substances (such as hydrocarbons and anesthetic drugs) can move across membranes at rapid rates and with little selectivity. It appears that these nonpolar molecules penetrate the membrane directly and randomly at some site, and are transported by means of diffusion. On the other hand, particular ions and polar molecules are transported more selectively across membranes, which seems to indicate there are relatively few sites on membranes that are specialized for the transport of ions and polar molecules.

The lipid layer of membranes behaves as if it were porous, allowing free passage of small polar molecules and regulated passage of ions. But most polar molecules do not pass across membranes except in association with "carriers" within the membrane, which are specific for various types of molecules needed by the cell. Such *carrier molecules* are not visible with the electron microscope, but their existence may be demonstrated chemically. Biochemists have isolated membrane proteins that can selectively bind the molecules to be transported. It is postulated that carrier molecules can move through the lipid portion of the membrane to deliver their bound substrates to the other side, as shown in Figure 4.16. Transport across a membrane is passive only if the flow is from a region of higher concentration to one of lower concentration. If carriers are used to effect this movement, the mechanism is referred to as *facilitated transport*.

Some substances of lower concentration, however, are moved across the membrane toward the side of *higher* concentration.

Figure 4.15 Diffusion of water or other solvent molecules across a semipermeable membrane is known as osmosis. In this demonstration of osmosis, a thistle tube is filled with a colored starch solution and immersed in distilled water. A semipermeable membrane separates the starch solution from the water. Through osmosis, water moves from an area of higher water concentration to an area of lower concentration in the thistle tube. As water molecules move across the membrane, the starch solution becomes more and more diluted, and inward pressure causes the solution to rise (lower diagram). Eventually the weight of the starch solution in the tube will exert just enough pressure to counterbalance the tendency of the water to enter the thistle tube. No more water will enter, and the system will be in a state of osmotic equilibrium.

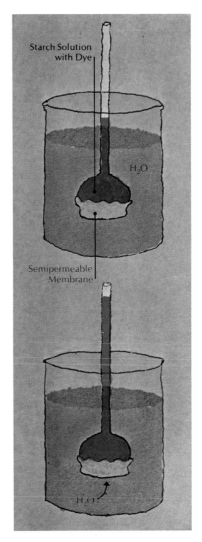

Starch Solution with Dye

H_2O

Semipermeable Membrane

H_2O

Figure 4.16 (left) In this model of facilitated transport, polar molecules require carriers. Part 1 shows a carrier protein within the membrane. An exposed terminal peptide joins with the solute and, through a change in shape followed by a dissociation of the solute-carrier complex, releases the solute. Part 2 shows a mobile carrier.

Figure 4.17 (right) In this model of active transport, as the solute is being transported against a concentration gradient, energy in the form of ATP breakdown is required. The mobile carrier-solute complex moves the solute into the cell.

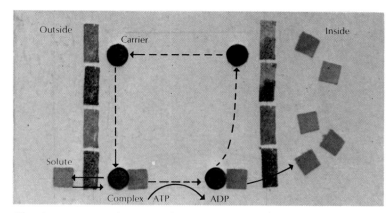

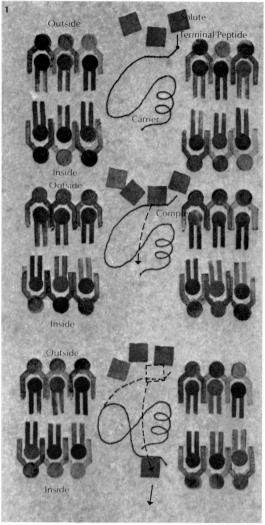

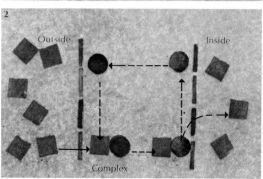

Clearly some mechanism of *active transport* must exist that uses chemical energy to move the molecules against the natural direction of diffusion. A model of such a mechanism is shown in Figure 4.17.

Endocytosis, another kind of transport mechanism, involves the passage of large molecules such as proteins or multimolecular particles such as viruses and bacteria into a cell. In this process, the membrane is enfolded to form a small pocket, or sac, within the cytoplasm. This sac contains materials from the exterior of the cell that were trapped as the membrane pinched in. The plasma membrane fuses across the opening, and the membrane sac detaches from this membrane and moves into the cytoplasm.

Endocytosis also occurs in cells that engulf bacteria and other relatively large particles. In such cases, the process is called *phagocytosis*, or "cell eating." The cell flows around the particle until it has completely engulfed it (Figure 4.18). The cell membrane then fuses and pinches off, and the membrane sac containing the particle moves into the cytoplasm where digestive enzymes are introduced into the sac. Certain cells, such as some white blood cells in animals, use phagocytosis to remove many of the bacteria, viruses, and other foreign particles that find their way into the intercellular spaces of the body.

Some cells take droplets of extracellular fluid into the cytoplasm. Although the tiny droplets can barely be seen in the light microscope, the mechanism appears to be similar to that of phagocytosis. It is called *pinocytosis*, or "cell drinking." Electron micrographs confirm that it involves the enfolding of minute sacs and the subsequent pinching off of those sacs into the cytoplasm, where large molecules contained within the membrane sac may be broken down by enzymes to units small enough to pass through the sac membrane into the cytoplasm. Alternatively, the sac membrane may be broken down so that large molecules may enter directly into the cytoplasm.

In many kinds of cells the plasma membrane is enfolded to form tiny projections called *microvilli* (Figure 4.19). These struc-

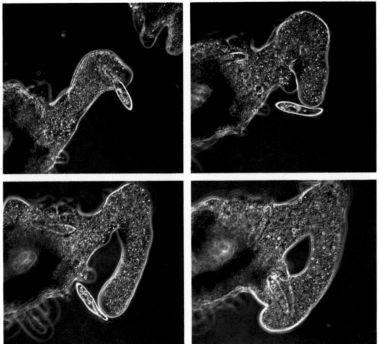

Figure 4.18 Phagocytosis in the amoeba *Chaos chaos*. To capture its prey, the amoeba extends pseudopodia (fingerlike extensions) to encircle and engulf it. The cell membrane then fuses, and the vacuole containing the captured prey moves into the cytoplasm, where digestion takes place.

tures are most abundant in cells that are specialized in the absorption of substances from the external environment—for example, some intestinal cells and some tubule cells in vertebrate kidneys. Microvilli increase the cell membrane surface area and therefore the absorptive capability of cells. They are nonmotile and have no specialized internal structure.

Working Components of the Cytoplasm

As described earlier, the cytoplasm is the region of the cell in which the primary metabolic activities are carried out. Within the cytoplasm are large numbers of different organelles and other components, which vary in configuration, size, and function.

ENDOPLASMIC RETICULUM The *endoplasmic reticulum* is a common feature of all eucaryotic cells and yet one of the most variable. It is a system of membranous sacs that may be flattened (like sheets) or inflated; it may be sparse or it may virtually fill the cytoplasm. Basically, endoplasmic reticulum is either smooth-surfaced (without ribosomes) or rough-surfaced (with ribosomes).

Most metabolic reactions proceed with the help of a sequential series of enzymes. For these reactions to be carried out most effectively, the enzymes must be spatially situated or aligned. Interfacial films or membrane sheets such as those of endoplasmic reticula provide a site for the orderly arrangement of enzymes in a metabolic pathway.

What determines the specific function of a given unit membrane? For the answer you have to look to the particular protein

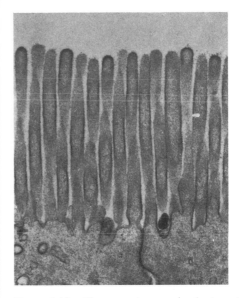

Figure 4.19 Electron micrograph of microvilli, the small projections of the plasma membrane in certain absorptive cells. Each cell has about 3,000 individual microvilli, and on one square millimeter of intestinal tissue there may be as many as 70,000 cells with microvilli.

Figure 4.20   Rosettes of ribosomes clustered on the endoplasmic reticulum. Studies indicate that each ribosome consists of two separate subunits, each containing ribosomal RNA. A complete ribosome is composed of about 60 percent RNA and 40 percent protein, and is approximately 170 angstroms in diameter. (× 54,000)

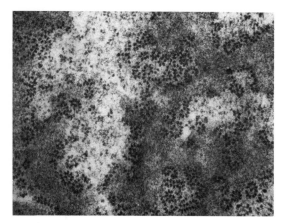

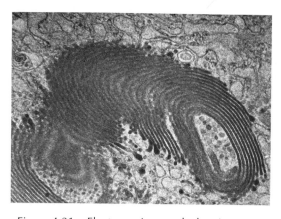

Figure 4.21   Electron micrograph showing a cross section of a Golgi apparatus, which is comprised of a stack of vesicles, or cisternae. One function of the Golgi apparatus is to package and transport materials to be secreted to the exterior of the cell. Substances to be eliminated accumulate in the vesicles of the Golgi apparatus. These vesicles enlarge, separate from the Golgi apparatus, and move into the plasma membrane. The membrane of the vesicle fuses with the plasma membrane, and the contents of the vesicle are discharged to the exterior of the cell. This process has been observed in the secretion of plant cell walls and in the secretion of enzymes and other substances by animal cells. (× 50,000)

molecules making up the surface layer or to the presence of carbohydrates of one kind or another on the surface layer. The membrane of the endoplasmic reticulum has a unit structure like that of the plasma and nuclear membranes. Some parts of the membrane are studded with ribosomes, which make it rough-surfaced. The degree of development of the rough-surfaced endoplasmic reticulum within a cell is an index of the cell's degree of activity in protein synthesis.

On the other hand, a smooth-surfaced endoplasmic reticulum lacks any specific ribosomal associations. It may be related to almost *any* kind of metabolic activity, such as lipid, steroid, carbohydrate, or vitamin synthesis. It may also be involved with conducting materials within the cell's cytoplasm.

RIBOSOMES   The RNA-containing granules that are found in cytoplasm, in organelles, or on the endoplasmic reticulum are called *ribosomes* (Figure 4.20). They are an essential link between the genetic information contained in the nucleus and the enzymes or other proteins that the cell synthesizes. They represent the "workbenches" where the genetic instructions are read and the corresponding enzymes are constructed. Moreover, they have what are analogous to jigs and vises to hold the components of enzymes in proper alignment while they are being assembled. How the ribosomes are able to carry out these accomplishments is only partly understood and will be discussed further in Chapter 5.

GOLGI COMPLEX   The *Golgi complex* appears in the cytoplasm as a tight cluster of one or, at most, a few flattened or rounded sacs similar to smooth-surfaced endoplasmic reticula (Figure 4.21). This well-developed system of cytoplasmic unit membrane sacs or channels is involved primarily with secretion. Although some products such as complex carbohydrates are probably synthesized in or on Golgi membranes, more typically it is believed that the Golgi complex concentrates substances that have been synthesized in the ribosomal regions and packages these substances in membrane sacs so they can then be transported to the plasma membrane, where they are released from the cell. This process has been observed in the secretions of plant cell walls and in the secretions of enzymes and other substances by animal cells.

There is little question today that endoplasmic reticula and the Golgi complex are slightly different forms of the same biological membrane structure, and that assigning them different names might imply a greater difference than actually exists. If a stick of wood is planted upright in the ground, we call it a post. If the same stick is secured in a horizontal position, we call it a rail or railing. If it is used to prop up a sagging clothesline, we call it a pole. Post, rail, or pole—it is still a stick of wood. Similarly, smooth-surfaced endoplasmic reticula, rough-surfaced endoplasmic reticula, Golgi complex, and many of the other organelles making up cytoplasmic membrane systems are undoubtedly the same structures put to different uses.

LYSOSOMES *Lysosomes* are small cytoplasmic particles that are enclosed in a single unit membrane (Figure 4.22). They are produced as a rule by the Golgi complex, but occasionally they are directly produced by endoplasmic reticula. Their internal appearance is variable, depending on the physiological stage they happen to be in when they are observed. When first produced, a lysosome is spherical and its interior is dense and finely granular. It then fuses with one or more bodies, such as ingested particles called "phagosomes," or with pinocytotic sacs or degenerating organelles, whereupon the lysosome becomes irregular in shape and varied in internal appearance.

The reason for this change in appearance becomes clear when the lysosome's function is understood. Lysosomes are the digestive cavities of the cell. They store a high concentration of digestive enzymes and make these enzymes available to the cell. For example, lysosomes isolated from mammalian liver contain enzymes that are capable of digesting DNA, RNA, nucleotides, polysaccharides, and protein. This battery of enzymes would utterly destroy all of the cytoplasmic components of the cell if they were released into it, but this does not happen because they are contained within their unit membrane. When a lysosome fuses with another particle, such as an engulfed bacterial cell, its enzymes are activated and it digests the particle. After the contents have been digested, the entire lysosome may be discharged from the cell.

How universal are lysosomes? They have been observed in almost every kind of animal cell in which they have been sought, so they are believed to be essential to the internal maintenance of every kind of cell. Some cells need more upkeep than others, however, so they vary from one kind of cell to another. When a cell dies, the limiting membrane of its lysosomes breaks down, the lysosome enzymes are released into the cytoplasm, and "self-digestion" takes place. When meat is aged to increase its tenderness, it is the lysosome enzymes that go to work and "tenderize" the meat by hydrolyzing the protein fibers. This time-honored and empirically devised procedure antedates all knowledge of the existence of lysosomes. But lysosomes were busy at work long before the art of butchery was developed.

VACUOLES Vacuoles are membrane-bound spaces found in all kinds of cells (Figure 4.23). They vary in size more than any other organelle and in fact may fall somewhere between true organelles and simple inclusions. Some vacuoles play an active role in cell processes such as phagocytosis or pinocytosis; others are merely storage sites.

Vacuoles in plant cells are filled with fluid under a pressure that helps maintain the cell's rigid shape. In mature plant cells, a single

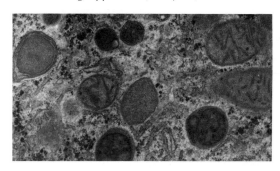

Figure 4.22 In this electron micrograph, the bodies having internal structure are mitochondria; the others are lysosomes. Lysosomes apparently act as the "disposal units" of the cytoplasm. They are believed to form by the pinching off of sacs from the Golgi apparatus. (× 50,000)

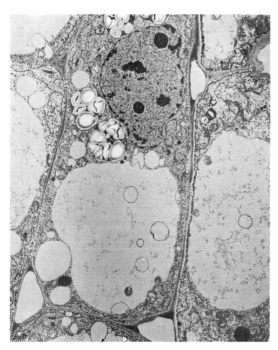

Figure 4.23 Electron micrograph showing vacuoles in separate plant cells. In this micrograph, the vacuoles appear as large, light-colored spaces. Just above the vacuole shown at center is the nucleus, which is surrounded by starch storage plastids. The two vacuoles near the bottom of the micrograph are separated by vacuolar membranes, plasma membranes, and cell walls. Although vacuoles are found in all kinds of cells, the largest ones appear in plant cells. In such cells, the cytoplasm, nucleus, and plastids may be pressed against the cell wall by the large central vacuole. (× 6,000)

vacuole may occupy 90 percent or more of the cell volume. That is why plants wilt: If the amount of fluid available is insufficient to maintain the fluid pressure of the vacuoles, the cells collapse.

Water makes up most of the fluid in the large vacuoles of plant cells. Dissolved in this water are salts, sugars, pigments, and other substances. The color of flowers is determined in part by the kind of pigments concentrated in the vacuoles of the flower petal cells. Lemons have their characteristic sour taste because in citrus fruit the contents of the vacuoles are acidic.

Vacuoles in microorganisms and animal cells show great variability in function and in size. For example, vacuoles in fresh-water protozoans such as amoebae and paramecia are contractile and they may be constantly changing in shape. Because the water outside the cell in these small fresh-water organisms is at a higher concentration than the water in the cytoplasm, water diffuses constantly through the plasma membrane into the cytoplasm. Contractile vacuoles gather up the excess water and pump it out periodically through a pore they form in the plasma membrane. The membrane is immediately repaired after the vacuole completes its contraction. Without contractile vacuoles, these organisms would burst from the accumulated water pressure.

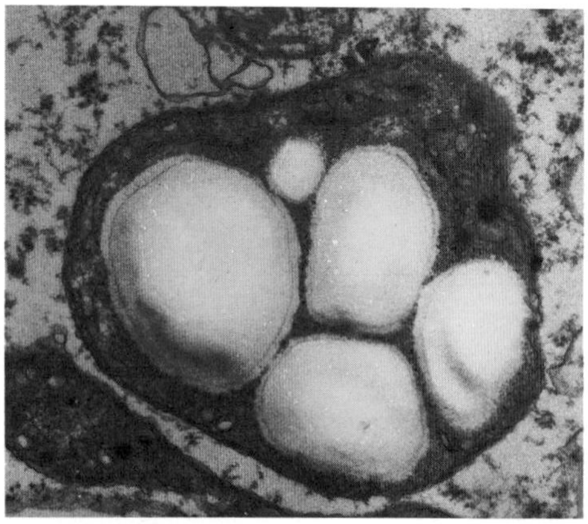

ENERGY-TRANSFERRING ORGANELLES Two organelles—chloroplasts (a plastid) and mitochondria—contain their own DNA and protein-synthesizing systems and perform important roles as "energy units" in plant and animal cells. Because these organelles possess such unique features, they are able to reproduce themselves without instructions from nuclear DNA. They may represent modern versions of the original symbiotic components of eucaryotic cells.

Plastids are present only in plant cells, and most are involved in the capture of the solar energy that is needed in the synthesis of energy-rich macromolecules. Most plastids are round or ovoid and are large, as cytoplasmic organelles go. They have a double membrane as well as a system of internal membranes. Their DNA differs significantly from that of the chromosomes.

Chromoplasts are plastids that contain pigments, and *leucoplasts* are plastids that are colorless (Figure 4.24). One kind of chromoplast is called a *chloroplast;* it contains the green pigment chlorophyll (Figure 4.25). It is in the chloroplasts that photosynthesis takes place (Chapter 3).

Mature chloroplasts are packed with a continuous membrane system that folds back on itself repeatedly to form interconnecting sacs called *grana.* The material surrounding the grana is called *stroma.* It contains dissolved salts, enzymes, more widely spaced membranes, ribosomes that are involved in chloroplast protein synthesis, and the DNA of the chloroplast.

Mitochondria are the cellular "power plants" where the cell's main energy source—ATP—is generated (Figure 4.26). Each of

Figure 4.24 Electron micrograph of a leucoplast, a colorless plastid found only in plant cells. Leucoplasts are the sites for conversion of glucose to starch and to lipids or proteins; these products are then stored in the leucoplasts. *Amyloplasts* are leucoplasts that are specialized to store starch; they are found in many fruits and vegetables. The whiteness of potatoes is due to the presence of amyloplasts. (× 86,000)

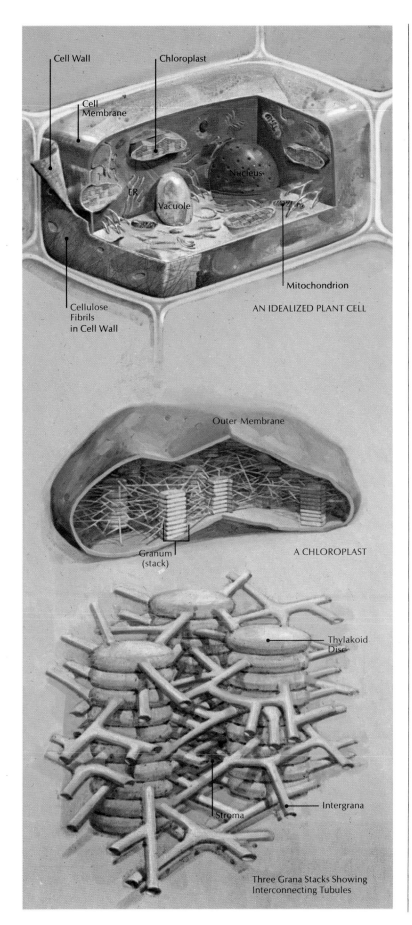

Cell Wall

Chloroplast

Cell Membrane

Nucleus

ER

Vacuole

Mitochondrion

Cellulose Fibrils in Cell Wall

AN IDEALIZED PLANT CELL

Outer Membrane

Granum (stack)

A CHLOROPLAST

Thylakoid Disc

Intergrana

Stroma

Three Grana Stacks Showing Interconnecting Tubules

Figure 4.25 These diagrams and electron micrographs show the highly complex ultrastructure of the chloroplast. This organelle has a double-membrane system. As seen in the diagrams at left, the convolutions of the inner membrane make up a vast network of stacked membranous discs. These discs, known as grana, are interconnected by tubules. Individual discs that make up the grana are called thylakoids. The electron micrographs at right show increasing magnification of a chloroplast structure. A complete chloroplast is shown at top ($\times$10,700); the middle micrograph shows two grana stacks and a portion of the outer membrane ($\times$ 58,000); at bottom is an enlarged region of a single granum ($\times$ 310,500). In this lower micrograph, the staining techniques are reversed from the one above—the loculus (L) is black and the inner membrane of the thylakoid is white.

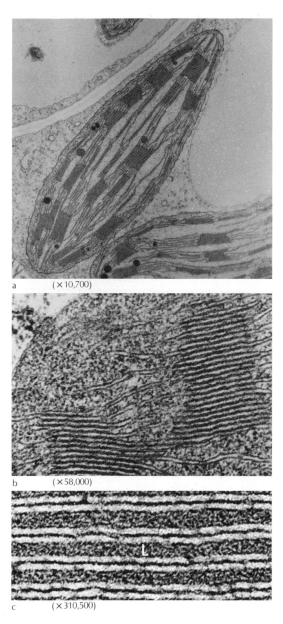

a   ($\times$10,700)

b   ($\times$58,000)

c   ($\times$310,500)

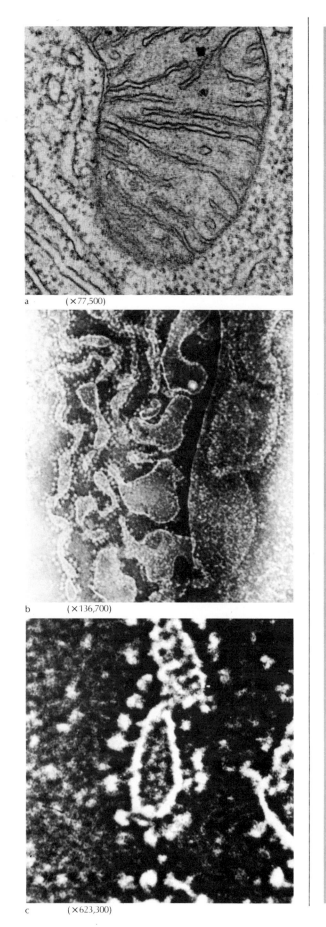

a   (×77,500)

b   (×136,700)

c   (×623,300)

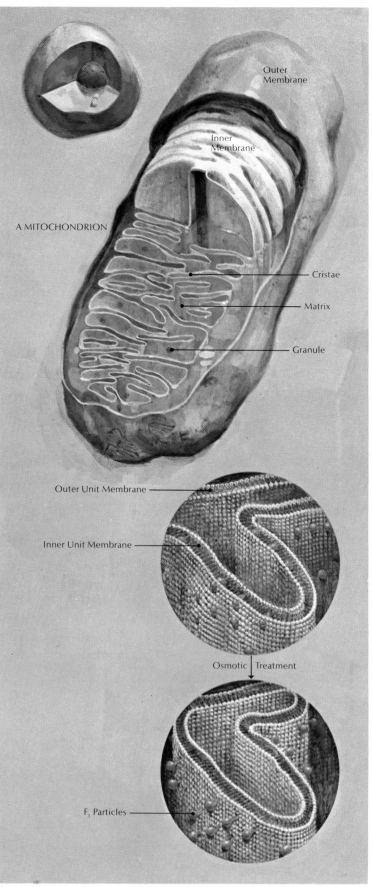

Outer
Membrane

Inner
Membrane

A MITOCHONDRION

Cristae

Matrix

Granule

Outer Unit Membrane

Inner Unit Membrane

Osmotic Treatment

F_1 Particles

the terminal phosphates of ATP can yield about 7,000 calories per mole in a form that is directly usable for nearly all energy-requiring processes in the cell, such as movement, muscular contraction, nerve conduction, glandular secretion, and protein synthesis. Mitochondria are present in all metabolizing eucaryotic cells, and the number of mitochondria in a given cell varies. Cells with high energy requirements have many large mitochondria; those with lower requirements have smaller and fewer mitochondria.

The evolutionary origin of mitochondria is controversial. They have about the same dimensions as bacteria; the functions of their membranes are similar to the energy-transforming functions of bacterial cell membranes; as self-reproducing bodies they contain their own DNA, RNA, and ribosomes—all of which suggests they may represent symbiotic procaryotic microorganisms.

Structurally, mitochondria consist of a smooth outer membrane; a rough-surfaced and convoluted inner membrane; and a heterogeneous matrix that contains RNA, ribosomes, and a minute quantity of DNA. The folds formed by the convolutions of the inner membrane are called *cristae*. The outer and inner membranes are separated by a narrow space that is occupied by a homogeneous fluid, which also fills the space between the cristae (Figure 4.26). The greater the demand placed upon a mitochondrion for ATP, the more richly convoluted its inner membrane becomes, for in this way the amount of inner membrane can be extensively increased without materially affecting the total volume of the mitochondrion.

The inner surface of the inner membrane is studded with granules supported by short stalks. The protein components of both the cristae and the outer membrane include a series of enzymes and hydrogen acceptors that effect the transport of metabolic substances into the mitochondria and their subsequent oxidative phosphorylation to synthesize high-energy ATP from ADP, as discussed in Chapter 3. These enzymes and hydrogen acceptors are arranged within the membrane in a precise order that corresponds with the sequence of reactions in the metabolic pathways that they mediate.

Such systematic "packaging" makes for great efficiency. By isolating mitochondria, fragmenting them, and analyzing the fragments for their metabolic characteristics, it has been shown that the outer membrane is concerned primarily with active transport, by which specific kinds of molecules produced by glycolysis are absorbed and transferred between the cytoplasm and the mitochondrion. The inner membrane, including its stalked granules, is concerned with the major role of oxidative phosphorylation. The matrix itself contains the enzymes that regulate the Krebs citric acid cycle.

Not only do mitochondria respond to the metabolic demands of a cell by adjusting their numbers and the numbers of their cristae,

Figure 4.26 Diagrams and electron micrographs of the mitochondrion, showing the ultrastructure of this double-membrane organelle. The diagrams at right show that the outer membrane of the mitochondrion envelops a highly convoluted inner membrane—both membranes are composed of a double layer of proteins separated by an area of lipid molecules. Further details of structure can be seen after an osmotic treatment that causes the mitochondrion to swell. New particles, previously within the membrane, now cover the cristae surface. Each particle has a polygonal head and is attached to the membrane by a stem. These particles have been variously termed F_1, or elementary, particles. They appear regularly dispersed along the cristae, with some estimates of up to 10,000 per mitochondrion. The electron micrographs at left show increasing magnifications of a mitochondrion. At top a partial mitochondrion is shown ($\times$ 77,500); the middle and lower micrographs were taken after the osmotic treatment ($\times$ 136,700 and $\times$ 623,000 respectively). Note the F_1 particles that are projecting from the cristae.

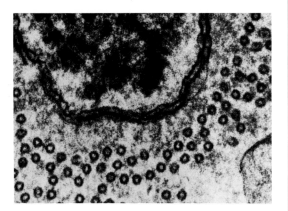

Figure 4.27 Electron micrograph of microtubules in cross section. The dark area in the nucleus is chromatin. (× 35,000)

they also respond in many cases to the particular *locality* or organelle in which the demands are being made. For example, in skeletal muscle fibers, the mitochondria are arranged in rows in intimate contact with the contractile fibrils. In the cells concerned with the rapid absorption or elimination of substances, as in salt-secreting glands or kidney tubules, the mitochondria are nestled in the richly folded cell membrane. In spermatozoa and other cells characterized by rapid movement, the mitochondria are arranged close to the region responsible for movement.

MICROTUBULES   *Microtubules* are tubelike structures that are common in the cytoplasm and in some cylindrical organelles of eucaryotic cells (Figure 4.27). In plant cells they are abundant in the cytoplasm near the plasma membrane and are also found extending along the streams of cytoplasm that move through the cell interior. In some cells they appear to serve as a "cytoskeleton" that gives cells structural integrity. Microtubules are also associated with cell movement, including the so-called amoeboid motion characteristic of amoebas, slime molds, and certain blood and tissue cells. Microtubules also are prominent features of nerve cells and their processes; they appear as spindle fibers in cell division (Chapter 6); and they serve as important parts of centrioles, basal bodies, cilia, and flagella.

Each microtubule is about 240 angstroms in diameter and is made up of thirteen protein filaments arranged in a circular pattern. Each filament is about 40 to 60 angstroms in diameter. Some microtubules are disaggregated by exposure to various chemicals or to low temperatures or to high pressures. For example, the microtubules in spindle fibers, when exposed to the drug colchicine,

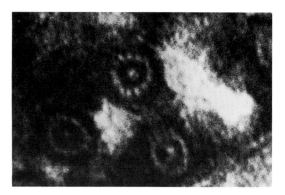

Figure 4.28   In this electron micrograph, which shows sections of microtubules from a plant root tip, the microtubular substructure is visible. Note the thirteen filaments arranged in a circular pattern around the core of the microtubule. This microtubular arrangement is similar for all organisms. (× 480,000)

Figure 4.29   Electron micrograph of cilia from the surface cells of a rat oviduct (in cross section). Note the arrangement of microtubules within each cilium: nine peripheral pairs of microtubules and one central pair of microtubules. Note that each cilium is surrounded by the plasma membrane of the cell. (× 150,000)

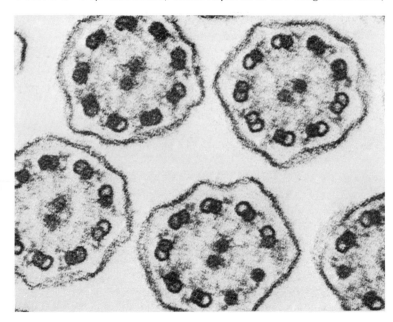

become deactivated and chromosomes do not move to poles of the cell during cell division (Chapter 6).

Centrioles are cylindrical bodies, about 1/5 micron in diameter and about 1/5 to 2 microns in length. They usually occur in pairs in the cytoplasm adjacent to the nucleus. The members of a pair typically lie at right angles to each other. Around the periphery of centrioles are nine sets of microtubules, usually with three microtubules, or a "triplet," in each set (Figure 4.30). The sets of microtubules are joined by fine, fiberlike connections like the spokes of a wheel.

Centrioles are found in all protistans and in all animal cells capable of cell division. These organelles are also present in the reproductive cells of primitive plants, but they are missing from the cells of higher plants. Centrioles have their own DNA and RNA and can be viewed as semiautonomous cell organelles. In this way they may be considered similar to mitochondria and chloroplasts. Little is known about the function they serve, however.

Basal bodies are produced initially in a cell by the centrioles, which they resemble closely. Thereafter, the basal bodies of a cell are self-reproducing and in some cases may number anywhere from 300 to 14,000. They consist of a cylindrical array of nine triplets of microtubules, just as in centrioles, which are joined by fiberlike connections to one another and to a central fibril. The axis of each triplet is tipped. Whereas centrioles appear to be concerned primarily with the mechanics of cell division, basal bodies are concerned with the production of cilia and flagella.

Cilia and *flagella* are motile projections of cells. They are bounded by an outfolding of the membrane. "Cilium" and "flagellum" are relative terms. "Flagellum" generally is used for longer structures and "cilium" for shorter ones, but the two kinds of appendages have identical microstructures. Cilia are commonly about 10 or 20 microns long, and flagella can be thousands of microns long.

Most cilia and flagella are capable of motion. Because of their length, flagella usually move in an undulating fashion, whereas cilia move with simple, oarlike strokes. Their activity propels the cell to which they are attached, or it moves things past a stationary cell. Some kinds of cells have hundreds of cilia, some have only a few, and many have none at all. Most flagellated cells have only a single flagellum, but in algae and fungi the flagella usually occur in pairs. Many unicellular organisms move by means of cilia or flagella. A sperm cell is propelled by a single, long flagellum. The meeting of sperm and egg is further facilitated in many organisms by the motion of cilia on cells that line the female reproductive tract. In lungs, cilia move foreign particles such as dust and soot out of the respiratory tract.

A flagellum is an extension of one of the centrioles of the flagellated cell (Figure 4.31). In a ciliated cell, many extra centrioles may

Figure 4.30   Electron micrograph and diagram of the structure of a centriole. In contrast to the structure of cilia or flagella, no central pair of microtubules is found. In addition, the centriole consists of nine parallel triplets of microtubules instead of doublets. (× 130,000)

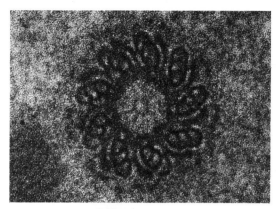

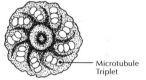

Microtubule Triplet

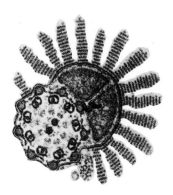

Figure 4.31   Electron micrograph of a sample of butterfly sperm with cross-sectional views of their flagella. Note the arrangement of microtubules within one flagellum—nine peripheral doublets of microtubules with a single pair of central microtubules. The fine structure of cilia and flagella are similar; flagella are usually longer and fewer occur per cell. (× 65,000)

Figure 4.32 These photographs show examples of ciliated and flagellated organisms. Each of the organisms is a unicellular eucaryote.
(a) *Vorticella,* a sessile ciliate protistan, creates a current of water with its cilia and directs food particles into its gullet. The recoil action of its contracile stalk is illustrated.
(Courtesy Carolina Biological Supply Company)
(b) In this marine ciliate, note the winglike extensions of the pellicle (surface membrane), which function as flotation devices.
(c) In *Euplotes,* an advanced ciliate, the ciliophora are structurally the most complex of the protozoa.
(d) *Euglena* is a flagellated autotroph in which the outer portion of the cytoplasm is rigid, holding the cell in its elongated shape.
(e) *Paramecium aurelia* is another representative ciliate protozoan.
(f) The stalked ciliate *Stentor* is noted for its complex gullet and elaborate ciliation pattern.
(g) *Trypanosoma gambiense* is a parasitic flagellate that causes African sleeping sickness.
(Courtesy Carolina Biological Supply Company)

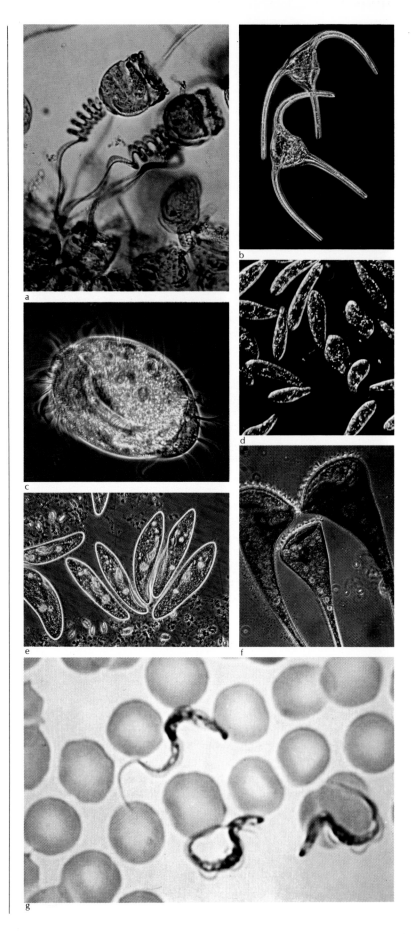

be formed and may serve as basal bodies from which cilia develop. Two of the microtubules of each centriolar triplet extend the full length of the flagellum or cilium. One of each pair of microtubules possesses enzyme molecules that are essential to the movement of the appendage. An additional pair of microtubules extends up the center of the flagellum or cilium.

Cytoplasmic Inclusions and Extracellular Structures

In the arbitrary category of *inclusions* are grouped all particles, droplets, storage granules, and other substances that are relatively inert with respect to the metabolic activities of the cell. They vary in size from glycogen granules (about 150 to 300 angstroms in diameter) to crystals of various sorts that are visible at low powers of the light microscope. In the intermediate range are lipid droplets, yolk granules, pigment granules, virus inclusions (often in crystalline form), and other crystals. Crystals of the organic base guanine are found in the surface cells of fish, amphibians, and lizards, and in the light-reflecting cells of the eyes of many nocturnal animals. The guanine crystals give a silvery luster to the tissues formed of the cells that contain them.

Many types of cells are embedded in a matrix of material that was produced by the cells themselves. They are referred to as *extracellular products*. Bone cells are interspersed within a matrix formed chiefly of calcium crystals. Ligaments and tendons, the connective tissues that connect bones and muscles (the "gristle" of meat), derive their toughness from an extracellular protein called *collagen*. The hard external skeleton of insects and other arthropods is composed chiefly of chitin, another cellular product that surrounds the cells that synthesize it. The cell wall of plant cells is a complex, multilayered construction containing the polysaccharides, cellulose, lignin and pectin, as well as waxes and silica. All of these materials are produced by the cells but are deposited outside the plasma membrane and are considered to be nonliving.

How does the cell regulate and integrate all the diverse functions of the many structures that have been described in this chapter? Complexity requires control. Over immense spans of time the cell has developed elegant control processes in order to remain alive and to remain a vital, integrated, and dynamic piece of the universe. These are the processes that will be discussed in the following chapter.

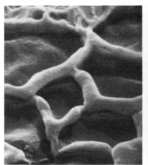 

Figure 4.33 Chitin is an extracellular polysaccharide found on the body surfaces of certain arthropods, fungi, and other organisms. Shown at left is a scanning electron micrograph of this material; the hard, external skeleton of the sowbug at right is composed of chitin.

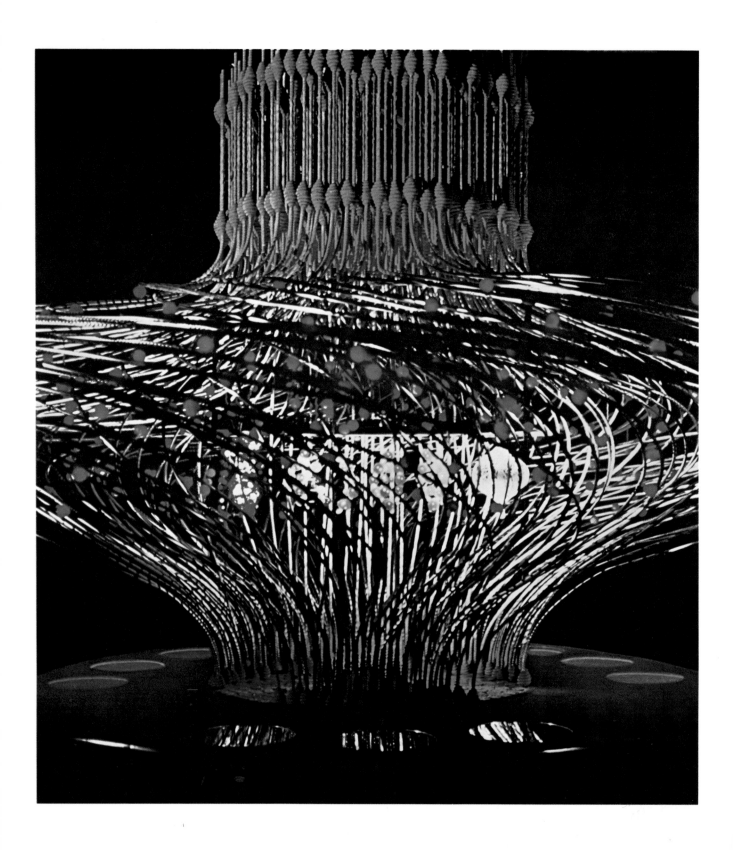

Information, in essence, is a message—a readable message. The message may be in various forms and it may be read in various ways; it may be stored in one form and delivered in another. But two characteristics never vary: Information implies *structure* and it implies *sequence*.

Consider a written message composed of paragraphs, which are made up of sentences, which are in turn made up of words, then letters. Understanding the message follows from understanding of the structure of the letters. But even if the message is not written out, there is structure just the same. Verbal messages, mathematical equations, musical notations, sculpture—these forms of information also have a structure. Alter the structure and the message may be altered; alter the structure sufficiently and the message may be lost entirely.

The coherence of messages set forth in linear form, such as writing, depends on the sequence of information. RUN does not carry the same message as URN does, and RNU is nonsense in English. And RUN read backward becomes NUR, also nonsense in English.

Consider, finally, the structure and sequence of a message that carries all the information needed to construct a certain specific enzyme. It is information that is used in reactions among molecules, so it is reasonable to assume that the information itself is stored in some sort of molecular structure. The kind of molecule needed to store this information depends on the complexity of the information to be stored. And the "blueprint" for any enzyme depends on the complexity of the enzyme molecule itself.

Every enzyme is a protein. And every protein is in turn a unique polymer of amino acids (each of which is a monomer). A blueprint for the construction of a polymer that is made up of only one kind of amino acid could be relatively simple. Such a plan would have to specify the monomer to be used, the way in which like monomers would be linked together, and the number of monomers to comprise the polymer. Because the monomers are identical, sequence is irrelevant. However, for a polymer such as an enzyme, which is made up of twenty *different* amino acid monomers, the blueprint must also specify the *sequence* in which the various amino acids are to occur in the chain.

The message—when read in the proper direction—must specify the sequence of monomers in the protein chain. Because the protein to be constructed is a linear molecule, the blueprint could be readily encoded in another linear molecule. The length of the molecule must be sufficient to include all the information needed to specify fully the structure of the protein.

The information should be stored in a linear macromolecule. This information molecule must be a polymer made up of different kinds of monomers, otherwise it could not carry the information needed to construct the proper sequence of different amino acids

5

The Cell As Information Center

in proteins. Proteins contain some twenty different amino acids, but the information molecule would not necessarily have to contain this many different monomers. Just as the twenty-six letters in our alphabet can be used to encode many different words, so groups of a few different monomers can be used to encode information about the twenty amino acids found in proteins.

The message molecule used by organisms has precisely these characteristics. It is called *ribonucleic acid* (RNA), a linear macromolecule made up of four different kinds of monomers (called ribonucleotides). In certain viruses (RNA viruses), RNA is the ultimate repository of genetic information. In other viruses and all other organisms, RNA is the working genetic information—in a sense it is the working copy of the blueprint taken to the construction site rather than the master copy kept in the office. The basic genetic information is stored in the master copy—deoxyribonucleic acid (DNA). DNA is also a linear polymer composed of four different kinds of monomers (called deoxyribonucleotides), differing slightly in structure but similar to the same four kinds as in RNA.

What is the blueprint for DNA? DNA is a polymer of *nucleotides*. Each nucleotide consists of a phosphate group, a sugar called deoxyribose, and one of four nitrogen-containing bases. A single strand of the DNA polymer can be represented by the following diagram:

. . . — phosphate — sugar — phosphate — sugar — phosphate — sugar. . .

| | |

base 1   base 2   base 3

Notice that the *backbone* of the polymer has repeated identical sugar-phosphate groups. But each *side-chain* base can be any of the four base units. The possibility for varied order of the bases is what makes DNA useful for storing information.

The four nitrogen-containing bases in DNA are *thymine* (T), *adenine* (A), *cytosine* (C), and *guanine* (G). DNA is usually a double-stranded molecule. The bases in one strand are attracted to the bases in the other strand because of hydrogen bonding between the bases. In DNA, the attraction involves only four different base combinations. The base A pairs to base T, and vice versa; base C pairs to base G, and vice versa.

Figure 5.1 depicts the two kinds of base pairing that can occur. This figure also shows the hydrogen bonds that give rise to them.

Once you know about this characteristic pairing of nucleic acid bases, you can easily see how a molecule of DNA might be replicated. The molecule consists of a double-stranded sugar-phosphate backbone, with one of the four bases extending from each sugar. Think of this molecule as the original blueprint. Now

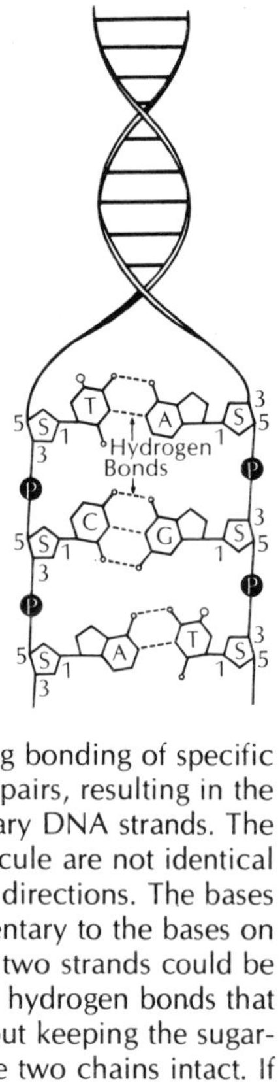

Figure 5.1 Diagram showing bonding of specific purine-pyrimidine base pairs, resulting in the production of complementary DNA strands. The two strands of the DNA molecule are not identical but are oriented in opposite directions. The bases on one strand are complementary to the bases on the other. Suppose that the two strands could be separated, breaking the weak hydrogen bonds that hold the strands together but keeping the sugar-phosphate backbones of the two chains intact. If the environment contains the necessary deoxyribonucleotides, the specific enzymes are present, and the two DNA strands are separated from each other, then each of the parent strands of DNA could build upon itself a new companion strand. Because of the base pairing restrictions, each of the new companion strands would be complementary to the strand on which it was built. Therefore, the final result would be two identical molecules, each with two complementary DNA strands.

bring up an assortment of base-sugar-phosphate groups—the nucleotides—and you will end up with something that looks like the structure shown below:

Original
Strand

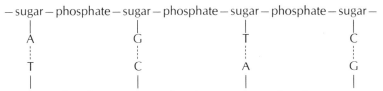

Each base will pair to its usual mate, as shown schematically above. As this pairing takes place, an enzyme called *DNA polymerase* forms the polymer bonds between sugar-phosphate groups on this second strand. The second strand is *complementary* to the first; in other words, it can act as a blueprint to reproduce the original. In order for this to happen, the complementary strand must separate from the original strand, and more nucleotides must be brought up to pair with the complementary strand. These nucleotides will then be polymerized into a strand that is a replica of the original.

How is genetic information encoded in DNA molecules? The code is actually in the sequence of nitrogen-containing bases. Each sequence of three bases on the DNA polymer is a code word—a *codon*—for one amino acid. It has been discovered, for example, that the codon TTT on DNA corresponds to the amino acid lysine. Each sequence of several hundred codons is a hereditary code message—a *gene*. It is this sequence that is transcribed into RNA, which in its turn is translated into a sequence of amino acids; in other words, into a protein. Much of the machinery of a living cell is devoted to translation of the base sequence of DNA molecules into proteins. Many of these proteins are the enzymes that catalyze all the chemical reactions of the cell. Therefore, the information that a new cell receives is primarily a "list" of the enzymes it should make. Which enzymes are present and the order in which they appear determine the details of the future history of the cell.

That, in outline form, is the way DNA stores and passes on information. And yet the simplicity of this picture belies the true elegance of the way in which the DNA molecule functions as the fundamental blueprint of heredity. All living organisms carry in their DNA a message of 3 billion years of evolutionary success. The evolutionary failures, with their DNA, are extinct; the succes-

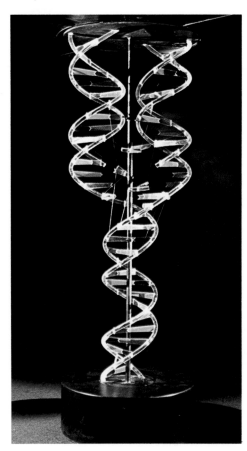

Figure 5.2   Model of the double-helix configuration of the DNA molecule in the process of replication.

ses live on, changing in a changing world. The workings of DNA have been subjected to eons of natural selection, so that by now it is a treasure of irreplaceable exactitude and detail. Bringing to light the intricacies of that treasure has been one of the greatest scientific adventures of this century.

THE SEARCH FOR THE GENETIC MOLECULE

The scientific discovery that laid the groundwork for the subsequent era of genetic research was made in the 1920s by Frederick Griffith as he worked with an organism that causes pneumonia, the bacterium *Diplococcus pneumoniae*. Griffith's intention was to develop a method for immunizing an individual against the disease, but he failed to achieve his objective. Instead he obtained experimental results that could not be explained by the knowledge of his time. Fortunately he recognized that his unexplained results might be significant and he published a report of the work in 1928. The results, which are illustrated in Figure 5.3, proved to be of enormous value.

Griffith worked with a highly virulent strain of the bacteria and a strain that was avirulent. (A virulent strain would be infectious and damaging to the host organism, whereas an avirulent strain would be noninfectious or harmless.) The virulent bacteria, when injected into a mouse, reproduced rapidly and killed the host animal within a few days. Avirulent bacteria injected into a mouse produced no ill effect and soon disappeared. Griffith anticipated both of these results. The virulent bacteria possess a polysaccharide coat, or capsule, at their surface that inhibits the ingestion and destruction of the bacteria by white blood cells. The avirulent bacteria are readily ingested by white blood cells and destroyed, which prevents the disease. The ability of a bacterial cell to encapsulate itself in a polysaccharide coat is an inherited, or genetic, property. The injection of killed bacteria of the virulent type caused no ill effects because the development of disease requires the proliferation of the bacteria within the animal. Griffith obtained his totally unexpected result after he had injected killed, virulent bacteria *and* living avirulent bacteria into mice: the mice rapidly developed disease and died. The cause of death was shown to be massive proliferation of virulent bacteria possessing the polysaccharide coat, or capsule. The results could mean that the killed, virulent bacteria in some way brought about a genetic change in the avirulent strain—a change that conferred the genetically determined ability to produce a polysaccharide capsule and thereby acquire virulence.

Griffith's experiments were later performed in a test tube. Virulent bacteria were disrupted and an extract of the broken bacteria was mixed with living, avirulent bacteria. The extract was also effective in heritable transformation of the living bacteria to the virulent type. Because the transformation was a genetic change, it had

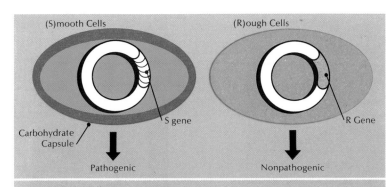

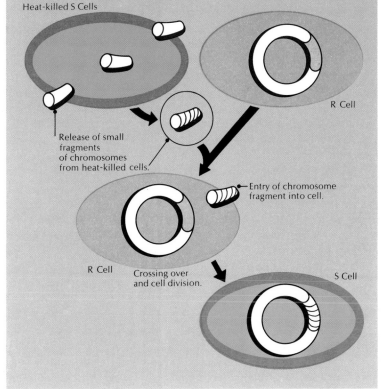

Figure 5.3 Diagram of Griffith's experiment. The genetic composition of the bacteria *Diplococcus pneumoniae* is transformed by the addition of heat-killed cells of a different strain. This experiment set the stage for the proof that DNA is the genetic material of the cell. (S = smooth cells, or the highly virulent strain of bacteria; R = rough cells, or the avirulent strain of bacteria.)
Printed by permission from J. D. Watson, Molecular Biology of the Gene, © *1965, J. D. Watson*

to represent the *transfer* of the genetic material of a single bacterium into another. Then, in 1944, O. Avery, C. MacLeod, and M. McCarty prepared DNA in highly purified form from virulent *D. pneumoniae*. They discovered that the purified material could cause the transformation of living avirulent bacteria. They strengthened their conclusion that the substance causing the transformation was DNA—not a trace of some unidentified contaminating material—by showing that cleavage of purified DNA with deoxyribonuclease destroyed the transforming activity of the preparation. Other macromolecules could not induce transformation.

In some plant, animal, and bacterial viruses, the genetic material has proved to be RNA instead of DNA, but in cellular organ-

Figure 5.4 Hydrogen bonding between the purine-pyrimidine base pairs of DNA. Guanine (G) is bonded to cytosine (C) by three hydrogen bonds; adenine (A) is bonded to thymine (T) by two hydrogen bonds. In order for the chains to fit together in the structure indicated by the x-ray diffraction studies, only those two specific pairings—G to C and A to T—are possible.

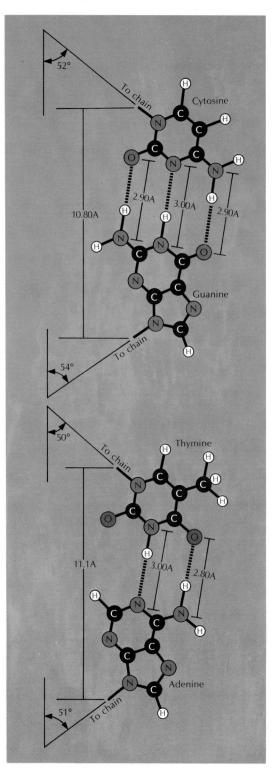

isms, the genetic material is always DNA. This assertion has been validated with extensive experimental evidence. Two examples of this evidence are the *stability* of the DNA molecules within the cell and the constancy in the *amount* of DNA present within the cells of a given species. The macromolecule comprising genetic material must be highly stable because breakdown of the genetic molecules would represent loss of genes and there would be no means of resynthesizing the destroyed genes. Consider what happens when radioactive substances such as thymidine (a precursor of DNA) are incorporated into DNA molecules at the time of DNA replication. Only minimal amounts of the radioactivity are lost from the DNA as long as the cell remains alive. Clearly DNA, once made, does not undergo appreciable breakdown and resynthesis; it shows instead the high degree of stability predicted for the genetic molecule.

The prediction that the macromolecule representing the genetic substance will be present in a constant and fixed amount in each cell is based on a wealth of genetic studies indicating that each gene is present in the nucleus of the cell in a fixed number of copies. This constancy in the number of genes should be reflected in a constancy in the amount per cell of the molecules that fill the role of the genetic substance. Many measurements on cells have shown that DNA meets the criterion of constancy, whereas the number for all other types of molecules in the cell varies over a wide range. Under certain circumstances, the amount of DNA may also vary, but such variations are attended by genetic alterations of the cell that are in some measure predictable.

From observations of this type and many others it has been established beyond doubt that DNA is the genetic substance in all cells—from the simplest bacterium to the highly evolved and complicated cells of multicellular organisms such as humans.

The forceful evidence obtained in the 1940s and early 1950s proving that DNA served the genetic role stimulated an intensive effort to determine the chemical structure of the DNA molecule. It had long been known that DNA was composed of the four different deoxyribonucleotides. The molecular structures of these four building blocks of DNA were also known, but the size of the DNA molecule and the number of building blocks had to be determined. In addition, somewhere within the order and arrangement of the four nucleotides resided the properties that endow DNA with the capacity for genetic function and self-replication.

THE STRUCTURE OF THE DNA MOLECULE

The first clue about DNA structure was provided by the work of Erwin Chargaff. In examining the DNA from a variety of organisms, he found that for each thymine base in DNA there was one of adenine; and for each cytosine base there was one of guanine. The implication was that thymine and adenine occurred in pairs, as did the other two nucleotides. Then, in 1952, James Watson and Francis Crick discovered the answer. Relying on results from x-ray diffraction analyses of DNA, they proposed a three-dimensional arrangement of atoms within the molecule.

The x-ray diffraction studies showed that the DNA molecule was *helical* and that at least two helices were present and bound to each other. Another key was the possibility that *hydrogen bonding* could occur between thymine and adenine and between cytosine and guanine (Figure 5.4). Hydrogen bonding in pairs of nucleotides in other combinations—for example, between adenine and guanine—could not readily occur, which again suggested that within the DNA molecule thymine + adenine and cytosine + guanine might be arranged in pairs.

Watson and Crick proposed that DNA was made up of two polynucleotide chains wound around each other. The nucleotides were known to be held one to another to form a chain by means of a bond in which the sugar molecules of sequential nucleotides are linked to one another by means of a phosphate molecule (Figure 5.6). They postulated that the two chains were held together by hydrogen bonding between adenine in one chain and thymine in the other and, similarly, between guanine and cytosine. For each thymine in a given chain, then, the *opposite* chain would contain an adenine base, and for each cytosine in one chain the *opposite* chain would contain a guanine base at that position. In order for the two nucleotide bases to be in the precise position for hydrogen bonding, the two chains would have to twist in a manner that produces a helical configuration in each chain. The net effect is a double helix in which the two chains remain in perfect registry while they twist around each other.

THE REPLICATION OF THE DNA MOLECULE

The picture of the complementary nature of the two intertwined chains of the double helix gave Watson and Crick insight into the mechanism by which the DNA might duplicate itself without jeopardizing the genetic information contained within its macromolecular structure. They proposed that the two chains were caused to *separate* from one another. Somehow the hydrogen bonds between the chains had to be disrupted to initiate an unwinding action. Just how the unwinding is achieved is not yet clearly understood, but it may occur by rotation of the duplex on its axis, ahead of the point at which the two chains are being separated from each other (Figure 5.7).

As the two chains separate from each other, each then becomes a template (or mold) by virtue of the specific hydrogen bonding properties of the nucleotides. As each chain becomes a template, *free nucleotides in the cell align with the chain according to the rules of base pairing.* As each free nucleotide falls into hydrogen bond pairing, it is joined to the preceding nucleotide just

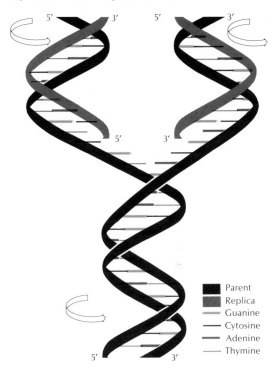

Figure 5.5   The DNA molecule in the process of replication according to the Watson-Crick model.

■ Parent
■ Replica
— Guanine
— Cytosine
— Adenine
— Thymine

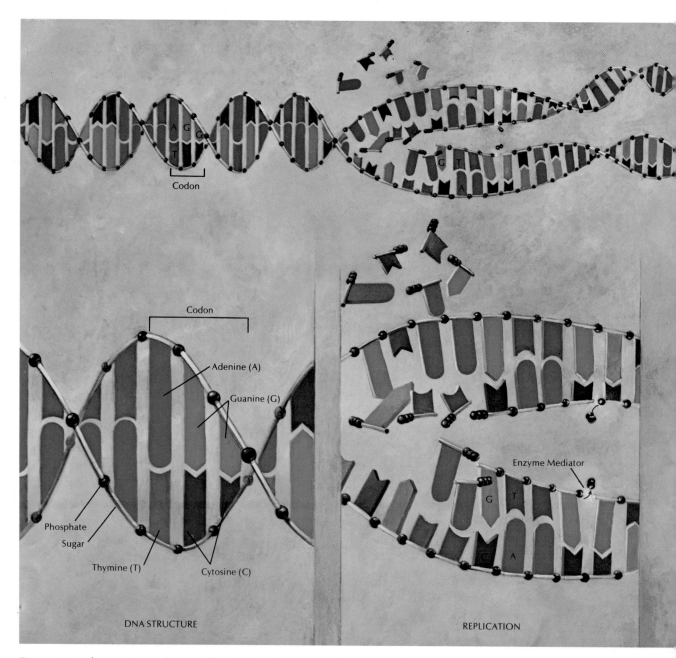

Figure 5.6 The structure of DNA. The DNA molecule, which takes the form of a double helical "ladder," is made up of sugars and phosphates to form the molecule's "supports," and of bases (cytosine, guanine, thymine, and adenine) to form its "rungs." Any group of three base pairs is called a codon.

Figure 5.7 The process of replication (DNA self-duplication). During replication, which occurs before a cell divides, the DNA molecule splits along its rungs, and each half of the ladder builds a new half (with the help of enzyme mediators) by adding on new components from the cytoplasm in exactly the same order as they were in the original strand.

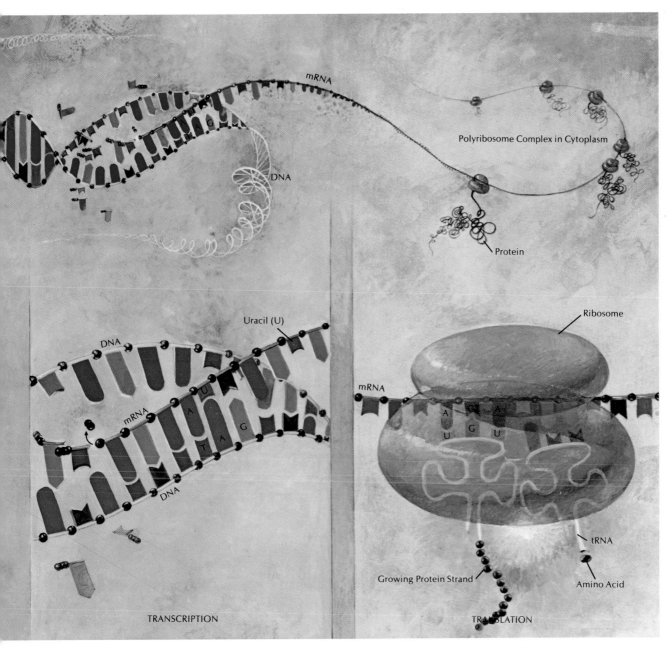

Figure 5.8   The process of transcription (the creation of RNA by DNA). The sequence of nucleotide bases (A, C, T, and G) on a strand of DNA forms a template for the formation of a complementary strand of messenger RNA (with the bases A, C, U, and G). The messenger RNA forms a "negative print" of the DNA (which is located in the nucleus) and proceeds to carry its message outside the nucleus.

Figure 5.9   The process of translation. The messenger RNA carries its message from the DNA outside the nucleus to the ribosomes, where, with the help of transfer RNA, it lines up a set of amino acids to form a protein molecule. Thus, the DNA in the nucleus, through a series of processes, ultimately controls the production of the proteins used in so many important ways by the body. Many ribosomes in a polyribosome complex may construct many identical protein molecules from the information in one mRNA molecule.

added to the growing chain by means of a phosphate molecule bond (Figure 5.7). In this way, each of the old chains guides the formation of a new chain of nucleotides that is complementary to itself. With the completion of the process, two identical duplexes of DNA will have been formed from a single duplex. This means of reproducing the DNA double helix is called *semiconservative replication* (to distinguish from other suggested schemes of replication) because each of the daughter duplexes contains half of the parental duplex in the form of one of the original nucleotide chains. Conclusive proof that semiconservative replication is the only mechanism by which DNA can replicate was provided by Matthew Meselson and Franklin Stahl in 1958 in their experiment with *E. coli* (Figure 5.10).

In sum, the general requirements for DNA synthesis are known. The preexisting polynucleotide chains must be present to act as templates, nucleotide precursors must be available, and an enzyme must be present to catalyze the formation of bonds between nucleotides as they become aligned on the templates. There are important details about DNA replication, however, that are not yet clear. Although the scheme summarized in Figure 5.10 is essentially correct, recent discoveries have shown that diagrams of this type seen so commonly in textbooks are an oversimplification of the actual molecular events.

One point not yet understood is the process by which the two chains of the duplex are caused to begin separation from one another so that replication may occur. Presumably there must be a molecule, possibly a protein, that can bring about the rupture of the hydrogen bonds holding the two chains together. Once the separation has been initiated at some point on the DNA molecule, the process of DNA replication may itself be sufficient to bring about the continued unzippering of the two chains.

Another unresolved detail is how initiation of DNA replication is controlled. DNA replication is the key event in cell reproduction. In order for a cell to divide it must have replicated its DNA so that each daughter cell will receive a copy of the genetic material. Cell reproduction is in general controlled by the controlled initiation of DNA replication. In cells that stop proliferating—for example, during the development of a multicellular organism—cell reproduction may cease because these cells are prevented from replicating their DNA. How that happens is not understood. The matter is of extreme importance, however, because the development and health of an organism depend on regulation of cell reproduction. In cancer, a defect is present in the regulatory process and cell reproduction becomes rampant.

THE ONE GENE—ONE ENZYME HYPOTHESIS

Insight into the relationship between the individual genes and the protein products of genes initially came about in the 1940s as the

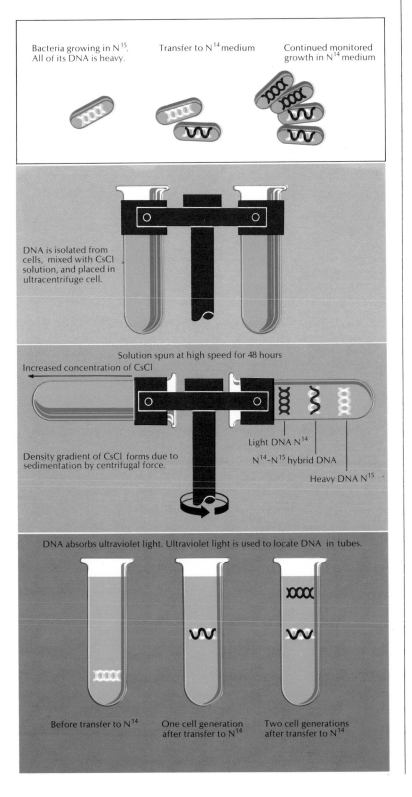

Bacteria growing in N^{15}. All of its DNA is heavy.

Transfer to N^{14} medium

Continued monitored growth in N^{14} medium

DNA is isolated from cells, mixed with CsCl solution, and placed in ultracentrifuge cell.

Solution spun at high speed for 48 hours

Increased concentration of CsCl

Density gradient of CsCl forms due to sedimentation by centrifugal force.

Light DNA N^{14}

N^{14}–N^{15} hybrid DNA

Heavy DNA N^{15}

DNA absorbs ultraviolet light. Ultraviolet light is used to locate DNA in tubes.

Before transfer to N^{14}

One cell generation after transfer to N^{14}

Two cell generations after transfer to N^{14}

Figure 5.10 Diagram of the Meselson-Stahl experiment, a demonstration of complementary strand separation during DNA replication by the use of the cesium chloride (CsCL) density-gradient technique. Bacteria were grown for many generations on a nutrient medium containing the heavy, nonradioactive isotope nitrogen-15 (N^{15}). During the DNA synthesis that precedes binary fission, nucleotides containing N^{15} were incorporated into the bacterial DNA. After several cycles of cell division, most of the normal N^{14} in the DNA had been replaced by the heavier N^{15}, making the molecules about 1 percent heavier and denser than normal. Using density-gradient centrifugation, Meselson and Stahl separated DNA containing N^{15} in one or both chains from DNA containing N^{14} in both chains. When DNA from the bacteria grown in the N^{15} medium (heavy DNA) was mixed with DNA from bacteria grown in a normal N^{14} medium (normal DNA) and the resulting mixture centrifuged, the two kinds of DNA formed distinct layers. The bacteria with heavy DNA were then transferred into a normal medium. After one cycle of cell division, each DNA double strand, or duplex, will have replicated, forming two daughter duplexes. According to the semiconservative replication scheme, each of the daughter duplexes should contain one of the original heavy chains and one light chain constructed from the precursors available in the normal N^{14} medium. As predicted, DNA extracted from these bacteria after a single division cycle in the normal medium formed an intermediate layer—halfway between the positions of the heavy and the light DNA duplex layers in the first test. Cells allowed to undergo two replications in the N^{14} nutrient should produce a population of cells whose DNA would consist of an equal mixture of light (N^{14}–N^{14}) duplexes and intermediate (N^{14}–N^{15}) duplexes. After three replications of the DNA, 75 percent should be of light intensity and 25 percent of intermediate density. These predictions were also confirmed in this experiment.

Printed by permission from J. D. Watson, Molecular Biology of the Gene, © 1965, J. D. Watson

Figure 5.11   Dr. George Wells Beadle, Nobel prize-winning geneticist noted for his work on *Neurospora crassa* leading to the one gene—one enzyme hypothesis.

result of George Beadle's and Edward Tatum's work with the red bread mold *Neurospora crassa* (Figure 5.12).

Neurospora has the enzymes necessary for the synthesis of all twenty amino acids that go into the formation of proteins. The synthesis of arginine, for example, begins with the synthesis of the ornithine molecule from a series of smaller precursor molecules. Ornithine is converted into citrulline and citrulline is converted into arginine. The pathway may be represented as ornithine →citrulline→arginine. Each step in arginine synthesis may be affected by a different mutation (an incorrect DNA message), which will result in the inability of the cells to perform that step. For example, mutations affecting the conversion of citrulline to arginine will prevent the synthesis of arginine, thereby forcing the cells of *Neurospora* to obtain arginine from the environment. A mutant that will grow when provided with either citrulline or arginine— but *not* when provided with ornithine—obviously has lost the ability to convert ornithine to citrulline. Therefore a given, single mutation affects only one of the steps in the pathway for the synthesis of arginine. These studies led to the hypothesis that the enzyme needed in each step of a biosynthetic pathway is provided for by a single gene.

In general, the concept of one gene—one enzyme has proven to be correct. It had to be modified, however, when it was discovered that many enzymes are composed of two (or even four) separate polypeptide chains, each of which is coded for by a separate gene. This discovery led to a restatement of the *one gene—one enzyme hypothesis* as the *one gene—one polypeptide hypothesis*. Actually it is no longer a hypothesis; its correctness has been extensively confirmed.

THE LINK BETWEEN DNA AND THE SYNTHESIS OF PROTEINS

The discovery that DNA is constructed of two polynucleotide chains held together in a helical configuration by hydrogen bond-

Figure 5.12   Photograph of the red bread mold *Neurospora crassa*. The single set of genes and relatively short life cycle (ten days between sexual generations) of this mold make it suitable for genetic research. Its ability to reproduce asexually enables a researcher to create a sizable sample of genetically identical individuals suitable for biochemical analysis.

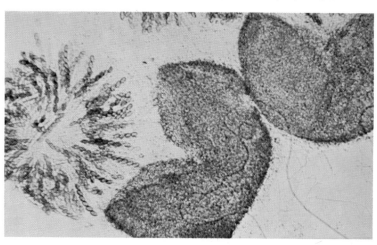

ing between matching nucleotides was a remarkable achievement. But the discovery in itself did not provide any direct or specific clue about the *localization* of genetic information in the molecule. Nevertheless, there was no reasonable alternative to the assumption that genetic information was stored in specific sequences of nucleotides. In some fashion, a given sequence of nucleotides was able to direct the sequence of amino acids during the synthesis of a protein molecule. A means by which nucleotides in DNA could act as a direct template, or a direct ordering force, in arranging the twenty different kinds of amino acids in protein molecules has never been found. Moreover, it had become apparent in the middle and late 1950s that most, if not all, protein synthesis is carried out in the cytoplasm of the cell (as discussed in Chapter 4), whereas DNA remains restricted to the nucleus.

At about the same time, evidence was accumulating that RNA synthesis requires the presence of DNA. It could be shown, for example, that removal of the nucleus from a cell by means of microsurgery results in the immediate cessation of all RNA synthesis in that cell. Furthermore, if a cell is given a radioactive precursor of RNA (for example, uridine that contains tritium atoms, known as H^3-uridine), the first RNA molecules to become radioactive are always found in the nucleus. The detection of the radioactive RNA was achieved by the technique of *radioautography*, in which a cell is covered by a thin film of photographic emulsion. Any region of the cell (in this case, the nucleus) containing radioactive material causes the exposure of the overlying photographic film (Figure 5.13).

Because the cytoplasm is rich in RNA, radioautographic experiments suggest that the RNA of the cytoplasm must come from the nucleus. Consider an extension of the experiment just described. A cell is fed on tritiated uridine for 5 minutes, as in Figure 5.13b, and then the radioactive uridine in the nutrient medium is replaced by nonradioactive (normal) uridine, thereby terminating the further incorporation of radioactivity into RNA. This cell is allowed to live for 60 additional minutes, during which time it continues to make RNA but, at this point, nonradioactive RNA. When the cell is killed and analyzed by radioautography, it is evident that a considerable amount of radioactive RNA has moved from the nucleus into the cytoplasm.

These and similar experiments led to the concept of *messenger RNA* (mRNA). Simply stated, during synthesis of RNA, the sequence of nucleotides in an RNA molecule is guided by a sequence of nucleotides in DNA in a template fashion, essentially in the same manner that the two chains of DNA act as templates for DNA replication. The four nitrogen-containing bases of RNA — *uracil* (U), *cytosine* (C), *guanine* (G), and *adenine* (A) — would be positioned by respective base pairing of the adenine, guanine, cytosine, and thymine of a DNA chain. In the case of RNA synthesis,

Figure 5.13 Series of radioautographs and electron micrographs depicting the progressive sequences involved in replication, transcription, and translation. (a) Radioautograph of DNA replication in *E. coli*. (b) In the upper radioautograph, a cell is fed on H^3-uridine for 5 minutes and then killed; all the RNA (black dots) is clustered in the nucleus. In the lower radioautograph, the cell has been fed as above, but the radioactive uridine has been replaced by normal uridine. The RNA (black dots) has moved from the nucleus into the cytoplasm. (c) Electron micrograph of a polyribosome with a thin strand of linking mRNA — the level of translation. (d) Radioautograph showing the production of protein with tritium-labeled leucine (an amino acid) in guinea pig pancreatic cells.

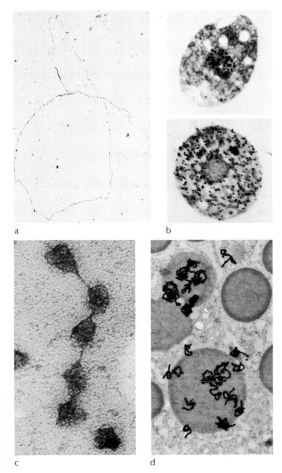

a

b

c

d

Figure 5.14 The chemical structures for the sugars, the nucleic acid bases, and the phosphate groups found in RNA and DNA. The DNA sugar deoxyribose is slightly different from the RNA sugar ribose in that it lacks one oxygen atom.

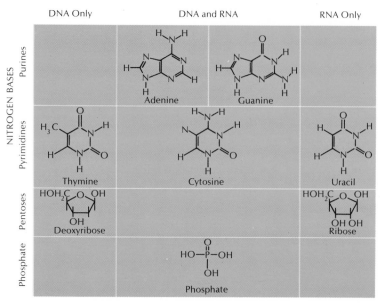

one or the other of the DNA chains acts as the template for the synthesis of an RNA molecule, which then migrates to the cytoplasm. In this process, the hydrogen bonds between the two DNA chains must be disrupted temporarily to allow templating for RNA synthesis. After dissociation of the newly synthesized RNA molecule from the DNA, the DNA duplex might re-form by regeneration of the hydrogen bonds; or the process of RNA synthesis might be repeated many times, producing many identical RNA molecules for export to the cytoplasm. The production of RNA molecules from DNA templates is called RNA transcription of DNA or, simply, *transcription* (Figure 5.15).

This concept of messenger RNA has proved to be essentially correct. The process has been directly observed in some detail in electron microscope studies. DNA supporting RNA synthesis was carefully isolated from a bacterial cell with partially completed RNA molecules still attached to and projected away from the DNA duplex. The enzyme molecules responsible for polymerization of the RNA molecules (RNA polymerase) are apparently still present in this preparation at the point where each partially completed RNA chain is attached to the DNA. Thus, the genetic information in DNA is transcribed into RNA molecules. These molecules then migrate into the cytoplasm where the information is translated into the sequences of amino acids during the process of protein synthesis.

Although this concept partially explains how genetic information is used in the cytoplasm, it does not indicate the chemical mechanisms by which the various genetically different RNA messages are translated in a way to produce the sequences of amino acids. The discussion of "translation," however, is better post-

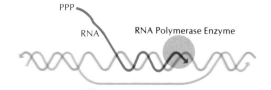

Figure 5.15 Transcription of a DNA template code to an RNA molecule. The enzyme RNA polymerase attaches to the DNA molecule, opening up a short section of the double helix for transcription. As RNA polymerase moves along the DNA template, the growing RNA strand peels off and attaches to a ribosome, while hydrogen bonds re-form the complementary DNA strands.
Printed by permission from J. D. Watson, Molecular Biology of the Gene, © 1970, *J. D. Watson*

poned until the nature of the genetic code in DNA and RNA has been considered.

THE GENETIC CODE

There are four different nucleotides in DNA (and RNA), but there are up to twenty different amino acids in proteins. If one nucleotide residue specified a single amino acid, then only four amino acids could be coded, using the nucleic acid alphabet. If two bases at a time were used, then sixteen amino acids could be specified. This number still is not enough, so it is most likely that at least three bases are required to specify one amino acid; then sixtyfour different combinations of three bases are possible, and this number is more than enough to specify the twenty amino acids found in proteins.

It is now known that the specification of nucleic acid to protein is in fact a *triplet* code. In the linear sequence of residues in an RNA molecule, three bases at a time are required to specify one amino acid residue in a protein chain. Each set of three bases is called a codon (as in DNA); each codon specifies an amino acid. There are sixty-four possible codons and only twenty amino acids, so it is clear that some amino acids may be specified by *more* than one codon (Table 5.1). All sixty-four possible codons are used in the genetic code. Three codons (UAA, UGA, and UAG) appear to act as end points, or "periods," in translation, giving the signal for termination of the polypeptide chain being formed.

The sequential addition of amino acids to a growing polypeptide chain takes place on small particles in the cytoplasm called ribosomes (see Chapter 4). The initial step in adding an amino acid to a growing polypeptide chain is a binding of the particular amino acid to the ribosome. (The second step is the joining of the amino acid to the chain.) Designation of which particular amino acid is bound is made by the mRNA that is directing the polypeptide synthesis. As the ribosome moves along the mRNA, the successive codons are translated. When the ribosome comes into con-

Figure 5.16 Structural formula for a section of ribonucleic acid (RNA). One of each of the nucleotide bases and associated sugar and phosphate groups found in RNA is shown. Because each of these bases contains nitrogen, they are sometimes called the nitrogenous bases.

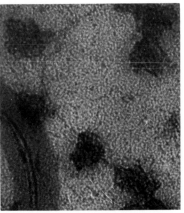

Figure 5.17 Electron micrographs of polyribosomes. Each ribosome consists of a large and a small subunit, each subunit containing ribosomal RNA molecules and a number of proteins. On the micrograph at right, the thin strand connecting the ribosomes is mRNA.

tact with a given codon, the corresponding amino acid becomes bound to the ribosome in preparation for its addition to the chain. The means by which mRNA causes specific amino acids to be bound to the ribosome is described in connection with protein synthesis. This situation was exploited to make the complete list of codon assignments. The various codons of nucleotides were synthesized and a given codon (for example, GCU) was added instead of mRNA to a cell-free system similar to that used to achieve protein synthesis. The presence of a particular codon caused the bind-

Table 5.1
Codons* in the Genetic Code

Codon	Message	Codon	Message
UUU	phenylalanine	AUU	isoleucine
UUC	phenylalanine	AUC	isoleucine
UUA	leucine	AUA	isoleucine
UUG	leucine	AUG	methionine
UCU	serine	ACU	threonine
UCC	serine	ACC	threonine
UCA	serine	ACA	threonine
UCG	serine	ACG	threonine
UAU	tyrosine	AAU	asparagine
UAC	tyrosine	AAC	asparagine
UAA	STOP	AAA	lysine
UAG	STOP	AAG	lysine
UGU	cysteine	AGU	serine
UGC	cysteine	AGC	serine
UGA	STOP	AGA	arginine
UGG	tryptophan	AGG	arginine
CUU	leucine	GUU	valine
CUC	leucine	GUC	valine
CUA	leucine	GUA	valine
CUG	leucine	GUG	valine
CCU	proline	GCU	alanine
CCC	proline	GCC	alanine
CCA	proline	GCA	alanine
CCG	proline	GCG	alanine
CAU	histidine	GAU	aspartic acid
CAC	histidine	GAC	aspartic acid
CAA	glutamine	GAA	glutamic acid
CAG	glutamine	GAG	glutamic acid
CGU	arginine	GGU	glycine
CGC	arginine	GGC	glycine
CGA	arginine	GGA	glycine
CGG	arginine	GGG	glycine

*These codons are sequences of nucleotides in mRNA. Each is represented by a letter symbolizing its base: U = uracil, C = cytosine, A = adenine, and G = guanine. Each codon causes the addition of a particular amino acid to the protein chain, except UAA, UAG, and UGA, which indicate the end of a protein chain.

ing of a particular amino acid to the ribosome present in the mixture. Testing the various codons with the different amino acids led to the complete deciphering of the triplet code.

According to the codon meanings in Table 5.1, mRNA with the base sequence AUGUUUCUCGCGGGG . . . will code for a polypeptide with the amino acid sequence methionine-phenylalanine-leucine-alanine-glycine. . . . This translation is based on a nonoverlapping code. Although it was expected that the code would be nonoverlapping, it was necessary to confirm this assumption. The mRNA message above could conceivably be read in a series of overlapping codons (Figure 5.18), yielding the code sequence AUG-UGU-GUU-UUU-UUC-UCU . . . and the amino acid sequence methionine-cysteine-valine-phenylalanine-phenylalanine-serine. . . . Such a reading of the code would occur if the ribosome advanced only one nucleotide each time it added an amino acid to the polypeptide chain. The genetic code was proven to be nonoverlapping by means of rather intricate genetic studies of a bacterial virus.

Implicit in the preceding discussion is the assumption that *the linear arrangement of nucleotides in a gene will translate ultimately into a linear arrangement of amino acids;* that is, the gene and its polypeptide product are *colinear.* Colinearity has been proved directly. The enzyme tryptophan synthetase in *E. coli* is composed of two polypeptide chains, which are coded for by two adjacent genes designated the A and B genes. A major part of the amino acid sequence of the polypeptide product of the A gene has been determined. Several known mutations in the A gene result in a single amino acid change in the A polypeptide. The single amino acid changes result in inactivation of the enzyme. By genetic mapping studies within the A gene, the position of the mutation in any given case has been shown to correspond with the position

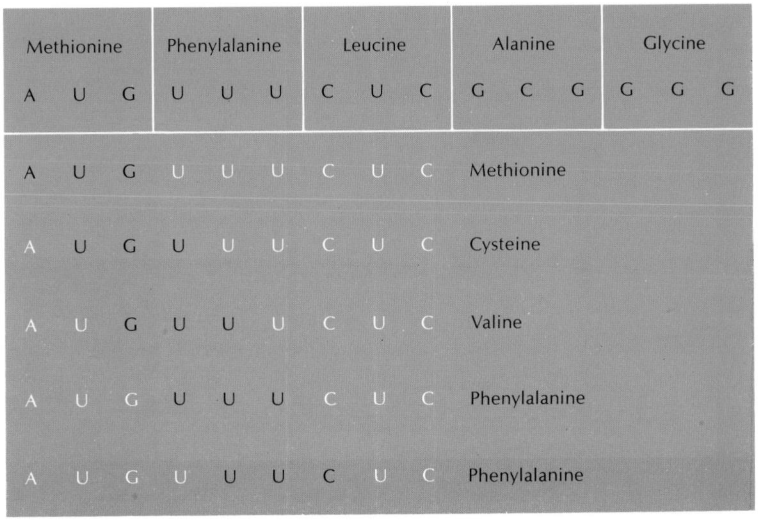

Methionine			Phenylalanine			Leucine			Alanine			Glycine		
A	U	G	U	U	U	C	U	C	G	C	G	G	G	G

A	U	G	U	U	U	C	U	C	Methionine
A	U	G	U	U	U	C	U	C	Cysteine
A	U	G	U	U	U	C	U	C	Valine
A	U	G	U	U	U	C	U	C	Phenylalanine
A	U	G	U	U	U	C	U	C	Phenylalanine

Figure 5.18 Comparison of the peptide products resulting from the same DNA base sequence being read as triplets in either an overlapping or nonoverlapping fashion. In the overlapping sequence, the reading frame is advanced one case at a time; in the nonoverlapping, three bases at a time. In an overlapping code, the amino acid sequence differs from that in a nonoverlapping code. The triplet code is known to be overlapping.

Figure 5.19 Mechanisms involved in protein synthesis. The left side of the illustration shows the activation of three different tRNAs, with specific anticodons for specific amino acids. ATP provides the energy for this reaction (the ATP's source being oxidative phosphorylation in the mitochondrion). The right side of the illustration depicts the translation process. Each ribosome has two tRNA binding sites. Through enzyme mediation, the polypeptide chain grows by the formation of a peptide bond between amino acid molecules (AA-1 and AA-2). After the bond has been formed, the tRNA is ejected from the amino acid-tRNA binding site. The movement of the mRNA over the ribosomal surface is still unknown, but the growing chain is translocated to a different site on the ribosome, allowing a different tRNA to bind and the transfer process to repeat itself. In this illustration, the ribosomal subunits attach themselves while the mRNA is moving from left to right. A given ribosome can read any mRNA molecule, and a single mRNA can flow through many ribosomes at the same time. Many identical protein molecules can be constructed by this polyribosome complex.

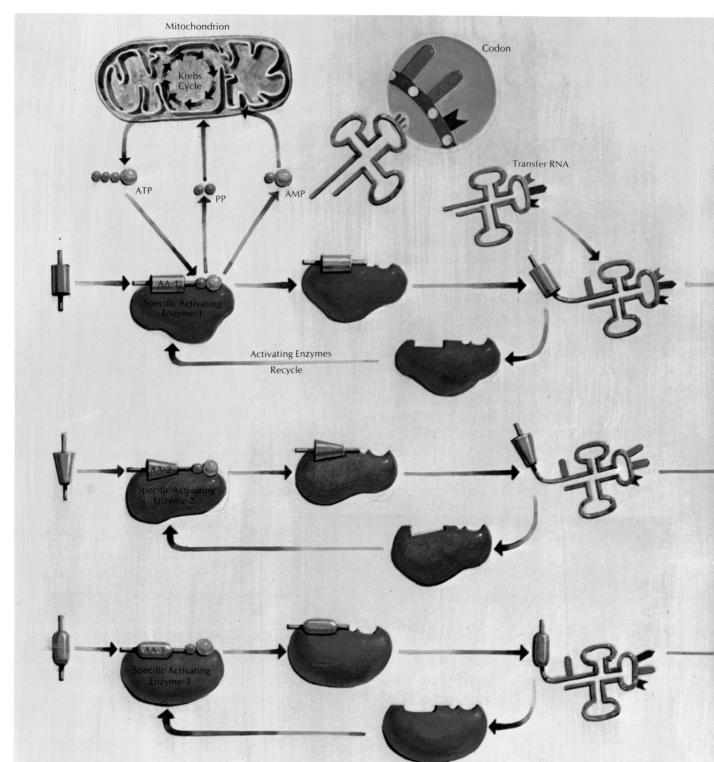

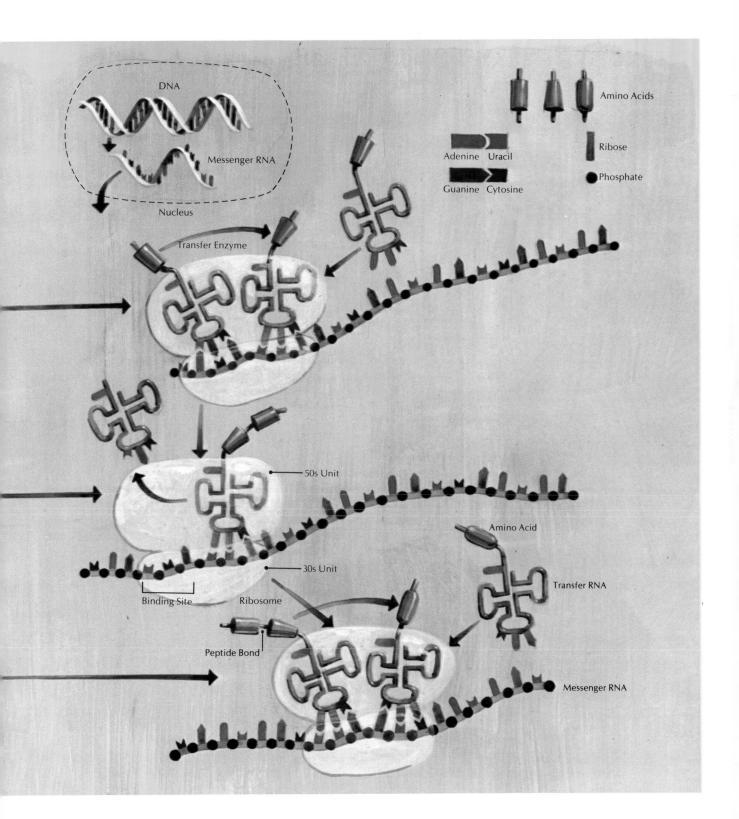

DNA

Messenger RNA

Nucleus

Amino Acids

Adenine   Uracil

Guanine   Cytosine

Ribose

Phosphate

Transfer Enzyme

50s Unit

30s Unit

Binding Site   Ribosome

Amino Acid

Transfer RNA

Peptide Bond

Messenger RNA

Figure 5.20   Transfer RNA (tRNA) delivering amino acids to build proteins. Of the three major types of RNA molecules found in all cells, the smallest are tRNA molecules. Transfer RNA performs the essential step in translating information from nucleic acid into protein; it is the "adapter" between an amino acid and its coded counterpart transcribed from DNA. It reads the coded information of the RNA molecule and inserts the corresponding amino acid into a growing polypeptide chain. It must be recognized by a specific enzyme, which equips the tRNA with the proper amino acid. It must also interact properly with the ribosome, the subcellular organelle in which protein synthesis occurs. Every cell contains at least one tRNA for each amino acid. Often tRNA also regulates transcription of the message itself. Specialized tRNA does not always serve as an amino acid donor but may be responsible for beginning or terminating synthesis of a polypeptide.

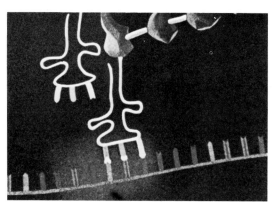

of the mutant amino acid in the peptide chain. Thus, the gene and the product it designates are colinear.

One of the important discoveries about the genetic code has been its universality. The triplet codon designations in Table 5.1 represent the same amino acids in all organisms—from viruses to highly evolved cells. The existence of one universal code implies a common ancestry for all biological systems now in existence.

THE TRANSLATION OF NUCLEIC ACID MESSAGES INTO PROTEINS

The central element in *translation* is the messenger RNA molecules derived by transcription of the DNA. The DNA in a chromosome (a threadlike body that appears in the nucleus of a cell during cell division) is probably one continuous molecule containing many genes in series. The place at which one gene ends and the next one begins has the property of a period at the end of a sentence. Presumably a series of nucleotides in the DNA contains the information that allows the identification of the beginning and end of a gene. Such identification is essential for the individualized control of the expression of single genes. In bacteria and perhaps in eucaryotic cells, certain genes with related functions are transcribed together as a unit with a single control point at one end of the gene series. Such a group of genes is called an *operon*.

In the cytoplasm, the mRNA combines with two pieces of machinery that will translate the message into a polypeptide chain. These two pieces of machinery are the ribosome and the *transfer RNA* (tRNA) molecules. The tRNA molecules are the decoding adapters that carry out the actual translation of the mRNA into a specific sequence of amino acids. For each triplet codon that stands for an amino acid, there is at least one different tRNA (Figure 5.20). Each of these tRNA molecules are composed of about seventy nucleotides. Hydrogen bond base pairing between uracil and adenine and between cytosine and guanine produce several regions with a double helical configuration, which results in three major loops and one stem protruding from the tRNA molecule. One of the loops contains a triplet of nucleotides that distinguishes it unequivocally from all the other kinds of tRNA molecules. This triplet is known as the *anticodon* because it is the complement to a given triplet codon in the mRNA. For example, the tRNAs containing the anticodon AAG will be a leucine tRNA, because AAG is the anticodon for CUU; and CUU is one of the codons for leucine.

The tRNAs are in turn matched by a series of amino acid activating enzymes that catalyze the addition of a given amino acid to the stem of the appropriate tRNA. The tRNA with the GAA (an anticodon for leucine) is recognized by one particular activating enzyme that will join leucine to that tRNA. In this way, each kind of tRNA becomes charged with the amino acid; the amino acid selected is the one for which the anticodon is present in that tRNA.

The various activating enzymes have the specificity to match the appropriate amino acid to the appropriate tRNA.

The next step in translation is the concurrence of the tRNA molecules (charged with amino acids) with the RNA message. Each ribosome is composed of a small and a large subunit, and each subunit is composed of RNA and a number of proteins. (The function of *ribosomal RNA* and some ribosomal proteins is not yet understood.) An RNA message binds at its beginning codon to the small subunit of the ribosome. This codon triplet is then matched by the corresponding anticodon of the appropriate tRNA. The tRNA molecule with its attached amino acid becomes bound to the large subunit of the ribosome with the codon-anticodon triplets in register. The second codon on the mRNA now directs the binding of a second tRNA by codon-anticodon matching to a second position on the ribosome. The two amino acids on the two bound tRNAs correspond to the first two codons of mRNA and these now become linked together by means of a peptide bond. This linkage is followed by release of the first tRNA from the ribosome and its replacement with a charged tRNA corresponding to the third codon of the mRNA. The amino acid of the third tRNA is now joined to the second amino acid by a peptide bond, and the second tRNA leaves the ribosome to be replaced by a fourth tRNA with its amino acid charge. The energy for these activities is supplied by ATP.

In sum, the tRNAs enter and leave the two ribosome binding sites in a reciprocating fashion under the specific guidance of the mRNA. Their movement is controlled in such a way that the appropriate amino acids are positioned for addition to the growing peptide chain. The peptide chain continues to grow until the codon UAA, UAG, or UGA is encountered on the message. This signals the termination of the gene, and the complete polypeptide is released from the tRNA that brought in the final amino acid.

The entire process of synthesizing many identical polypeptide molecules, each containing 200 amino acids, is accomplished in a matter of seconds. The genetic fidelity of the translation mechanism appears to be remarkably mistake-free. To date it is the only known mechanism by which cells synthesize proteins.

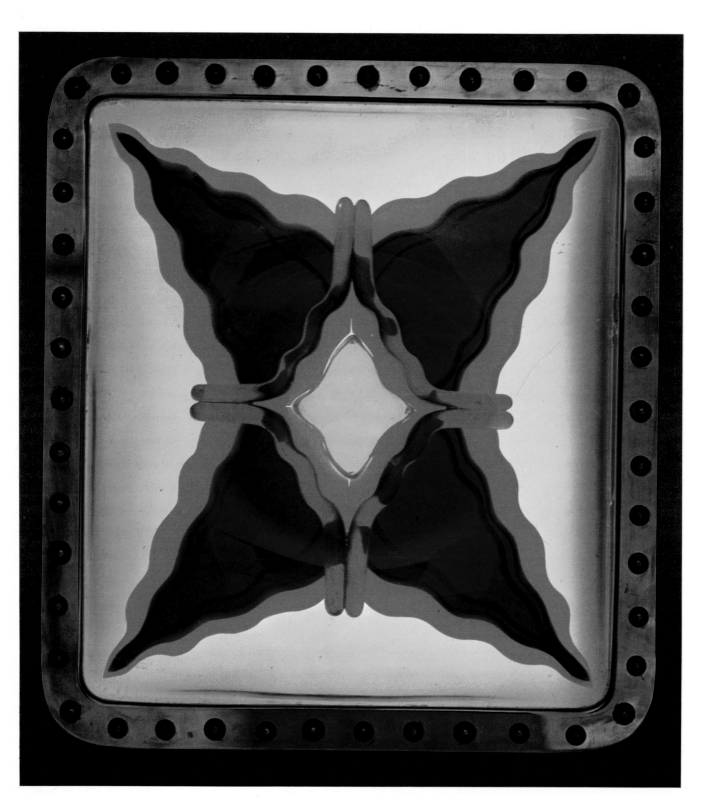

One of the characteristics of living systems is the ability to reproduce. Every cell and every organism alive today came from preexisting cells. The body of a 150-pound adult human contains about 100 trillion cells, all of which can be traced directly to the divisions of a fertilized egg called a *zygote*. The cell nuclei of an adult organism have the same amount of genetic information as the zygote and twice that of an egg or sperm nuclei.

Cells do more than just reproduce identical copies of themselves. After they go through cell division, cells grow and gradually change their structures and functions through the processes of "differentiation," which will be described further in Chapter 9. In multicellular organisms, such as humans and flowering plants, differentiation acts to make many different kinds of specialized cells with different functions.

Procaryotic cells—bacteria and blue-green algae—undergo a unique kind of cellular reproduction called "fission" (see Chapter 9). Shortly before the cell is to divide, a single circular molecule of DNA replicates and one molecule goes to each of the two cells produced at fission. Electron micrographs reveal that procaryotic DNA is generally localized in the nuclear area of the cell but is not segregated from the other cellular contents by a membrane. The DNA in eucaryotes is found within the membrane-bound nucleus in the more complex nuclear bodies called *chromosomes*. Chromosomes in eucaryotic cells are threadlike bodies that also contain RNA and protein in addition to DNA. For any particular species chromosomes are usually constant in size, shape, and number (Figure 6.1).

Chromosomes, which can be seen only at cell division, have at least one constriction called a *centromere*. This narrow region is where the spindle fibers, or microtubules, attach. The two structures on each side of the centromere are called "arms." The overall length of a chromosome and the ratio of its arm lengths are identifying features. In any normal human cell, such as a white blood cell, there are twenty-three pairs of chromosomes. The two chromosomes in each pair are called *homologous chromosomes*, and each chromosome of a pair can be referred to simply as a *homolog*. In sexual reproduction, one set of homologs is inherited from each original "parent" cell. In most stages of life cycles of advanced multicellular organisms, cells contain *two sets* of homologs.

Three different life cycles in eucaryotic organisms are depicted in Figure 6.2. Perhaps the term "life spiral" is more accurate than "life cycle" because the chromosomes and genes in each succeeding generation are somewhat different due to genetic recombination in sexual reproduction. The three life cycles have several features in common. First, they all have *haploid* stages during which each cell has only *one* set of homologs. (In such a haploid cell, the number of sets of homologs it contains is called the hap-

6

Cell
Division

Figure 6.1 Chromosome numbers of various organisms. In this list, the first number represents the diploid number (2n) of chromosomes. For example, the alligator possesses a total of thirty-two chromosomes, or sixteen pairs of chromosomes. In birds (such as the chicken, pheasant, pigeon, and turkey), male members have one more chromosome than females. Some species of plants have polyploid chromosome numbers, which are multiples of the basic diploid number (note the magnolia, onion, rose, and white ash).

Alligator	32
Amoeba	50
Brown Bat	44
Bullfrog	26
Carrot	18
Cat	32
Cattle	60
Chicken	78, 77
Chimpanzee	48
Corn	20
Dog	78
Dogfish	62
Earthworm	36
Eel	36
English Holly	40
Fruit Fly	8
Garden Pea	14
Goldfish	94
Grasshopper	24
Guinea Pig	64
Horse	64
House Fly	12
Human Being	46
Hydra	32
Lettuce	18
Lily	24
Magnolia	38, 76, 114
Marijuana	20
Onion	16, 32
Opossum	22
Penicillium	2
Pheasant	82, 81
Pigeon	80, 79
Planaria	16
Redwood	22
Rhesus Monkey	42
Rose	14, 21, 28
Sand Dollar	52
Sea Urchin	40
Starfish	36
Tobacco	48
Turkey	82, 81
White Ash	46, 92, 138
Whitefish	80
Yucca	60

loid number (n) for the species.) Second, all the life cycles have *diploid* stages during which the cells have *two* sets of homologs (2n). Third, they all undergo two kinds of cell division: mitosis and meiosis. In *mitosis*, the "daughter" cells produced have the same number of chromosomes as in the parent cell (2n), and in *meiosis*, the cells are produced with *half* the number of chromosomes (n) as their parent cell. The fourth feature that all eucaryotic life cycles share is the process of *fertilization*, where haploid cells (n) are joined to form a diploid *zygote* (2n). In addition, many organisms (especially among plants and lower animals) may have some form of reproduction that does not immediately require fertilization.

Reproduction, thus, produces new organisms that embark on their own journeys of cell division, growth, and differentiation, which lead, eventually, to their own capacity to reproduce. Or, as the nineteenth-century novelist Samuel Butler put it, "A hen is only an egg's way of making another egg."

We all have a general idea of the cyclic nature of life—we have watched plants and animals come into being, grow, mature, reproduce, and eventually die; we have observed the same broad sequence of events among human beings. This cyclic nature of life is also present at the cellular level, where the story of the continuity of life begins.

Figure 6.3 shows the cell cycle of a typical eucaryotic cell. The long period called "interphase" is the active metabolic stage when most or all the components of a cell are synthesized. The reproductive period in which cell division (mitosis or meiosis) takes place is, by comparison, usually short. The DNA molecules of the chromosomes are replicated during interphase so that when cell division occurs the two sets of DNA can be equally divided between daughter cells in mitosis. The other parts of the parent cells are divided less equitably, and the daughter cells may differ in size and in cytoplasmic components.

A technique called radioautography (described in Chapter 5) has become a useful tool for the study of DNA synthesis during interphase. It was found, using this technique, that there is an interval, or gap (called the G_1 period), after the end of cell division. This period may last for several hours during which much RNA and protein are synthesized but no new DNA synthesis occurs. The G_1 phase is followed by the *S-phase* of DNA synthesis, which lasts from a few minutes in rapidly dividing embryonic cells to 6 to 8 hours in bean root cells used in the original radioautography studies. After the S-phase, there is another gap—the G_2 period—of variable length during which no new DNA synthesis occurs. Generally the G_2 period is not as metabolically active as the G_1 period. However, the G_2 period is extremely active in those cells that will produce large yolky eggs in the process of meiosis.

Studies of DNA synthesis in many kinds of plant and animal cells have shown the same general cycle of activity. However, in procaryotic cells, it was found that DNA synthesis proceeds almost continuously throughout the short life cycle.

MITOSIS

For descriptive purposes, the process of mitosis has been divided into phases (Figure 6.4). In living cells, however, mitosis proceeds without sharp changes from one phase to the next; the stages are thus artificial descriptive aids, not inherent discontinuities in the process.

With the aid of a light microscope we can observe the nuclei of cells during the nondividing *interphase* stage of the cell cycle. In this stage, the DNA material appears in granules or threads called *chromatin*, the nucleolus is visible, and the cell is preparing for division.

The first signs of mitosis appear during *prophase*, when the chromosomes become distinctly visible. The chromatin appears to condense into highly twisted chromosomes. Each chromosome is made of two coiled threads, called *chromatids*, which are joined at a point called the *centromere*. A chromatid, then, is a structure that *shares* a centromere, and two joined chromatids compose a double-stranded chromosome. Early in prophase, the cell becomes rounder; the cytoplasm becomes denser and more viscous. As prophase progresses, the chromatids become shorter and thicker, the centromeres become more pronounced, and the chromosomes move toward the nuclear membrane. The *spindle*—a structure made up of thin, microtubular filaments stretching across the cell

Figure 6.2   Examples of three types of life cycles: (A) a primitive plant—such as the green alga *Ulothrix*; (B) an intermediate plant—such as a fern or a moss (higher plants have a reduced multicellular haploid stage); (C) an animal—such as a man or some other multicellular animal.

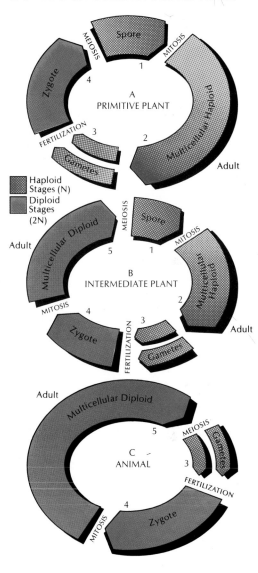

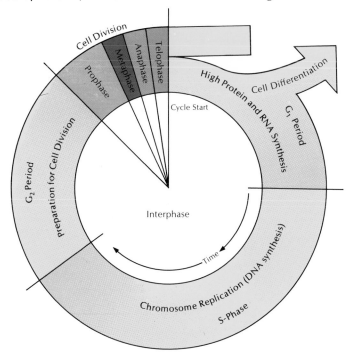

Figure 6.3   Diagram illustrating the life cycle of a generalized cell. Note the time difference between interphase and the cell division stages. The total time required to complete one cycle varies from one cell type to another and is influenced by environmental conditions, such as temperature.

137

Figure 6.4   Sequence of light microscope photographs of mitotic stages in the division of an onion root cell. Mitosis is a continuous, dynamic process, and this sequence must be viewed in that context. The body cells of onion plants each have sixteen chromosomes. The diagrams depict six of these chromosomes, which are undergoing movements characteristic of mitosis.

Courtesy Carolina Biological Supply Company

1. Interphase. Note the darkly stained nucleus, containing what appears to be granular chromatin. Often called the "resting stage", this is actually a period of high metabolic activity within the cell nucleus.
2. Prophase.   Chromosomes are quite distinct; as depicted in the drawing, each is composed of two coiled, threadlike filaments called chromatids.
3. Early metaphase.   Chromosomes are beginning to become oriented on the equatorial plate with their "arms" pointing toward the poles.
4. Late metaphase.   Chromosomes have become noticeably oriented on the equatorial plate (lengthwise to each other but not in pairs).
5. Early anaphase.   Single-stranded chromosomes have formed, with division of the centromere that joined the two chromatids of each chromosome.
6. Late anaphase.   Single-stranded chromosomes move toward the poles. The spindle filaments attached to each chromosome appear to be pulling them from the equatorial plate.
7. Early telophase.   The clumping of daughter chromosomes at the poles can be seen. Spindle filaments joining chromosomes across the equatorial plate are also visible. These spindle filaments apparently pull the chromosomes apart during their movement across the cell.
8. Late telophase.   Cytoplasmic division is visible in this photograph, with a cell plate forming along the equatorial plate. Individual chromosomes have lost their definition, and the formation of daughter nuclei is in progress.
9. Interphase (daughter cells).   Two new identical daughter cells have formed. These are conspicuous from surrounding cells at this time by their reduced size.

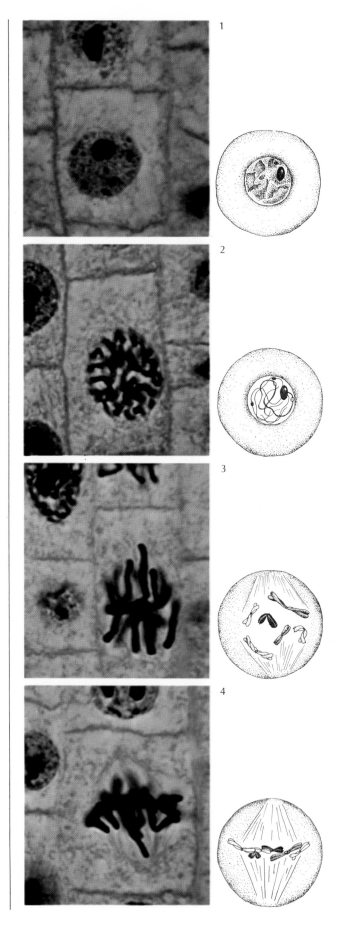

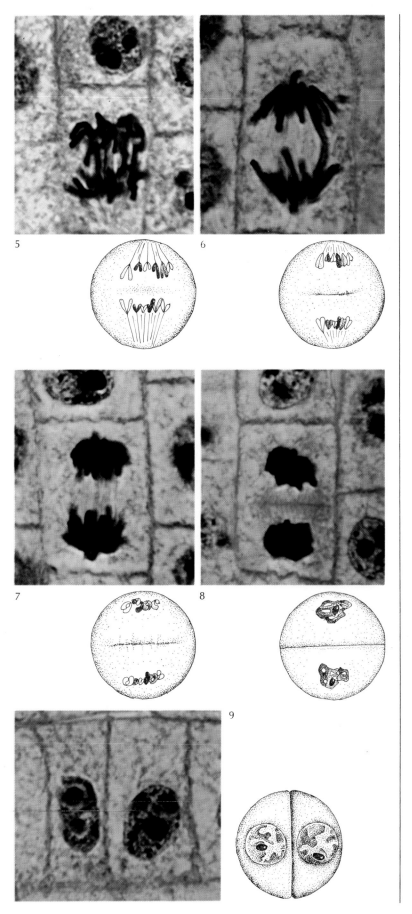

5

6

7

8

9

5. Early anaphase. Chromatids are beginning to separate, with division of the centromere joining the two chromatids of each chromosome.
6. Late anaphase. Separation of chromatids continues with movement occurring toward the poles. The spindle filaments attached to each chromatid appear to be pulling them from the equatorial plate.
7. Early telophase. The clumping of daughter chromatids at the poles can be seen. Spindle filaments joining chromatids across the equatorial plate are also visible. These spindle filaments apparently push the chromatids apart during their movement across the cell.
8. Late telophase. Cytoplasmic division is visible in this photograph, with a cleavage plate forming along the equatorial plate. Individual chromosomes have lost their definition, and the formation of daughter nuclei is in progress.
9. Interphase (daughter cells). Two new identical daughter cells have formed. These are conspicuous from surrounding cells by their reduced size.

Figure 6.5   Radioautographs showing untagged chromosomes of the lily plant *Bellevalia* (2n = 8) in metaphase. Chromosome replication has been extensively studied through the use of tritiated-thymidine, or H³-thymidine, labeling. This substance is known to be used almost exclusively in DNA synthesis. Root cells of several plants were the first test subjects for H³-thymidine labeling studies. (a) Because *Bellevalia* has only eight large chromosomes in each cell, particularly clear results were obtained from studies of its root cells. Bulbs and roots of the plant were immersed in a solution containing H³-thymidine for one to two hours. Some of the roots then were fixed and squashed on glass slides, while the others were put into a nonradioactive solution to continue growing. The radioautographs showed that about one-third of the cells had incorporated the labeled thymidine into their DNA. This observation was consistent with the expectation that about one-third of the cells would be in S-phase at any given time.
(b) Radioautographs of labeled cells reaching metaphase showed that all of the chromosomes were about equally labeled and that both chromatids of each pair had incorporated the H³-thymidine into their DNA. This observation confirmed the model of DNA replication suggested by Watson and Crick. Each chromosome entering S-phase contains a double strand of DNA. During DNA replication, this strand separates and a new complementary strand is built on each of the original strands. Thus, each chromatid would be expected to contain one original unlabeled strand and one labeled strand built during immersion in H³-thymidine solution. Next, samples were taken from roots that had grown in nonradioactive solutions for 36 hours after immersion in H³-thymidine solution. Because the complete cell cycle was known to take about 24 hours, these cells would be expected to have passed through one S-phase in the unlabeled solution. The chromosomes of each labeled cell would have entered that S-phase with a double strand of DNA in which one strand was labeled and the other unlabeled. When the strands separated, each would acquire an unlabeled complementary strand. Thus, cells in metaphase after 36 hours would be expected to have chromosomes in which one chromatid was labeled and its sister chromatid was unlabeled. (c) Some cells did show the expected pattern, but some segments had broken and had been exchanged between sister chromatids.

between two poles—begins to form. The nuclear membrane and the nucleolus disintegrates, and the chromosomes move toward the center of the cell, where they arrange themselves in a plane perpendicular to the spindle axis about midway between the two poles.

When the chromosomes have become oriented within the plane, the cell is in *metaphase*. Some of the filaments of the spindle have become attached to the centromeres of the individual chromosomes by this time.

The next phase, *anaphase*, begins with the separation of the chromatids in each chromosome. The centromere appears to split in two, producing two single-stranded chromosomes from each double-stranded one. The newly formed chromosomes move to opposite poles of the spindle, where they gather into compact groups. During this movement, the chromosomes look as if they are being dragged across the cell by the spindle filaments attached to the centromere. The filaments between the two groups of single-stranded chromosomes look as if they are being stretched. The indentation of the cell membrane or the laying down of a cell wall along with the gathering of the daughter chromosomes at the two poles marks the beginning of *telophase*. During this phase, the chromosomes become longer and thinner, nuclear membranes form around the two groups of chromosomes, nucleoli reappear, and the cytoplasm usually is divided to form two daughter cells.

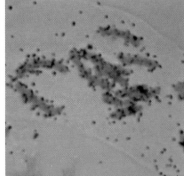
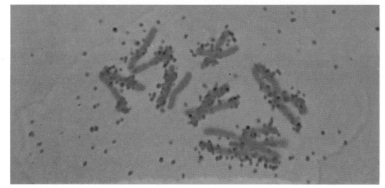

The two daughter cells, each containing one of the newly formed nuclei, now return to the stage of interphase.

The Spindle

During metaphase, a transparent spindle forms that is made up of fibers running from pole to pole and from chromosome to pole. The structure of the spindle is not yet well understood. In many kinds of cells, filaments or fibers radiate into the cytoplasm from the poles, forming *asters*. The spindle, the asters, and the centrioles (in the cells that possess them) make up the *mitotic apparatus*. The mitotic apparatus has been isolated from some kinds of cells during division (Figure 6.7). The main component is a protein; water, RNA, and ATP are also present in significant amounts. As much as 15 percent of the protein in the cytoplasm may be used in the formation of the mitotic apparatus.

There is still considerable debate about how the chromosomes are separated in the spindle. The protein fibers of the spindle appear in electron micrographs as microtubules. Visual observation in the light microscope suggests that the spindle fibers contract between the chromosomes and the poles, and then elongate between the pairs of chromosomes, with the fibers pushing or pulling the chromosomes apart. There is some evidence that ATP may be necessary for the contraction and elongation of the spindle fibers. Electron micrographs confirm that spindle fibers attach to the chromosomes at the centromeres. It has been postulated that the fibers leading from centromere to pole begin to be formed at the centromere and extend toward the pole as they grow. Some investigators have suggested that the fibers contract with expenditure of ATP energy. Others have suggested that the fibers between centromere and pole decrease in length by loss of protein molecules, whereas those between centromeres grow longer by addition of protein molecules. The mechanism of chromosome movement during anaphase, however, remains a mystery.

The Chromosomes

Synthesis of the major components of the chromosomes occurs during the middle of interphase at the S-phase of the cell cycle, although some events essential to chromosomal reproduction may occur during the G_2 period or in prophase. In addition to DNA replication, the synthesis of the basic proteins associated with chromosomal DNA—called *histones*—occurs during the S-phase. RNA and other protein components of the nucleus are synthesized throughout interphase.

The arrangement of DNA in chromosomes (whether single- or double-stranded) is still a controversial matter, but it seems clear that long chains are folded to form nucleoprotein fibrils about 250 angstroms in diameter. The DNA molecules, which are only about 20 angstroms in diameter, are folded or coiled in some unknown

Figure 6.6 Proposed diagrammatic representation of chromosome organization and replication that interprets the radioautographic results of Figure 6.5. Solid lines represent nonlabeled units, and those units in dashed lines are labeled. The dots represent grains in the radioautographs.

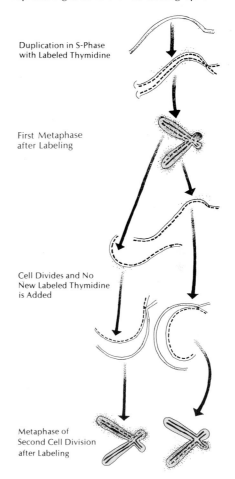

Duplication in S-Phase with Labeled Thymidine

First Metaphase after Labeling

Cell Divides and No New Labeled Thymidine is Added

Metaphase of Second Cell Division after Labeling

Figure 6.7 Phase-microscope photograph of an isolated mitotic apparatus from a fertilized sea urchin egg (magnified 1,000 diameters). In the center of the photograph is the spindle (which is composed of two sets of chromosomes) and the spindle fibers (which have asters at either end). The light area in the center of the asters represents the mitotic poles.

Figure 6.8   Photomicrograph depicting the giant salivary cell chromosomes of *Chironomus tentans*. Note the distinct banding—dark bands contain DNA and green bands contain RNA. Also note the two distinct chromosome "puffs." These are called *Balbiani rings* and may be areas of gene activity where RNA transcription is taking place.

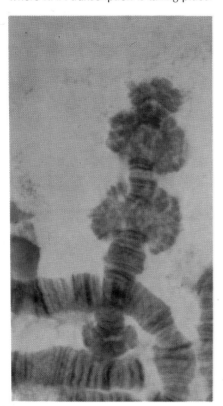

way in the larger fibrils. The fibrils, in turn, appear to extend as rather loose loops from the central axis of the chromosome.

In certain tissues, such as the salivary glands of the larvae of flies and mosquitoes, giant chromosomes are formed (Figure 6.8). These giant chromosomes are more than 1,000 times the volume of the corresponding normal chromosomes. They are thought to be formed by repeated replications of the DNA molecules, forming perhaps 1,000 copies of the DNA molecule within a single giant chromosome. Because the giant chromosome is about the same length as the corresponding normal, uncoiled interphase chromosome, the giant chromosome is thought to be made up of DNA strands lying side by side with little or no coiling. For this reason, the giant chromosomes are called *polytene* (many-stranded) chromosomes. Regular patterns of light and dark bands appear on the polytene chromosomes. These patterns have been mapped and have been shown to be consistently the same for a given kind of chromosome from a given species.

The Centrioles

All eucaryotic cells except those of higher plants have centrioles, and these structures appear to play a major role in the formation of the mitotic spindle. Two pairs of centrioles exist in the interphase cell, with the centrioles of each pair at right angles to each other. As prophase begins, the pairs of centrioles separate. Each pair lies at the center of an aster, and the spindle fibers appear to be spun from the centrioles as they separate. Electron micrographs, however, have shown that the spindle fibers do not actually contact the centrioles.

During telophase, as the two daughter nuclei form, each of the two pairs of centrioles replicates to give two pairs of centrioles to each daughter cell. Although the centrioles appear to play a major role in cell division in most eucaryotic cells, the higher plant cells that lack centrioles are able to construct a mitotic apparatus almost identical to that of the cells having centrioles.

Cytokinesis

The final step in mitotic cell division, occurring at the end of telophase, is *cytokinesis*, or cell cleavage. In this process, the nuclei and cytoplasm of the daughter cells are separated by the plasma membrane to form two complete and independent cells. In some cells, mitosis may occur without cytokinesis, thus forming cells with more than one nucleus.

The mechanism of cytokinesis is markedly different in animal and in plant cells. In plant cells, a structure called the cell plate forms along the plane of the spindle during telophase. This cell plate appears to be composed of membranes from the Golgi complexes, which gradually fuse to form new plasma membranes that separate the two daughter cells. The growth of the new plasma membranes probably begins near the center of the cell and proceeds outward, until the membranes fuse with the membrane of the parent cell.

In animal cells, the cell membrane constricts or furrows, gradually closing in until the two daughter cells are separated. The mechanism by which this furrowing is carried out is unknown but probably involves contractile microfilaments. The fact that cell cleavage has been observed in cells from which the mitotic apparatus has been removed rules out the hypothesis that the cell membrane is pulled inward by the spindle or aster fibers. Hydrolysis of ATP is apparently involved in the process of furrowing, and there is some evidence that a contractile protein is involved in the process, obtaining energy for its contraction from ATP.

MEIOSIS

Mitosis is more than a process of cell reproduction that occurs in multicellular organisms. It is *the* mode of organismal reproduction for a sizable number of unicellular organisms. As such, mitosis is an efficient asexual process in terms of time and material. Only one parent is required and that parent acquires a kind of "immortality" by being converted into two offspring that are identical to it. However, this advantage of efficiency goes hand in hand with the major disadvantage of mitosis as a process by which organismal reproduction is accomplished.

It is evident that mitosis maintains the constancy of chromosome number and kind from one cell generation to the next. But it does not provide any mechanism for "shuffling" genetic information; it cannot rearrange genes that are already in the cell into new combinations. Meiosis solves this dilemma by producing reproductive cells, the *gametes*, which contain one-half the chromosome number (n) instead of $2n$, thereby making cell "fusion" feasible (Figure 6.9). This fusion process is called *fertilization* and will be described further in Chapter 9.

Sexual reproduction is customarily identified almost exclusively with diploid organisms and is a rare occurrence among haploid organisms. The evolutionary advantage of genetic reshuffling among early haploid creatures required the introduction of meiosis as a means of restoring their traditional haploidy. Thus diploidy was at first an intervening condition of sexual reproduction, with no more significance than just that. But the adaptive advantages of diploidy proved to be so enormous that roles were reversed in the evolution of higher organisms, with diploidy becoming the dominant phase and haploidy the intermediate one. Diploidy made possible the almost explosive evolution of eucaryotic organisms into the spectacular array of higher animals and plants that now populate the earth. Evolution can only proceed in the presence of genetic variation. Diploid organisms have two methods of producing variation: mutation to produce new genes and genetic shuf-

Figure 6.9   Sequence of meiotic stages in the
division of the lily *Lilium michiganense*. The body
cells of the lily have twenty-four chromosomes, but
only six are shown undergoing meiosis
in these diagrams.

Courtesy Carolina Biological Supply Company

1. Early prophase.   The chromosomes are
becoming visible.
2. Late prophase.   The double-stranded chromosomes lie close to the periphery of the
nucleus in pairs.
3. Metaphase I.   The chromosomal pairs lie side
by side along the equatorial plate.
4. Anaphase I.   The individual double-stranded
chromosomes of each pair separate and move
toward the poles, possibly pulled by the
contracting spindle fibers.
5. Telophase I.   The double-stranded chromosomes have reached the poles and are
bunched up tightly. They are separating into
the chromatin network of the daughter nuclei,
and a new membrane is forming.
6. Interphase.   Two nuclei are visible. Note the
remnants of the spindle fibers.
7. Prophase II.   Chromosomes are condensing,
and the spindle is forming as the nuclei
prepare for the second division.
8. Metaphase II.   Spindle fibers have appeared,
and the chromosomes are lining up along the
equatorial plate.
9. Anaphase II. The centromeres have replicated,
and single-stranded chromosomes are moving to
the opposite poles of the spindle.
10. Telophase II. Nuclear membranes are forming
around the haploid sets of chromosomes. The
plasma membranes and cell walls will separate
the daughter cells.
11. Spores.   The diploid cell has become four
haploid spores.

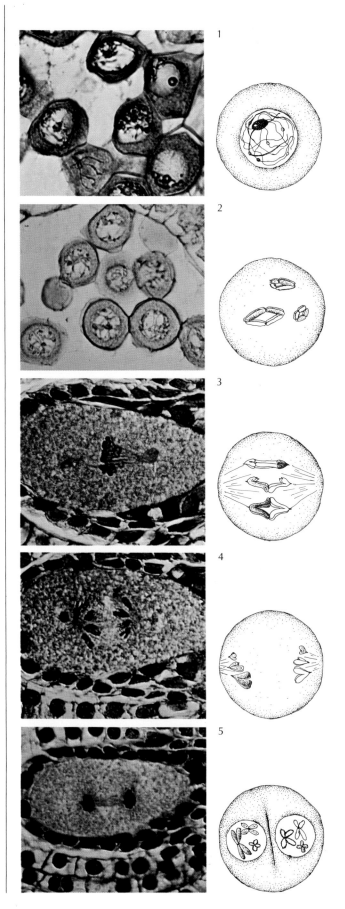

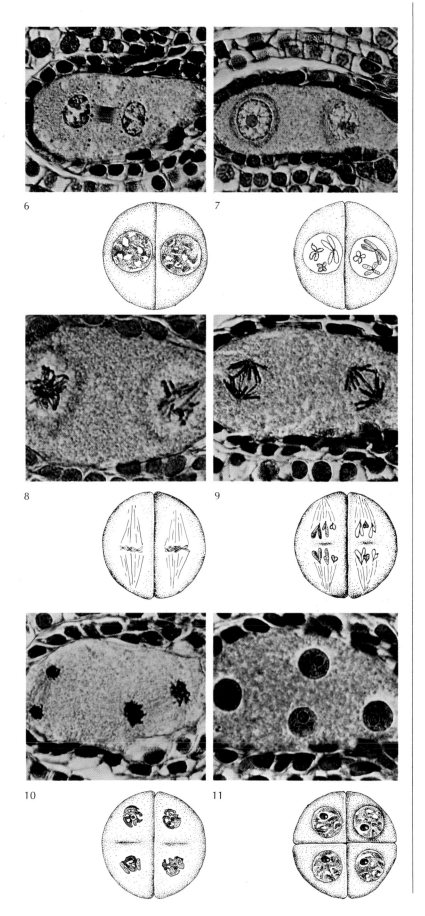

Figure 6.10 This series of diagrams shows stages in animal meiosis. Primary sex cells (spermatogonia or oogonia) can produce many different variations of chromosomal material in their gametes by means of crossing over (which occurs at Prophase I) and independent assortment (which occurs at Metaphase I). Meiosis also enables organisms to reduce the chromosome number in their gametes so that fertilization will restore the diploid complement. Only the nuclear areas are shown on these diagrams.

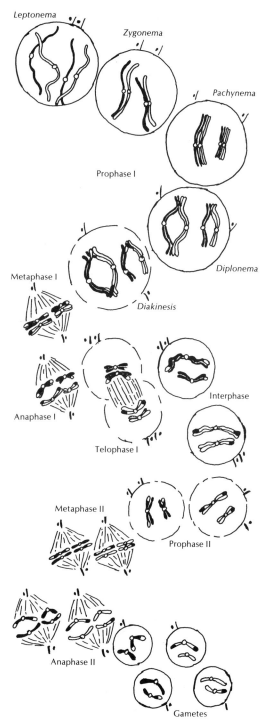

fling, or recombination, to produce new combinations of genes that can act in new ways.

Meiosis in diploid organisms involves two successive cell divisions preceded by DNA replication (Figure 6.10). In the first meiotic division (meiosis I), pairs of homologous chromosomes are separated but, unlike mitosis, each chromosome retains *both* chromatids. In the second meiotic division (meiosis II), which follows a brief interphase, centromeres replicate, and the chromatids are separated. Mitosis produces two diploid daughter cells from a single diploid parent, and meiosis produces four haploid daughter cells from a single diploid parent.

Meiosis I

Meiosis I is characterized by a long and complex *prophase I*. As in other prophases, the first visible change is a gathering of chromatin into long, tangled, threadlike chromosomes less than 1 micron in diameter. Because prophase I of meiosis is so complex, it has been subdivided into several stages. The first stage is called the *leptotene* stage. The leptotene chromosomes appear longer and thinner than the chromosomes of mitotic prophase. They are distinguished by the presence of a series of dark granules, or *chromomeres*, along the chromosomes. Although studies of DNA synthesis indicate that each chromosome must already contain two double strands of DNA at the beginning of the leptotene stage, the chromosomes are not visibly double as they are in mitosis.

In the *zygotene* stage, each chromosome pairs with its homolog in a regular fashion. Each chromomere matches up with the corresponding chromomere on the homolog. The chromosomes continue to shorten and thicken. The joining of homologous chromosome pairs is called *synapsis*. Homologous chromosomes always synapse in pairs even in "polyploid" nuclei, where several copies of each homolog are present. At the close of the zygotene stage, the nucleus appears to have only the haploid number of chromosomes, but each apparent chromosome is actually a pair of homologs closely bound together.

In the *pachytene* stage, the chromosomes shorten and thicken even more. Segments are interchanged between pairs of homologous chromatids. This process apparently involves breakage of the chromatids at corresponding points, interchange of the two segments, and rejoining of the chromatids.

In the *diplotene* stage, the homologous chromosomes begin to separate, and the pair of chromatids making up each chromosome is visible. The chromosomes do not separate entirely but are joined together at their ends and cross each other at points called *chiasmata* (Figure 6.11).

Each chiasma, which is also known as a *cross-over area*, represents the approximate point at which an exchange of chromatid segments between the two chromosomes occurs. One chiasma is formed for each pair of homologous chromosomes, and longer pairs generally form several chiasmata.

In the next stage, *diakinesis*, the chromosomes become maximally shortened and thickened. The chiasmata appear to move toward the ends of the paired homologs until each is held together only by chiasmata at its ends. Diakinesis is the final stage of prophase I, and it is accompanied by disintegration of the nucleolus and nuclear membrane and the formation of a spindle.

Metaphase I has one major difference when compared to metaphase in mitosis. Chromosome pairs are side by side at the equatorial plate rather than lengthwise to each other as in mitosis. In metaphase I of meiosis, the centromeres of each paired homolog usually are separated and lie off the plane toward the poles.

In *anaphase I*, the chromosomes move toward the poles of the spindle. At this stage, each chromosome is made up of a pair of chromatids, joined by the centromere. As the two centromeres of the pairs of homologs begin to move toward opposite poles, any remaining chiasmata slide apart, and the pairs of homologous chromosomes are separated from each other. Anaphase I ends with a complete haploid set of chromosomes clustered at each of the poles. But these chromosomes are not quite the same as the chromosomes that existed at the beginning of prophase I. Each chromosome contains one of its original chromatids and one chromatid that is a mixture of segments from its own original chromatid and from a chromatid of its homolog. This is due to chiasmata, or cross-over, phenomena.

In *telophase I*, nuclear membranes may form around the two sets of chromosomes, the chromosomes uncoil, and a cell membrane is formed between the two nuclei.

After a relatively brief interphase, the two haploid cells enter meiosis II. In some species, there is no noticeable interphase, and meiosis II may proceed without any formation of nuclear or cell membranes and without uncoiling of the chromosomes. In any case, the chromosomes pass from meiosis I to meiosis II without further chromosomal replication.

Meiosis II

Prophase II is relatively brief, particularly in species characterized by minimal telophase I and interphase. It is marked by gathering of the chromosomes and formation of the spindle. In *metaphase II*, the chromosomes become aligned in the plane of the spindle, and the centromeres become attached to the spindle fibers. In *anaphase II*, the daughter centromeres separate and the individual single-stranded chromosomes move to opposite poles. During *telophase II*, the nuclei are formed around the resulting haploid sets of chromosomes, and cell membranes separate the daughter cells.

The second meiotic division is similar to mitosis, but it begins with a haploid number of chromosomes and ends with the forma-

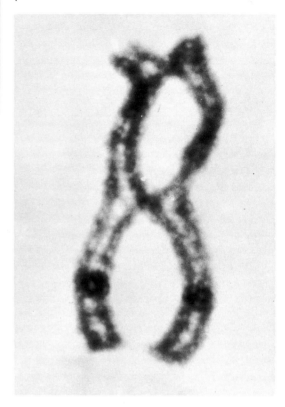

Figure 6.11   Micrograph of a pair of homologous chromosomes. The chromatids cross each other at points called chiasmata.

147

Figure 6.12   Successive stages in the embryological development of the frog. The number of cells is increased by mitosis, and the form of the embryo is changed by differentiation. Growth is restricted until such time as the digestive tract is complete and the tadpole larvae begin to feed. From left to right, the following sequences are depicted: 1–6, early cleavage (mitosis); 7–9, blastula (hollow ball of cells); 10–11, gastrula (digestive tract starting to form and beginning of organogenesis); 12–15, formation of the neural tube.
Courtesy Carolina Biological Supply Company

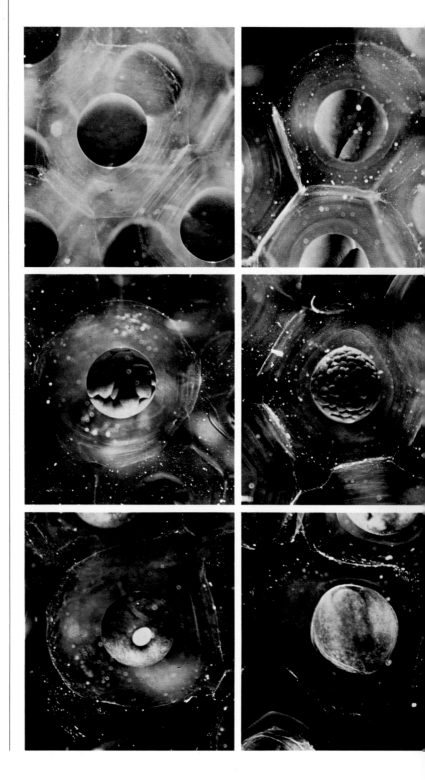

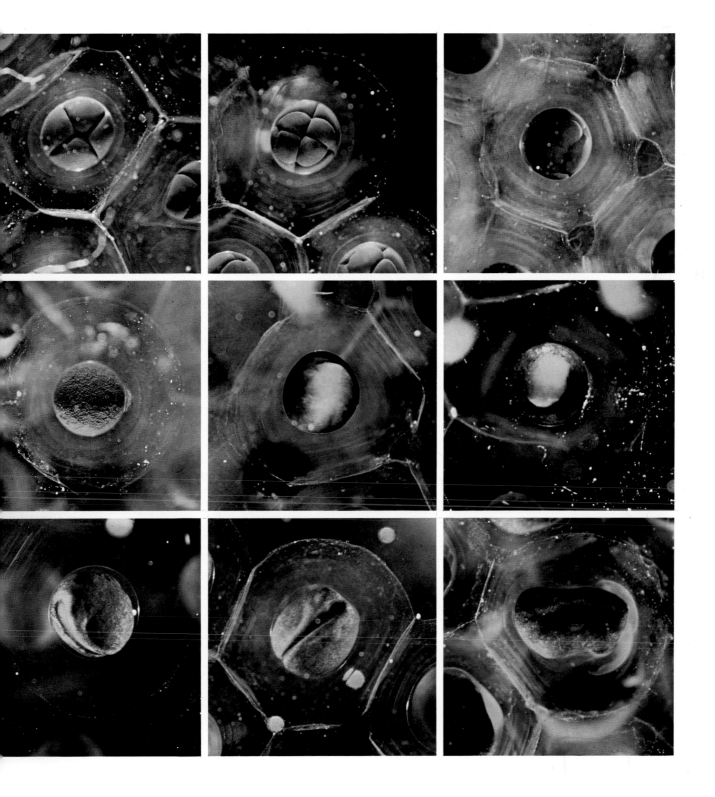

tion of daughter cells that contain a haploid set of chromosomes. Each of the haploid cells produced by meiosis could fuse with a haploid cell from another individual to form a normal diploid cell—a 2n zygote. This diploid cell will contain a complete set of chromosomes from each of the parent organisms. In many plant life cycles, meiosis produces haploid spores which, through the processes of mitosis, growth, and differentiation, can produce a haploid adult.

Primary Reproductive Cells

Meiosis occurs only in the *primary reproductive cells* of sexually reproducing plants and animals.

In the male animal, the primary reproductive cell gives rise by repeated mitotic divisions to a large number of cells called *gonocytes*, which become specialized to *spermatogonia*. Each spermatogonium divides twice mitotically to form four primary *spermatocytes*. Each of the spermatocytes then undergoes meiosis to form

Figure 6.13 Comparison of spermatogenesis in a multicellular male animal (left) and oogenesis in a multicellular female animal (right). In both the male and the female, meiotic divisions produce four haploid cells. In the male, all four cells will become functional gametes; in the female, however, only one cell will form a functional egg. At birth, a female human being has all the oogonia that she will need for life. Mitosis is therefore halted before birth, and meiosis occurs from puberty to menopause. In a male human being, new spermatogonia are produced by mitosis and sperm are produced by meiosis throughout his adult sex life.

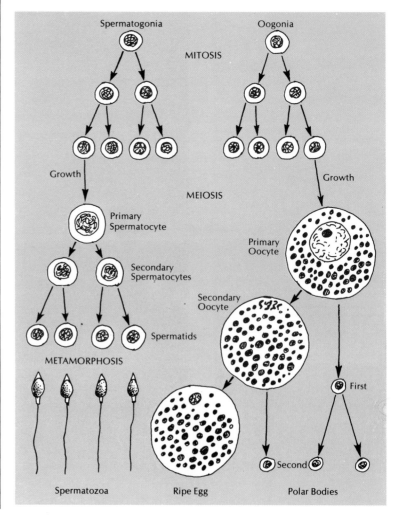

four haploid *spermatids*. These cells are then transformed into the specialized structures called *spermatozoa*, or *sperm*.

In the female animal, the gonocytes become specialized to form *oogonia*. Each oogonium divides twice mitotically to form four primary *oocytes*. Each oocyte undergoes meiosis to form four haploid *ootids* — one *ovum*, or *egg*, and three polar bodies. The polar bodies allow reduction and sorting of genetic material without significantly reducing the amount of cytoplasm in the egg.

In plants, meiosis occurs at various times during the life cycle. In some cases, male and female sex cells similar to the sperm and ova of animals are formed. In other cases, the products of meiosis are asexual spores. In some plants, the haploid stage involves a significant portion of the life cycle of the organism.

The result of meiosis and sexual reproduction is to give each diploid cell a reshuffled combination of genetic information. Each zygote obtains a complete haploid set of chromosomes from each of its parents. The structures and functions involved in all forms of reproduction will be described further in Chapter 9.

POLYPLOIDY

The presence of one or more complete extra sets of chromosomes is a condition called *polyploidy*. Cells with three complete sets of chromosomes are called triploid, those with four sets are called tetraploid, and so on.

Polyploid species of some plants occur naturally; they can also be produced in the laboratory. In one type of experiment, dividing plant cells are treated with the chemical colchicine, an alkaloid that blocks mitosis at metaphase. Although chromosomes have finished duplicating by the end of metaphase, (centromeres have replicated), colchicine acts to prevent chromosome separation in microtubules. Cytokinesis will not occur. If these cells are removed from the presence of colchicine, they will complete the next cell cycle normally except that they will have a polyploid (4n) number of chromosomes.

In 1972 Peter Carlson and his colleagues performed an experiment using diploid cells from two different species of the tobacco plant. Enzymes were used to remove cell walls from the leaf cells; the naked cells were then joined together so that new cells with a polyploid number of chromosomes (2n from each species) were formed. These newly-formed cells can reproduce and form a new polyploid species (which can reproduce sexually).

A new polyploid species, whether formed by chemical treatment or by somatic cell hybridization, may be endowed with such features as increased resistance to disease, insects, or drought. Plants of a polyploid species may also show an improved yield or an increased protein content.

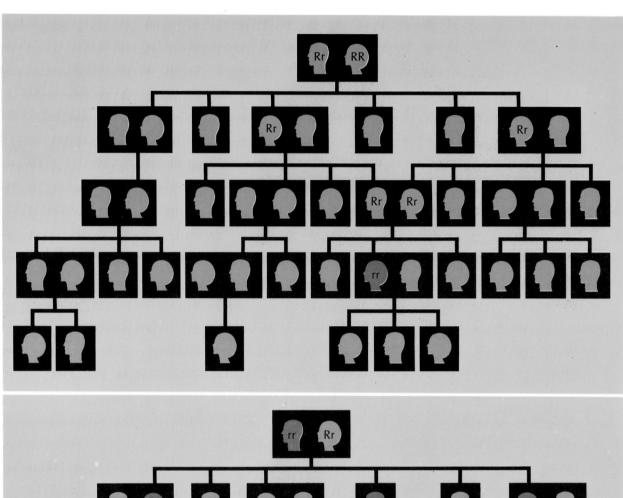
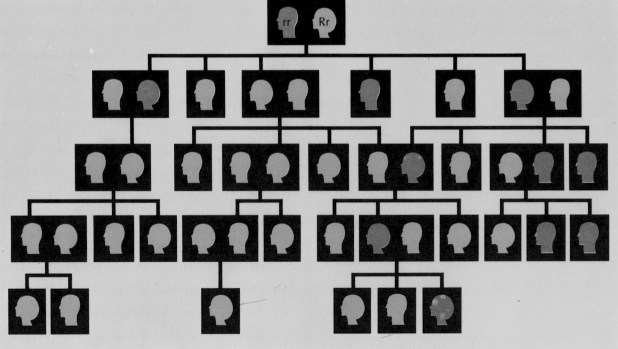

The general sequence of events in mitosis and meiosis had been described by the early 1900s. Scientists had not yet discovered that chromosomes exist in pairs, but they did know that meiosis results in a halving of the chromosome number. Acting on what was known about cell division, most biologists assumed that hereditary information is carried in the nucleus. And by observing the regular assortment of chromosomes between daughter cells, several biologists were able to speculate that it is the chromosomes which are involved in the transmission of hereditary information. It was at this point in time that biologists suddenly realized the significance of a paper on hereditary factors by Johann Gregor Mendel, which had been gathering dust for more than four decades.

MENDEL'S EXPERIMENTS

Mendel was an Austrian monk who experimented with plant hybridization in order to study the transmission of characteristics from one generation to the next. In his paper, which he read to the Brünn Natural Science Society in 1865, he wrote of three fundamental principles governing heredity: dominance, segregation, and independent assortment. These principles may be summarized as follows: (1) When parents differ in one characteristic, their hybrid offspring resemble one of the parents, not a blend of both—the principle of *dominance*. (2) When a hybrid reproduces, its reproductive cells are of two kinds—half transmitting the dominant character of one parent, and the other half transmitting the recessive character of the other parent—the principle of *segregation*. (3) When parents differ in two or more pairs of characters, each pair shows dominance and segregation independently of the other pairs, so that all possible combinations of the various pairs occur in their chance frequencies in the reproductive cells of the hybrid—the principle of *independent assortment*, or *recombination*.

These fundamental principles were not the conceptualizations of a naïve priest puttering with some peas. Mendel received a comprehensive education in science and applied mathematics from the University of Vienna. It is possible that he first became interested in the mutability of species and the history of the plant kingdom while he was at the university; both issues were raging at the time. Perhaps he undertook his studies into heredity in the hope of shedding light on these issues. In any event, within a few months of his return from the university Mendel had established some thirty-four pure strains of peas; he obviously was planning for some kind of hybridization experiments. He was not the first hybridizer to try discovering hereditary mechanisms, but he was the first to study large numbers of crosses in order to theorize about statistical distributions among successive generations.

Many observable characteristics of pea plants are influenced by heredity—the length and color of the stem; the size and form of leaves; the position, color, and size of flowers; the form and size

7

Patterns of Inheritance

(Opposite) A typical family pattern of a recessive trait (above) and a dominant trait (below). The orange represents the appearance of the dominant trait, magenta the recessive.

Figure 7.1 A monohybrid cross between pea plants differing in seed types. One parent is homozygous dominant for round seeds (RR), and the other parent is homozygous recessive for wrinkled seeds (rr). The phenotype of the F$_1$ offspring is round, but note that the genotype is Rr, or heterozygous. If two of these round F$_1$ plants are mated, their offspring (F$_2$) will show a 3:1 phenotypic ratio and a 1:2:1 genotypic ratio.

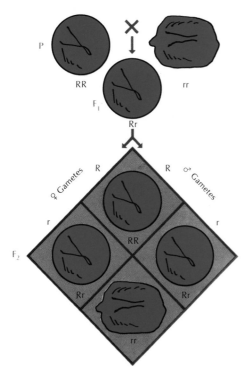

of seeds; and the color of seed coats and seed contents. Mendel chose to study seven pairs of characteristics:

1. *The form of ripe seeds.* The seeds may be *round* (or roundish, with only shallow surface depressions), or they may be irregularly angular and deeply *wrinkled*.
2. *The color of seed contents.* The contents of ripe seeds may be *yellow* to orange, or they may have a *green* tint.
3. *The color of the seed coat.* The seed coat may be *white* (such a seed produces a plant with white flowers), or it may be *gray* to brown, with or without violet spots (such a seed produces a plant with reddish-violet flowers).
4. *The form of ripe pods.* The ripe pods may be *inflated* with a smooth surface, or they may be deeply *constricted* between the seeds and more-or-less wrinkled.
5. *The color of unripe pods.* The unripe pods may be light to dark *green*, or they may be a vivid *yellow*.
6. *The position of flowers.* The flowers may be *axial* (distributed along the main stem), or they may be *terminal* (bunched at the top of the stem).
7. *The length of the stem.* Stem length varies greatly in different strains of pea plants, but Mendel chose one pure-breeding strain with *short* stems (9 to 18 inches) and another with *long* stems (6 to 7 feet).

For each pair of characteristics, Mendel obtained two pure-breeding strains, differing only in that single pair of characteristics. He did so because in each pair the two contrasting characteristics are always distinct; no intermediate characteristics appear in crosses between the two pure-breeding strains.

He began his experiments by cross-pollinating plants for each pair of strains differing in a single characteristic.

Mendel carefully collected the seeds that were produced on the original parent-generation (P) plants, recorded their characteristics, planted them, and recorded the characteristics of the resulting adult plants (Figure 7.1). These seeds and the plants into which they develop are the individuals of the first filial generation (F$_1$). Mendel called them hybrids, but modern geneticists prefer to use this term only for individuals produced by cross-breeding between two different species. In each cross, Mendel found that all the F$_1$ individuals resembled one of the contrasting parental characteristics. For example, all the F$_1$ seeds collected after the cross of wrinkled-seed and round-seed plants show the round-seed characteristic. Mendel called round seeds a *dominant* characteristic and wrinkled seeds a *recessive* characteristic. In a cross between pure-breeding strains with dominant and recessive characteristics, all the F$_1$ individuals show the dominant characteristic. The recessive characteristic is not expressed in the F$_1$ generation, but it must be present in some form because it is passed on to the offspring, or the second filial

| | Male Gametes | |
	R	r
R	RR (round seed)	Rr (round seed)
r	Rr (round seed)	rr (wrinkled seed)

Female Gametes labels the left column rows R and r.

Figure 7.2 The same F$_1$ cross—diagrammed in a Punnett square—shows in more detail the origin of the three round to one wrinkled seed phenotypic ratio and the underlying 1:2:1 genotypic ratio—that is, one homozygous round seed (RR), two heterozygous round seeds (Rr), and one homozygous wrinkled seed (rr).

generation (F$_2$). Mendel found the following characteristics to be dominant: round seeds, yellow seed contents, gray seed coats, inflated pods, green pods, axial flower distribution, and long stems.

The F$_2$ generation for each cross was produced by self-fertilization of the F$_1$ plants. In each case, some of the F$_2$ individuals showed the recessive character that had disappeared in the F$_1$ generation. For each cross, the ratio of individuals showing the dominant character to individuals showing the recessive characteristic averaged about 3:1. None of the F$_2$ individuals showed characteristics intermediate between those of the original parental strains as shown in Table 7.1.

Table 7.1
Ratios of Characters in Second Generations of Mendel's Seven Crossing Experiments

Dominant Characteristic	Number of F$_2$ Individuals	Recessive Characteristic	Number of F$_2$ Individuals	Ratio
Round seed	5,474	Wrinkled seed	1,850	2.96:1
Yellow seeds	6,022	Green seeds	2,001	3.01:1
Gray seed coats	705	White seed coats	224	3.15:1
Inflated pods	882	Constricted pods	299	2.95:1
Green pods	428	Yellow pods	152	2.82:1
Axial flowers	651	Terminal flowers	207	2.14:1
Long stems	787	Short stems	277	2.84:1

The next year a third generation (F$_3$) was produced by self-fertilization of the F$_2$ plants. The F$_2$ individuals that showed recessive characters produced only F$_3$ individuals showing recessive characters. However, the F$_2$ individuals that showed dominant characters proved to be of two kinds. One-third of these dominant-characteristic individuals produced only dominant-characteristic individuals of the third generation. The other two-thirds produced F$_3$ individuals with both dominant *and* recessive characters in the ratio 3:1. In other words, all of the recessive-character and one-third of the dominant-character F$_2$ individuals proved to be pure-breeding types. The other two-thirds of the dominant-character F$_2$ individuals showed the same pattern of offspring as the first generation had shown.

The distribution of characteristics in the third generation makes it clear that the 3:1 ratio of dominant to recessive characters in the second generation actually reflects a 2:1:1 ratio. Among the offspring of the F$_1$ hybrids, one-half were F$_2$ hybrids showing the dominant character, one-quarter were pure-breeding dominants, and one-quarter were pure-breeding recessives. At the time he wrote his paper, Mendel had carried most of his crosses through four or five generations and found that the 2:1:1 ratio continues to appear among the offspring of hybrids in each generation,

Figure 7.3 The microscope that Gregor Mendel used in his plant-hybridization experiments. In eight years of intensive research prior to the publication of his famous paper on the principles of heredity, Mendel counted over 10,000 plants.

Figure 7.4   A dihybrid cross, illustrating independent assortment, between a round, yellow pea plant (RRYY) and a wrinkled, green pea plant (rryy) carried through the F$_2$ generation. Those members of the F$_1$ generation are completely heterozygous (RrYy) for both traits and, when mated, yield nine possible genetic combinations.

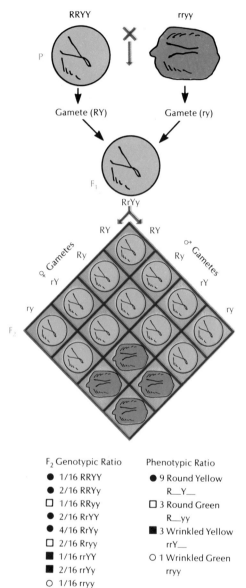

F$_2$ Genotypic Ratio	Phenotypic Ratio
● 1/16 RRYY	● 9 Round Yellow
● 2/16 RRYy	R__Y__
▢ 1/16 RRyy	▢ 3 Round Green
● 2/16 RrYY	R__yy
● 4/16 RrYy	■ 3 Wrinkled Yellow
▢ 2/16 Rryy	rrY__
■ 1/16 rrYY	O 1 Wrinkled Green
■ 2/16 rrYy	rryy
O 1/16 rryy	

whereas the pure-breeding lines continue to produce offspring like their parents.

Mendel explained his observations by assuming that each seed contains two hereditary factors affecting a particular characteristic. One of these factors is obtained from each parent. For example, in the characters related to seed form, let *R* represent a factor for the round-seed characteristic and *r* represent a factor for the wrinkled-seed characteristic. The pure-breeding plants of the original parent generation can be represented as *RR* and *rr*. Cross-breeding produces a seed that obtains one factor from each parent; it is thus *Rr*. The seed becomes round because the dominant *R* is expressed despite the presence of the recessive *r*.

When the *Rr* seed grows into a plant, it will produce, in equal numbers, reproductive cells carrying *R* and cells carrying *r* factors. So four combinations are equally likely to be found among the F$_2$ seeds produced by self-fertilization: *RR, rR, Rr,* and *rr*. The second generation will consist of pure-breeding dominants, hybrids, and pure-breeding recessives in the ratio 1 : 2 : 1, just as Mendel had found experimentally. Because one-half of the F$_2$ individuals have precisely the same *Rr* factors that the first generation possessed, further self-fertilizations of these individuals should produce the same ratios of offspring.

Next, Mendel carried out crosses involving two or more pairs of characteristics (Figure 7.4). In one experiment, for example, pure-breeding plants with round yellow seeds were crossed with pure-breeding plants with wrinkled green seeds. As expected, all the resulting seeds were round and yellow. The plants of the original P generation may be represented as *RRYY* and *rryy*. All of the reproductive cells from one strain carry *ry*. Therefore, all the F$_1$ plants will possess the factors *RrYy*. If the factors are assorted randomly in the production of reproductive cells (assuming that each reproductive cell gets only one of each kind of factor), four kinds of reproductive cells are possible: *RY, Ry, rY,* and *ry*. The checkerboard diagram shows that the F$_2$ individuals will have nine different genetic combinations of characters in the ratio 1:2:1:2:4:2:1:2:1 (Figure 7.4).

Crosses involving more than two independently assorting factor pairs can be diagrammed in the same way, but the results rapidly get more complex. With only twenty-three different, independently assorting pairs of factors (as in humans), the offspring of a cross between multihybrid individuals could be of about 8 million different observable types. It is easy to see why Mendel was impressed with the ability of his simple hereditary factors to account for variability in nature.

Mendel counted over 10,000 plants in the 8 years he gathered data. Unfortunately, the biologists of his time who read his paper were skeptical of his results and theories, and Mendel was unable to get them to repeat his laborious experimental work in order to check his conclusions. Nevertheless, his principles gave rise in the early years of this century to many of the basic ideas of *genetics*, as the study of inheritance then came to be called.

Other useful terms were introduced early in the 1900s. Individuals showing the same observable characteristics are said to be of the same *phenotype;* those possessing the same set of hereditary factors, or *genes*, are said to be of the same *genotype*. Pea plants of different genotypes (*RR* and *Rr*) both display the same round-seed phenotype. An individual whose genotype is made up of identical factors (*RR* or *rr*) is called *homozygous*, whereas an individual with the hybrid genotype *Rr* is called *heterozygous*. The word "gene" was not introduced until 1909 by the Danish botanist Wilhelm Ludwig Johannsen. Although it has meant various things at various times, one of its original uses was for the factors within cells that produce the characteristics studied by Mendel.

CLASSICAL GENETICS

For the first two decades of this century, students of heredity extended Mendel's principles to explain the many apparent exceptions they had discovered and attempted to discover the *physical basis* for Mendel's hypothetical hereditary factors. Most investigators had assumed that all chromosomes in a cell are essentially equivalent. The mechanism of meiosis was viewed simply as a device that put half of the chromosomes into each daughter cell. The American cytologist T. Montgomery, Jr. showed in 1901 that the cell contains pairs of distinctive kinds of chromosomes. He worked with cells from grasshoppers because the chromosomes in the cells of that insect are unusually visible and fewer in number (twelve pairs) than in humans. Another American working with grasshoppers suggested that the chromosomes play a role in determining the sex of an individual. C. McClung described the existence of two kinds of sperm cells: one with an extra, or accessory, chromosome that is not present in the other kind of sperm cell. McClung concluded that eggs fertilized by sperm having the accessory chromosome develop into males, whereas eggs fertilized by sperm lacking the accessory chromosome develop into females (the *reverse* of the actual situation, as it later turned out).

W. Sutton, too, studied grasshopper cells and found that the twelve chromosomes in a sperm or egg cell can be distinguished by size and that a fertilized egg cell or body cell contains a pair of each kind of chromosome. In meiosis, one member of each pair goes into each reproductive cell, as Montgomery had observed. At first, Sutton thought—as Montgomery had—that the maternal chromosome set is separated from the paternal set in meiosis. However, Sutton strongly suspected that the chromosomes are the carriers of the hereditary factors. If so, the chromosomes had to be assorted independently among the reproductive cells, as are the hereditary factors in Mendelian theory. Although he was unable to

157

prove it, Sutton's microscopic observations convinced him that gametes do receive a mixture of maternal and paternal chromosomes.

Sutton concluded that the hereditary factors are closely associated with the chromosomes. He pointed out that there are far more hereditary factors than there are chromosomes in any species, so *a number of different factors must be associated with a single chromosome.* A variety of recessive and dominant factors might be expected to remain together on a single chromosome rather than assorting independently, as assumed by Mendel.

Thomas Hunt Morgan and the Fruit Fly

Thomas Hunt Morgan, who had specialized in embryology before becoming interested in genetics, was convinced that sudden changes (mutations) play a more important role in evolution than do the hypothetical recombinations of hereditary factors described by the Mendelists. Like many other researchers exploring heredity and the chromosomes, Morgan turned to insects as experimental subjects. It is easy to maintain large populations of small insects. They reproduce and grow rapidly, and their cells and chromosomes can be viewed easily under the microscope. Morgan had heard of some experiments being done with *Drosophila melanogaster,* a small fruit fly. These little flies thrive on a diet of mashed fruit or yeast, can be kept by the hundreds in half-pint milk bottles, and require only about 12 days to reach maturity—thus providing some thirty generations each year for genetic studies. Furthermore, each *Drosophila* cell has only four chromosome pairs, making this an ideal organism for study in the search for simple relationships between heredity and chromosomes (Figure 7.7).

Morgan subjected his fruit flies to heat, cold, x-rays, radioactivity, and various chemicals, but he was unable to detect any mutations produced by these treatments. Then, in 1910, Morgan discovered a single, white-eyed male fly in a bottle of normal, red-eyed flies. No other white-eyed flies had appeared in the dozens of generations through which Morgan had observed this population, so he was sure that the white-eyed male was a mutation.

Morgan mated the white-eyed male with wild-type, red-eyed females from the same generation (Figure 7.8). The first generation produced by this cross was entirely red-eyed, as Mendelian principles would predict if the white eyes are produced by a recessive gene. Next, Morgan allowed the first generation to interbreed. The resulting generation contained 3,470 red-eyed and 782 white-eyed flies. This result is far from the 3:1 ratio predicted by Mendel's laws, but the reappearance of the white-eyed characteristic in this generation showed that Morgan had indeed located a new heritable characteristic.

One property of the second generation was incompatible with Mendel's laws. All of the 782 white-eyed flies were males. The red-eyed flies included 2,459 females and 1,011 males. At first,

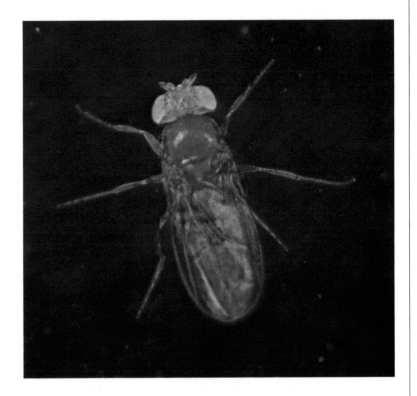

Figure 7.5 The fruit fly *Drosophila melanogaster*, an experimental insect used in numerous classical genetics studies.

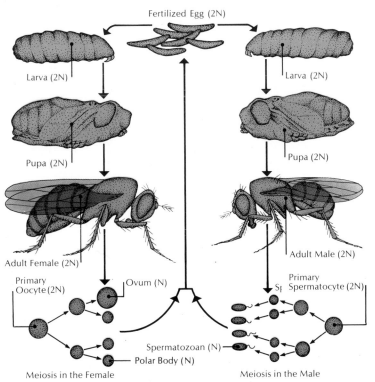

Fertilized Egg (2N)

Larva (2N)

Larva (2N)

Pupa (2N)

Pupa (2N)

Adult Female (2N)

Adult Male (2N)

Primary Oocyte (2N)

Ovum (N)

Primary Spermatocyte (2N)

Spermatozoan (N)

Polar Body (N)

Meiosis in the Female

Meiosis in the Male

Figure 7.6 (left) Life cycle of *Drosophila*. Because these fruit flies require only about 12 days to reach maturity, they can provide some thirty generations each year for genetic research.

Figure 7.7 (below) Chromosomes of *Drosophila*. There are four pairs of chromosomes in each cell. Note the differences in the sex determiner chromosomes of males and females.

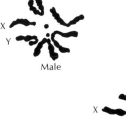

X

Y

Male

X

X

Female

Figure 7.8   The inheritance of white eye color, a sex-linked recessive trait, in *Drosophila*. (A) A wild-type (red-eye) crossed with a white-eye produces an F$_1$ generation in which all flies are red-eyed but the females are carriers (XWXW). The F$_1$ flies are crossed among themselves to produce an F$_2$ generation, three-quarters of which are red-eyed males and females and one-quarter of which are white-eyed males. (B) A homozygous recessive white-eye crossed with a wild-type (red-eye) produces an F$_1$ generation in which all females are red-eyed and all males are white-eyed. The F$_1$ flies are crossed among themselves to produce an F$_2$ generation of one-half red-eyed males and females and one-half white-eyed males and females.

Morgan thought that the white-eyed characteristic might somehow be restricted to males, but when the original white-eyed male was crossed with some of its red-eyed daughters from the first generation, 129 red-eyed females, 132 red-eyed males, 86 white-eyed males, and 88 white-eyed females emerged. White-eyed females had become as common as white-eyed males.

Clearly the inheritance of the white-eyed characteristic was somehow related to the hereditary determination of sex. In his first report on the white-eyed mutant, Morgan used Mendelian principles to explain the results of his crosses by assuming that the eye-color factor and the sex-determining factor are *linked together* rather than assorting independently. Morgan knew that one of *Drosophila's* four chromosome pairs is responsible for sex determination. In the cells of females, this pair is made up of two rod-shaped chromosomes—now called X chromosomes—whereas in

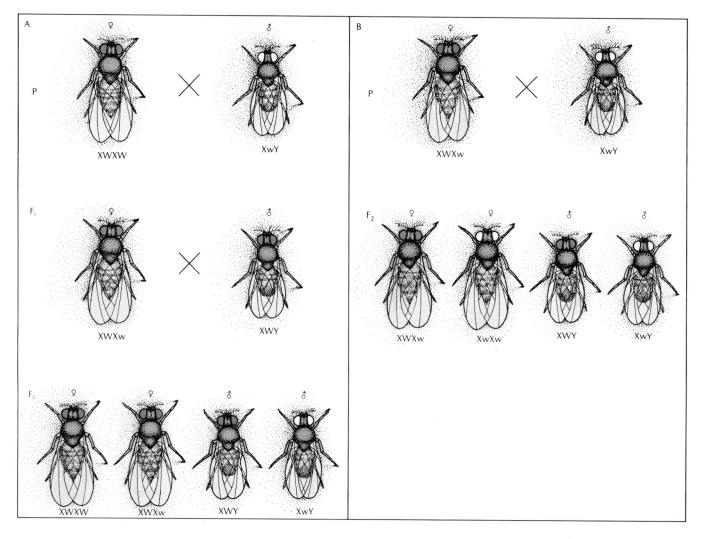

the male the pair consists of one X chromosome and one Y chromosome (a J-shaped chromosome). Later, Morgan showed that the eye-color gene is carried only on the X chromosome—that is, eye color is a *sex-linked trait*. The chromosomes of the remaining pairs are called *autosomes*.

Because the white-eyed factor is recessive, the female will have white eyes only if she carries the white-eyed factor on both X chromosomes. In a male, on the other hand, the presence of the white-eyed factor on the single X chromosome will lead to development of white eyes.

By 1915 Morgan and his students had identified nearly a hundred different mutant characters in *Drosophila*. More than twenty are sex-linked and are controlled by factors carried on the X chromosome. The remaining characteristics fall into three groups, with the characters of each group tending to remain linked together. The four linkage groups correspond well with the four chromosome pairs of *Drosophila*, and one of the groups contains only a few characteristics, as might be expected from the fact that one of *Drosophila*'s chromosome types is little more than a small dot.

Further evidence that genes are carried by the chromosomes was soon to come from Morgan's laboratory. His important work with *Drosophila* led many other geneticists to begin experimenting with this insect, and it became the most common organism for genetic research. In fact, because this inconspicuous little fly is of minor importance to man as a pest or otherwise, someone once remarked that God must have created *Drosophila* just for Morgan.

Gene Linkage and Crossing Over

In his early work with *Drosophila* mutants, Morgan found some mutant characters that appear only as a result of the *combined* effects of two recessive characteristics. For example, certain wing defects do not occur if the individual has a dominant wild-type allele for either of two different genes. (An *allele* is one of two or more alternative forms of a gene found at the same position, or locus, on homologous chromosomes.) This example of *incomplete dominance* can be explained by a relatively minor modification of Mendel's original laws—merely by recognizing that in some cases the heterozygous phenotype may be intermediate between the homozygous phenotypes. Gene linkage can be explained by adopting Sutton's hypothesis that the genes are linked together on chromosomes and therefore cannot assort independently.

Among the early mutant genes Morgan discovered in the fruit fly are those for miniature wings and vermilion eye color. Both are sex-linked genes carried by the X chromosome, so the two genes may be expected to be linked to each other. In both cases, the mutant allele is recessive to the wild-type allele. In an early experiment, one of Morgan's students, A. Sturtevant, crossed a long-

Figure 7.9 (above)   Variations in eye structure and color in the fruit fly *Drosophila melanogaster.*

Figure 7.10 (below)   Inheritance of the sex-linked recessive trait bar-eye in the fruit fly. In a male that has the recessive gene and a female that is homozygous recessive for the trait, the eye form is a narrow bar. A female that is heterozygous for the trait has a wide-bar phenotype—an example of incomplete dominance.

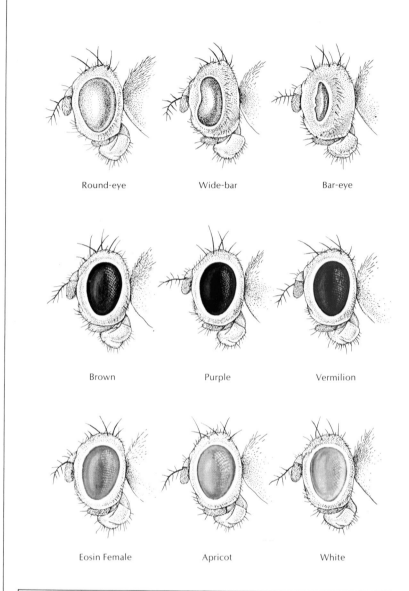

Round-eye   Wide-bar   Bar-eye

Brown   Purple   Vermilion

Eosin Female   Apricot   White

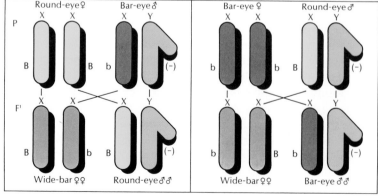

winged, vermilion-eyed female with a miniature-winged, red-eyed male. The female was homozygous for both genes. The genotype of the male was a recessive mutant allele for miniature wings and a dominant wild-type allele for vermilion eyes.

A group of second generation (F_2) individuals was then produced by interbreeding members of the first generation (F_1). Recombination of these gametes yields four genotypes, each of which corresponds to a different phenotype. Thus, the expected F_2 generation would contain equal numbers of long-winged, red-eyed females; long-winged, vermilion-eyed males; long-winged, vermilion-eyed females; and miniature-winged, red-eyed males. Table 7.2 compares the predicted and observed phenotypes of the F_2 generation.

Table 7.2
Observed and Predicted Phenotypes of Second-Generation (F_2) *Drosophila*

Phenotype	Observed *Drosophila*		Predicted Percentage
	Number	Percentage	
Long-winged, red-eyed males	8	1.7	0.0
Long-winged, red-eyed females	138	29.4	25.0
Long-winged, vermilion-eyed males	117	24.9	25.0
Long-winged, vermilion-eyed females	110	23.4	25.0
Miniature-winged, red-eyed males	97	20.5	25.0
Miniature-winged, red-eyed females	0	0.0	0.0
Miniature-winged, vermilion-eyed males	1	0.2	0.0
Miniature-winged, vermilion-eyed females	0	0.0	0.0

Source: A. H. Sturtevant, "The linear arrangement of six sex-linked factors in *Drosophila*, as shown by their mode of association," *Journal of Experimental Zoology*, 14 (1913): 43–59.

There were two unexpected results. First, although three of the four phenotypes were present in about the predicted proportions, there were fewer miniature-winged, red-eyed males than were expected. Morgan and Sturtevant found that miniature-winged phenotypes always occur in much smaller numbers than predicted in any cross. They concluded that the phenotype has a low viability. In other words, zygotes developing into miniature-winged adults tend to die before reaching maturity and do not get counted in research results.

The second unexpected result was the presence in the F_2 generation of small numbers of two phenotypes that were *not* expected to appear. Morgan suggested that their appearance might be due to an exchange, or *crossing over*, of alleles between the homologous chromosomes in the female during meiosis (Figure 7.11). Crossing over cannot occur in the male because there is no second X chromosome with which alleles can be exchanged. (It was later

Figure 7.11 Schematic model of chromosome cross-over. (1) Two homologous double-stranded chromosomes, one bearing the linked alleles A and B and the other bearing the linked alleles a and b are joined in synapsis. (2) Corresponding breaks occur in one chromatid of each pair and the fragments are exchanged. (3) After crossing over, one chromatid of the first chromosome bears alleles A and b and one chromatid of the second chromosome bears alleles a and B.

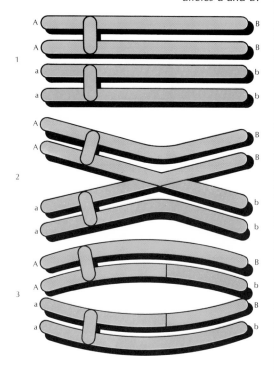

Figure 7.12 Pedigree chart tracing the inheritance of hemophilia, a sex-linked recessive gene, through the royal families of Europe. Queen Victoria can be traced as the original carrier of the hemophilic gene.

discovered that crossing over never occurs on any of the chromosomes of the male *Drosophila*, but this species is unusual in this respect.)

Morgan began to search for a physical explanation of crossing over and found a clue in the work of F. Janssens, a Belgian cytologist who described the process of chiasmata formation at the beginning of meiosis (see Chapter 6). When homologous pairs of chromosomes come together in synapsis, some material appears to be interchanged between chromatids of the two chromosomes. Janssens suggested that the chromatids break at corresponding places and interchange equal segments. Then, when the chromosomes are pulled apart in anaphase, each of the separating chromatids contains a number of segments derived from the other chromosome of the pair.

Morgan and his colleagues quickly realized that this physical crossing over provided exactly the mechanism needed to account for the crossing over of genes indicated by their experiments with *Drosophila*. Chiasmata do not form between the X and Y chromosomes during meiosis in the male. Furthermore, the chromatids always seem to exchange exactly corresponding segments. These observations are consistent with the genetic observations that crossing over does not occur with sex-linked genes in the male and that genes are exchanged but never lost during crossing over.

Morgan recognized another important implication of this mechanism for crossing over. He had found that the percentage of crossing over remains roughly constant for any given pair of genes but varies greatly among different pairs of genes. Some pairs of genes are completely linked so that crossing over is never observed between them. With other pairs, the percentage of crossing over reaches about 50 percent. Morgan suggested that the genes fall at

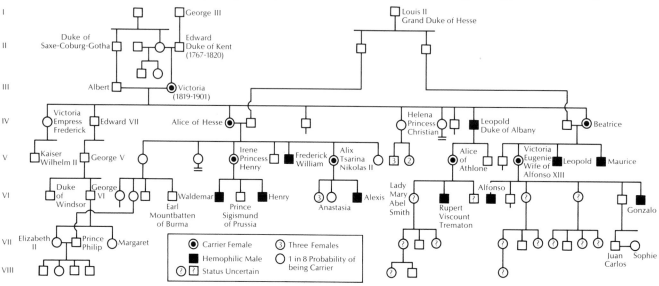

particular locations along the chromosomes. If two genes happen to be located close together, they will rarely be separated during chiasmata formation; if crossing over occurs, they will cross over together, and the crossing over will not be detected by breeding experiments. On the other hand, if the two genes are located near *opposite* ends of the chromosome, crossing over would be expected almost every time that chiasmata are formed. The fact that cross-over frequency seldom rises above 50 percent is probably due to multiple crossing over, which could not always be detected in breeding experiments. Morgan pointed out that cross-over frequencies should provide a means of mapping the relative locations of genes on the chromosomes.

Chromosome Mapping

Morgan and Sturtevant soon were able to show the relative positions of a number of genes on the X chromosome of *Drosophila*. Within a few years they had prepared chromosome maps for each of the four chromosomes of this species, showing the relative locations of each of the eighty-five mutant genes then known.

The consistent results of the mapping convinced Morgan and most other biologists that the genes are indeed located in linear order along the chromosomes. They found no evidence of branching or parallel chains of genes on a single chromosome. Every known gene could be assigned a consistent location on one of the chromosomes. Occasionally, two genes were found to have the same apparent location. In some cases, further experiments with larger numbers of flies revealed a very low frequency of crossing over, so that the two genes could be assigned locations near to each other. In other cases, no crossing over could be found. For example, at a particular point near one end of the X chromosome, several mutant genes for different eye colors have been assigned the same location. It appears that all of these are different mutant alleles of the same gene, for they are all recessive to the wild-type characteristic.

Morgan, who had begun his career in genetics as an opponent of the Mendelists, eventually became one of the outstanding proponents of a modified, chromosomal version of Mendelian genetics. He and his colleagues became the recognized leaders of genetic research and theory.

POPULATION GENETICS

With the development of classical genetics in the early years of this century, quantitative estimates could be made of the rates at which the characteristics of a population will change as a result of selection or of other factors. Changes in the distribution of particular genetic characteristics in a population could now be studied over time. The discipline known as population genetics came into being as an attempt to understand quantitatively the evolutionary

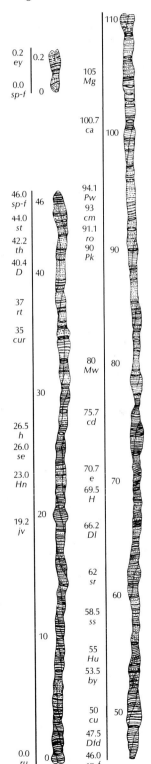

Figure 7.13   Map depicting two arms of a single chromosome.

process in terms of the genes, the chromosomes, the mutations, and other concepts as had been developed by classical and molecular geneticists.

Population genetics emerged as a statistical science concerned with groups of organisms in terms of the *gene pool*—the total collection of all the allelic forms of all the genes in all the gametes of the population.

The basic unit of study in this particular field is known as the *deme*, a population of freely interbreeding individuals. The gene pool of a deme can be described as the *gene frequencies;* in other words, the allele frequencies. The gene frequency is a mathematical expression of the proportion of one allelic form of a gene compared to other allelic forms of the same gene in the gene pool. Also of importance are the *genotype frequencies* and *phenotype frequencies* of the population, both of which depend on the gene frequencies.

Mendel observed the proportions of various phenotypes (such as long-stemmed and short-stemmed plants) among his pea plants. By using various controlled breeding experiments, he was able to calculate the proportions of various genotypes among his pea plants. It was on the basis of this information that a geneticist could calculate the gene frequencies of the various alleles involved.

Genetic Equilibrium

In order to describe the ways in which a gene pool can change from generation to generation, it is necessary to make some simplifying assumptions, many of which may be somewhat unrealistic for most natural populations. For example, the simplest situation for theoretical analysis would be one in which mating occurs in a random manner, where any sperm cell is likely to be combined with any egg in the population. Such *panmictic populations* are undoubtedly rare or nonexistent in nature. If nothing else, an organism tends to mate with its geographic neighbors. In most natural populations, mating behaviors depend on other nonrandom factors such as a tendency to mate with individuals of similar or different phenotype, with related individuals, or with individuals high in the dominance hierarchy. For example, individuals in human populations are more likely to mate with other individuals from the same geographic area, race, religion, or ethnic background.

A model of a population that will remain in genetic equilibrium for generation after generation *must include* the following factors: (1) it will be a large, sexually-reproducing population; (2) all genotypes will show an equal response to natural selection; (3) all members of the population will breed randomly; (4) no mutations will occur; and (5) no individuals will leave or enter the population (there is no gene flow). This concept of genetic equilibrium was developed independently in 1908 by the English mathematician G. Hardy and the German physician W. Weinberg and is known as the *Hardy-Weinberg* law.

In natural populations, however, several factors disturb this equilibrium and cause changes in the gene pool. Various behavioral and physiological factors can lead to nonrandom production and combination of gametes during mating. Mutation directly alters the gene pool by substituting one allele for another in a particular gamete. Selection results from different survival and reproduction rates for various genotypes, thus causing particular genotypes to contribute either more or less heavily to the next generation than they would under random conditions. Gene flow through immigration and emigration can cause significant changes in the gene pool. In small populations, chance effects can lead to changes in the gene pool because random fluctuations make the gene frequencies among gametes or zygotes significantly different from the theoretically expected values. Such random fluctuations lead to genetic drift of gene frequencies over time.

The equation for the Hardy-Weinberg law may be stated simply:

$$(P + Q)^2 = 1$$

If we expand this equation as follows:

$$(P + Q) \times (P + Q) = 1$$

and multiply, we find the following expression:

$$P^2 + 2PQ + Q^2 = 1$$

If we substitute the above symbols into our model population, which is in genetic equilibrium, the following factors will be true:

P = frequency of the dominant allele (such as the allele *A* or *B*)

Q = frequency of the recessive allele (such as the allele *a* or *b*)

P^2 = frequency of the homozygous, dominant genotype (such as the genotype *AA* or *BB*)

2PQ = frequency of the heterozygous genotype (such as the genotype *Aa* or *Bb*)

Q^2 = frequency of the homozygous, recessive genotype (such as the genotype *aa* or *bb*)

1 = one population in genetic equilibrium

The Hardy-Weinberg equation is useful for application in modern medical science. For example, if the numbers of babies born each year and the frequencies of certain genetic diseases in these babies are known, then the approximate number of heterozygous *carriers* — those individuals who carry one recessive allele — can be calculated.

Albinism is an absence of pigmentation and it has been noted in individuals of many species, including man. In some species, such as a certain butterfly, the term signifies only a partial lack of pigment; in birds and mammals, a total lack of pigment is shown by pink eyes. In human beings, albinism is caused by a recessive autosomal gene. The normal body and eye color allele *A* is dominant to albino allele *a*. Surveys have shown that the frequency of albinism (with a genotype of *aa*) in human populations is about 1 in 20,000 individuals. Assuming genetic equilibrium in human populations, what proportion of the population would you expect to be heterozygous for albinism (with a genotype of *Aa*)? How many individ-

Figure 7.14 Diagram showing how phenylalanine in milk is normally metabolized, compared with what occurs in phenylketonuria (PKU).

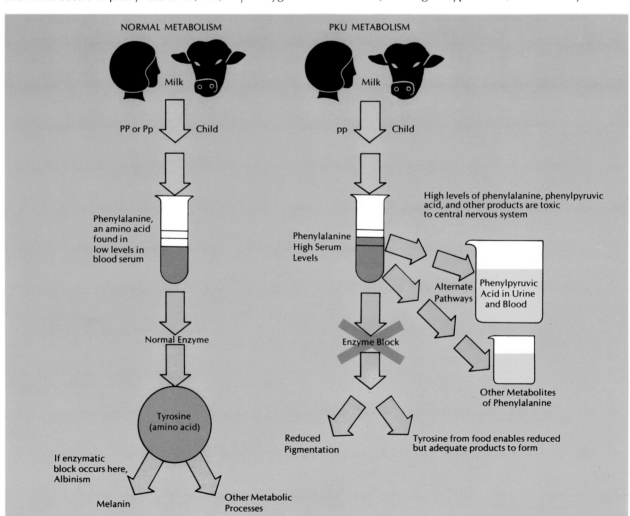

uals in a population of 20,000 would you expect to be heterozygous? Here is the solution:

Q^2 = frequency of *aa* genotype = 1/20,000

Q = frequency of *a* allele = $\sqrt{1/20,000}$ = 1/141

P = frequency of *A* allele = 1 − Q = 1 − 1/141 = 140/141

2PQ = frequency of *Aa* genotype = 2 × 140/141 × 1/140 = 1/70 = 1.4 percent

1/70 × 20,000 = 286 individuals who are heterozygous for the albino allele *a* in a population of 20,000

The incidence of individuals with the disease *phenylketonuria* (with a genotype of *pp*), is about 1 in 15,000 in human populations. If similar calculations are made to those in the albinism problem above, it can be determined that the frequency of heterozygotes (with the genotype *Pp*) in human populations is about 1.6 percent.

Sickle cell anemia is also a recessive autosomal disease. In black populations of African origin, the incidence of sickle cell anemia (*ss*) is about 1 in 400. The heterozygous carrier condition (*Ss*) would be expected to occur in about 9.5 percent in these black populations.

Genetic Drift

In a large population, the observed gene frequency fluctuates slightly about the value predicted by the Hardy-Weinberg law if no disturbing factors are operating. In a small population, the fluctuations may be much greater. In a population of less than a hundred individuals, then, the loss of alleles from the gene pool through genetic drift is not uncommon.

Genetic drift may be of evolutionary importance in these small populations. Some alleles may have their frequencies reduced or be lost from the gene pool in a population reduction, whereas others may become more common among the surviving population than they were before the population reduction (Figure 7.15).

Mutations

The rate at which mutations, or changes in genetic information, occur in single cells can be readily measured only in unicellular organisms such as bacteria. But even in these organisms it is difficult to determine whether a given phenotypic alteration is the result of a mutation of a single gene or of a number of genes. The rate at which spontaneous mutations arise in a given gene in microorganisms varies between about once in a million and once in a billion divisions.

In cultures of animal cells, mutations of any particular gene are observed about once in a million cell divisions. In a multicellular organism, mutations of most body cells affect only the particular

Figure 7.15 In a population in genetic equilibrium, random variations will occur in the frequencies of allelic forms of a gene. These alleles may be dominant or recessive, or they may be expressed equally. When population numbers are large (over 100 individuals), random variations in the frequencies of allelic forms will probably be slight. If numbers are small, however, then allelic frequencies may vary greatly. The graph demonstrates this random genetic drift in allelic frequencies when a population size is large (left) or much smaller (right). At point A, allele 1 is permanently eliminated from the population, and at point B, allele 2 is permanently fixed in the population.

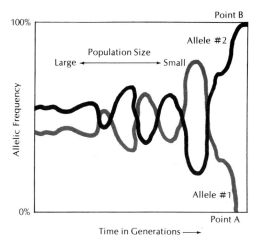

Figure 7.16 (opposite page)   In the photograph above, *Biston betularia*, the peppered moth, and its black form (*B. carbonaria*) are shown at rest on a soot-covered oak trunk; the same two moths are shown below on a lichened tree trunk in the countryside. One of the earliest and best-studied examples of the phenomenon of directional selection was this species of moth, which lives in industrial areas of Britain. As the terrain and vegetation have become darkened by the effects of smoke and other industrial pollution, dark-colored phenotypes have become predominant in many species of moths and butterflies where light-colored phenotypes formerly predominated. The typical phenotype of *B. betularia* was light-colored with dark speckling. About 1845 a dark, or melanic, form of this species was first reported near Manchester, England. Because the melanic form was seldom seen by collectors, it is estimated that this phenotype made up less than 1 percent of the population at that time. By 1895 the population of *B. betularia* near Manchester was about 99 percent melanic individuals. In most cases studied, melanism, or dark coloring, is due to the effect of a dominant allele at a single gene locus. In populations made up entirely of typical individuals (recessive homozygotes), melanic individuals do appear occasionally, apparently as a result of mutations. However, the mutation rate is far too low to account for the rapid increase in the proportion of melanic individuals in industrialized areas. Presumably, the melanic genotype is introduced into a typical population by mutation, but some form of selection must act in industrial areas to rapidly increase the proportion of melanic phenotypes in the population. The selection in the moth population was explained by the hypothesis that predators notice and kill moths that are colored conspicuously differently from the surface on which they rest. In nonindustrialized areas, typical individuals are almost invisible on lichen-covered tree trunks, whereas melanic individuals are very conspicuous. Industrial pollution leads to a reduction of the amount of lichen on tree trunks and a deposit of dark soot on the surface. On these polluted trunks, the melanic form is inconspicuous and the typical form is highly visible.

individual. Only mutations of gametes or their precursor cells alter the gene pool of the population. A single mutation can lead to the production of zero, one, or more mutated gametes, depending on the particular cell affected and the stage of development in which the mutation occurs.

For a population otherwise in genetic equilibrium, mutation shifts the equilibrium frequencies of particular alleles. Mutation may maintain a low frequency of a particular allele against other factors that might tend to make that allele disappear from the gene pool. However, because the observed mutations occur at extremely low rates, mutation cannot account for the rapid rate of evolution in a great number of natural populations.

Selection

In a population at Hardy-Weinberg equilibrium, all genotypes are considered to have an equal chance of surviving and reproducing. In fact, any two differing genotypes are likely to have slightly differing probabilities of producing gametes. Certain alleles are likely to become more common because individuals possessing those alleles have greater success in survival and reproduction. Thus, selection causes gene frequencies to shift away from the constant values expected in a population at genetic equilibrium.

The relative survival value and reproductive capability of a particular genotype vary under different environmental conditions. The *fitness*, or adaptive value, of a genotype is defined on a range of values from zero (for a genotype that contributes no gametes to the next generation) to one (for the genotype that proportionately contributes the most gametes to the next generation). Defined in this way, fitness can be determined only through observation of the changes in the gene pool. If one genotype is more successful in reproducing itself than is another, it is more fit than the other. Fitness, then, is a measure of the degree to which a genotype succeeds in having its alleles reproduced in the next generation. A genotype may have a high fitness as a result of any combination of a number of different factors—a longer reproductive period, an increased number of offspring, increased efficiency in mating, resistance to disease or environmental stresses, and so on.

Multiple Factors

In any real population of individuals, the gene frequencies at any time are likely to result from the combined effects of mutation, gene flow, genetic drift, and selection. Furthermore, in real organisms, genes are not assorted independently but are linked together on chromosomes. Therefore, the frequency of alleles of one gene can be influenced by the factors affecting the frequencies of alleles of a linked gene.

The effects of recombination through cross-overs must then be considered. For example, selection may favor a particular chromosome containing a number of relatively fit linked alleles, even though individual alleles in this group may not be the most fit possible for their particular genes.

Attempts to develop complete mathematical representations of genetic changes in realistic populations are still in a primitive state. However, combinations of simplified models make it possible to draw some general conclusions about the factors acting together on natural populations.

Mutation is a significant influence on the gene pool only when the resulting mutant allele is rare in the gene pool, because mutation rates are too low to alter significantly large gene frequencies. The effects of selection are strongest on relatively common alleles in the gene pool. Selection acts slowly on rare alleles — particularly if they are recessive — either to eliminate an unfavorable allele or to increase the frequency of an advantageous one. Genetic drift is a significant factor only in a very small population, where it may account for the random loss of alleles from the gene pool. Gene flow is of significance in establishing equilibrium among semi-isolated populations, but its effects on a particular gene pool become significant only if the migrants make up a relatively large part of the population or if the gene pool of the migrants is different from that of the population.

The combined effects of mutation, gene flow, selection, and genetic drift can significantly alter the gene pool of a population and can therefore account for evolutionary changes in species. Under most conditions, however, selection is apt to be the most important factor. Can these effects account for the evolutionary changes that are known to have occurred in nature? Is this model of the evolutionary mechanism adequate to account for the evolution of life on earth? These questions have not been answered to the satisfaction of all. Although many population geneticists now feel that selection is the major factor in evolution, some put more emphasis on random events. All agree that known mechanisms of population genetics can account for the evolution of modern life in the time that has passed since the origin of the earth.

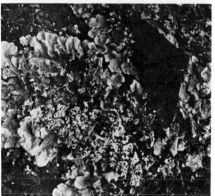

During the last several decades, the explosion of knowledge in many of the natural sciences has focused on the problems of human disease. The result has been a remarkable increase in our understanding of the way in which the health of organisms is controlled and how breakdowns in the control mechanisms lead to illness. It is safe to say that more has been learned about medical science during the twentieth century than during all of previous human history.

A great deal of our modern effort in understanding human biology and its derangements began with the pioneering concepts of Sir Archibald Garrod, an English physician who studied several kinds of human genetic diseases of amino acid metabolism. In 1908 Garrod described his concept of "inborn errors of metabolism." These errors are, in a sense, part of the variability occurring in nature: One defective step, or block, in a series of interconnected metabolic steps leads to a build-up of materials just before the block or a deficiency just after the block. These materials are not abnormal in themselves but are toxic if allowed to accumulate in large amounts or if insufficient quantities are produced. The result is genetic disease.

Garrod postulated that the block in a step from, say, $A \rightarrow B$ of a metabolic pathway is due to the absence of an enzyme that is required to convert A to B (Figure 8.1). He further postulated that such an enzyme deficiency could be inherited according to the genetic laws of Mendel, which were just then being rediscovered.

In the decades following Garrod's work, scientists have come to a better understanding of how these enzyme deficiencies arise. In order to follow the development of this understanding, it is useful to remember some features of enzymes — the biological catalysts that are required to drive most or all of the metabolic reactions in all organisms.

ENZYMES AND PROTEIN STRUCTURE

Enzymes are proteins that have the property of binding to another reactant molecule, or substrate, and changing it, either by cleaving the substrate at a specific site into a number of smaller constituents, or by joining smaller components into larger molecules. The overall structure of enzymes, and some other proteins, is complicated but specific. The amino acid chain in a protein is coiled and convoluted into a three-dimensional, globular structure known as the "tertiary structure." Parts of the chain are folded back over other parts, forming specific grooves, crevices, and interacting regions. One such region in many enzymes, which is called the *active site*, is the part of the molecule that recognizes and interacts with the substrate.

The way in which the sequence of amino acids in the protein is converted to the folded, three-dimensional molecule has now been shown to be directed by the amino acid sequence alone. We

8

Mutation and the Genetic Code

Figure 8.1   Model of a biochemical pathway in a living system. In order for compound A to be converted to compound B, a specific enzyme, E_1, whose production is controlled by gene G_1, must be present. This kind of mechanism must also exist to effect the conversion of compounds B to C, C to D, and D to F. If an enzyme block occurs in the conversion of compound C to D—that is, if enzyme E_3 does not form—compound C will accumulate or degrade, using an alternate pathway, and compound D will not form. As a result of the block, the D to F reaction will not occur. The amount of compound F in a system may regulate the activity or production of the first enzyme in the pathway (E_1) in a feedback control loop.

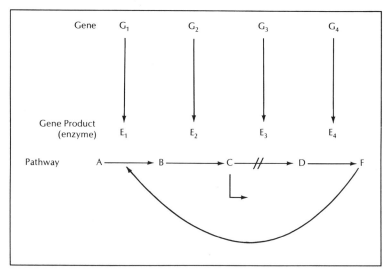

know from the work of Fred Sanger that the sequence of amino acids is unique for any given protein. Other protein chemists have shown that this unique sequence causes all like protein molecules to fold up into a form with a unique tertiary structure. If the folding is incorrect, then important regions such as the active site might be incorrectly constructed, therefore making an enzyme inactive. Because the folded structure is dictated by the amino acid sequence, could an error of even *one* amino acid cause an incorrect tertiary structure and an inactive enzyme?

This question was partly answered in 1949 when L. Pauling, H. Itano, J. Singer, and I. Wells showed that hemoglobin in patients with the disease sickle cell anemia is physically and structurally different from normal hemoglobin. Then, in 1956, Vernon Ingram showed that sickle cell hemoglobin, or hemoglobin S, is different from normal hemoglobin in only one position of the molecule; the defect is due to the presence of the neutral amino acid valine instead of the negatively-charged amino acid glutamic acid. The result is an abnormal folding, or incorrect tertiary structure, for hemoglobin S, which leads in some unknown way to an abnormal shape for the red blood cells. These abnormal red blood cells, which are shaped like "sickles" or "bananas," become trapped in the small

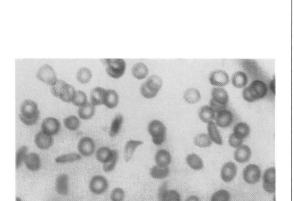

Figure 8.2 (above)   The characteristic form taken by red blood cells in an individual who has sickle cell anemia. The hemoglobin molecules in the round cells have normal activity.

Figure 8.3 (right)   Generalized schematic diagram showing an enzyme catalyzing the conversion of a complementary-shaped substrate molecule to products. The enzyme molecule remains unchanged during the reaction.

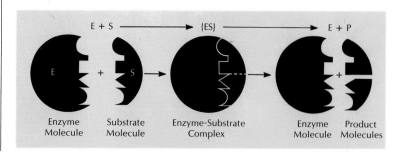

blood vessels and lead to tissue destruction in the patient, as well as cause a number of other complications.

Similarly, single amino acid changes in many enzymes and other proteins are now known to cause abnormal, and therefore inactive, three-dimensional structures. An absence of such enzyme activities then might lead to disease according to Garrod's concepts.

MUTATIONS

As described in Chapter 5, all genetic information, including the information for the amino acid sequence of proteins, is contained in the DNA of an organism. This fact, which is taken for granted today, was discovered as recently as 1944 when O. Avery and his colleagues showed that they could permanently convert one genetic characteristic of a microorganism into another characteristic by exposing it to DNA. During the same time, G. Beadle's and E. Tatum's work with the fruit fly *Drosophila* and with the bread mold *Neurospora* was beginning to show that single regions of the DNA—the genes—code for single proteins. Genes were known to undergo changes which render the product of a gene inactive. It was also known that some chemical and physical agents, such as x-rays, caused these changes in DNA to occur with greater frequency.

Changes in the genetic information that lead to defective genes are known as mutations. The nature of these changes could not be clarified biochemically until after the epochal discovery of the physical and chemical structure of DNA in the late 1940s and 1950s. DNA was shown to be composed of two intertwined strands, each consisting of organic ring compounds called nucleo-

Figure 8.4 Levels of organization in a protein molecule. In this diagram, the hemoglobin molecule is analyzed to show primary, secondary, and tertiary structure. Primary structure is concerned with how amino acids join together by means of peptide bonds. Secondary structure shows the helical nature of the amino acid sequence. Tertiary structure shows how an amino acid chain (polypeptide) folds in a three-dimensional shape. These three-dimensional polypeptide chains make up a complete hemoglobin molecule (as seen in Figure 11.9). A substitution of only one amino acid in the hundreds that make up a polypeptide chain can change the primary, secondary, and tertiary structures of a protein molecule.

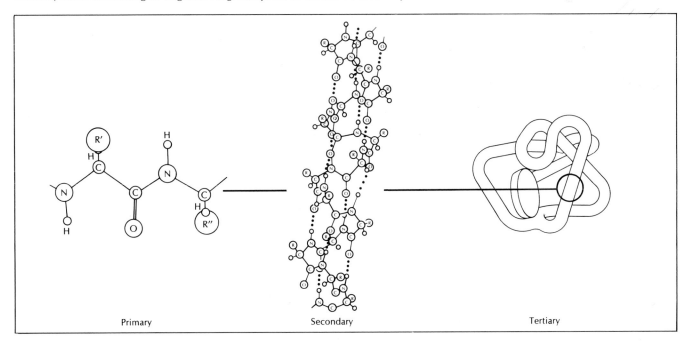

| Primary | Secondary | Tertiary |

tides stacked one on top of another, interacting with one another by hydrogen bonds and held in place by a backbone of sugars and phosphate groups. Molecular biologists, including Marshall Nirenberg and H. Khorana in the United States and François Jacob and Jacques Monod in Paris, soon worked out the details of the mechanisms by which the genetic information in DNA is converted to functional protein gene products.

Nirenberg had been studying the ability of ribosomes from the bacterium *E. coli* to incorporate amino acids into proteins. In one of the simplest yet most profound experiments in science, he discovered that small stretches of uracil-containing nucleotides that are joined together dramatically increase the ability of the ribosomes to incorporate the amino acid phenylalanine into protein. Nirenberg and his associates then tested all potential polynucleotide combinations of the four bases uracil, cytosine, adenine, and guanine to find codes for each of the twenty amino acids. They also studied the ability of base combinations to increase the binding of molecules of transfer RNA to ribosomes.

On the basis of their experiments, it soon became clear that the code in nucleic acids that is translated into protein is one in which

Figure 8.5   Genetic mutation. There are several ways that errors can occur during DNA replication. For example, the positions of one or more bases may be switched causing altered copies to be formed. These altered molecules will then transmit the mutation to later generations.

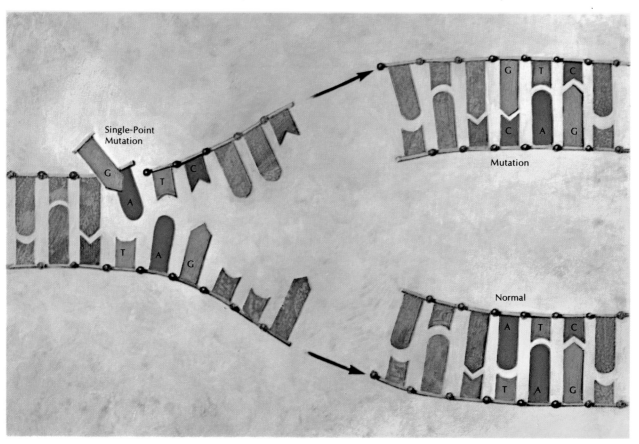

each combination of three bases codes for a single amino acid. (Each three-base combination is known as a codon.) The enzyme RNA polymerase transcribes the chain of codons of DNA language into a complementary chain of RNA still carrying the transcribed message for protein. This messenger RNA is then translated from the language of bases into the language of amino acids by interaction of the codons with parts of the transfer RNA that carry the amino acids to the ribosomes for polymerization into protein.

ERRORS IN PROTEIN SYNTHESIS

There are several ways in which the scheme of protein synthesis can go wrong. If a base is put into DNA in place of one that normally is present, the codon "meaning" clearly changes. The messenger RNA is "wrong" as a result, and the protein will be made "wrong" with an amino acid substitution. If the substitution is such that the subsequent folding of the protein is incorrect, then an inactive or structurally abnormal protein could result.

Some substitutions do not interfere with folding. For instance, if the middle base T of the DNA codon TTT coding for the amino acid lysine is mutated to a C, the resulting DNA codon TCT would code for arginine, an amino acid quite similar to lysine and usually located in similar positions on the folded structure of proteins. The resulting "error" is not likely to interfere markedly with protein function. But if the first T of the lysine triplet is mutated to a C, the resulting codon CTT is the code for glutamic acid, which is of opposite charge to lysine. That change is likely to result in abnormal folding and a functionally abnormal protein.

Such single changes in the base composition of DNA are called *point mutations*. If they lead to an amino acid change, the resulting mutation is called a *mis-sense mutation*. Base changes are caused by natural cosmic radiation, by industrial and manmade chemicals of many kinds, and probably by the radiation spilled into our environment by products of atomic industries and weapons testing. In another kind of mutation, one or more bases are inserted into a gene; in still another, one or more bases are deleted. The consequence of either of these two events is known as a *frame-shift mutation*. If one or two bases are inserted or deleted, the entire "frame" of codon reading will be put out of phase, so the entire amino acid sequence after the mutation will be wrong. If combinations of various codons are inserted or deleted, the reading frame will remain unchanged but an amino acid will be added or removed.

HUMAN DISEASES

Since the writings of Garrod at the beginning of this century, the concept that mutant genes lead to abnormal products which in turn lead to human disease has now been extended to include almost 2,000 human disorders or malformations. Of these 2,000

Figure 8.6 A frameshift mutation, in which the codons are read out of phase. Addition or deletion of a base still results in the reading of codons as triplets, but now different triplets are read. This alteration causes the insertion of the wrong amino acids in the polypeptide or may even result in the premature termination of the protein.
Printed by permission from J. D. Watson, Molecular Biology of the Gene, © *1970, J. D. Watson*

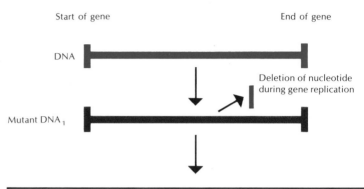

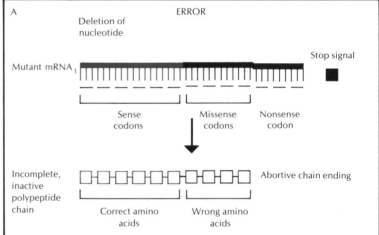

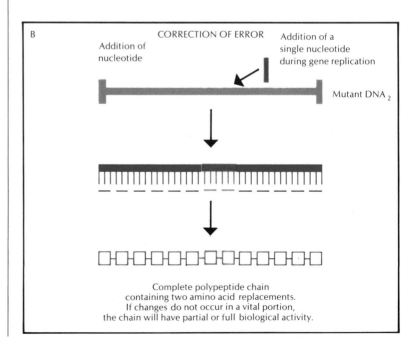

conditions, which were compiled recently by Victor McKusick of Johns Hopkins University, at least 103 are diseases traceable to an enzyme deficiency. And in every case in which the biochemical defect has been thoroughly worked out, the protein abnormality has involved a mutation of a *structural gene* — a gene that codes for a protein. In addition, in almost all cases, the defect is in *one* amino acid position in the protein, which presumably reflects a point mutation in the DNA of the affected individual.

The number of identified enzyme defects is growing rapidly, and it is almost certain that many more human disorders will be shown to involve similar defects. But it is just as likely that some human disorders will be shown to involve a defect not in a structural gene, but rather in one whose product must interact with other genes and cause their expression to be changed. These *regulatory genes* have been identified in many control points in the metabolism of microorganisms. Although it is often tempting to look at some human diseases as being of the regulator gene type, there is no proof of their involvement in any human disorder at this time.

Many of the enzyme deficiencies in human diseases were discovered in patients by analyses of their blood, urine, or other body products. During the past several decades, however, our ability to study these and many other diseases has been improved by remarkable techniques involving the removal of small pieces of skin or other tissues from whole animals or humans. These pieces are minced up into single cells and sheets of these cells are then grown for long periods of time in synthetic nutrient fluids in the laboratory. Cells cultured in this way continue to grow and divide, not in the form of tissues or organs but rather in the form of single cells. These cells carry (and may express) the genetic errors that cause the disease.

CELL DIFFERENTIATION

Why are tissue culture cells important in the study of human genetic disease? All cells in an organism contain the same total amount of genetic information in their nucleic acids, but different organs and different cells express selected portions of the genetic information. In other words, only liver cells make the protein albumin, only blood-forming cells make the proteins of hemoglobin, and so on, despite the fact that *all* cells contain the genetic information in their DNA for all protein gene products of the whole organism. The process by which different cells express different parts of their information is called "differentiation," and will be described further in Chapter 9.

In many cases, cells removed from an animal and placed into culture cease performing many specialized, or differentiated, functions. It is fortunate, however, that in some cases the enzymes and other proteins made in the tissues of whole animals continue to be made when the cells of that tissue are removed and grown in cell culture. This remarkable fact has enabled scientists for the first time to study genetic diseases extensively and effectively in the laboratory, because characteristic enzyme deficiencies in some diseases continue to be expressed in cells in culture.

DEFECTS IN THE AMOUNT OF GENETIC INFORMATION

In contrast to the class of human diseases caused by mutations in the quality of genetic information are diseases caused by an abnormality in the amount of genetic information.

In 1956, J. Tjio and A. Levan for the first time correctly identified the twenty-three pairs of human chromosomes that constitute the normal total genetic information. Normal males carry twenty-two pairs of chromosomes called *autosomes*, and one pair of *sex chromosomes* (an X chromosome and a Y chromosome). Normal human females carry twenty-two pairs of autosomes and one pair of X chromosomes (and no Y chromosomes).

In some situations, the pairing and separation of chromosomes during meiosis and egg or sperm formation (oogenesis or spermatogenesis) is aberrant. As a result, eggs or sperm may not end up

Figure 8.7 Chromosome transmission in gametogenesis and fertilization. The sex cells (gametes) have only half the chromosome number of the somatic cells, or twenty-three instead of a total of forty-six. When egg and sperm fuse together, the newly formed zygote acquires chromosomes from both parental gametes, and the full number is restored. The complete human being that will evolve from the zygote will therefore draw on the genetic characteristics of each parent. In this way, every human being receives one allele of each pair of alleles from his father and the other from his mother.

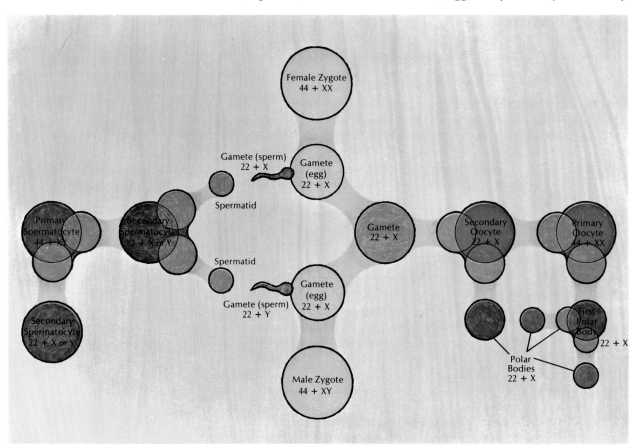

carrying one copy of each chromosome (as usually happens) but may carry two chromosomes of a given type. It is also possible that a chromosome of a given type will not be present at all. An incorrect chromosome pattern due to this abnormal separation is therefore said to be caused by chromosome *nondisjunction*.

If the extra (or deleted) whole chromosome is one of the autosomes, the result is usually a nonviable embryo that cannot survive for more than a short time *in utero*. However, such a defect does become manifest in one common condition. In some women, especially those over 35 years of age, the twenty-first chromosome pair segregates incorrectly and some eggs are formed with two chromosomes for set number 21. When such an egg is fertilized by a normal sperm, which contains one chromosome for set number 21, the resulting embryo will have *three* chromosomes instead of the two chromosomes that normally make up that set. Presumably all the genetic information is normal and no mutations cause the defect. Nevertheless, the result of excess normal genetic information is the devastating condition known as *Down's syndrome*, otherwise called Mongolism or Trisomy-21. Affected individuals suffer severe mental retardation and often require hospitalization and institutionalization.

Other conditions are known in which the wrong amount of genetic information is due to a deficiency or an excess of the sex chromosomes X or Y. These conditions are relatively common and are less devastating than the autosomal defects of chromosome number. For instance, a female with only a single X chromosome (XO instead of XX) may be of normal intelligence but she is likely to be short and sterile. Similarly, males with an extra Y chromosome (XYY instead of XY) are usually of normal intelligence, but some geneticists believe that such men are likely to develop some antisocial behavior and perhaps even criminal tendencies. This correlation has not yet been proved.

PRENATAL DIAGNOSIS

One of the most useful new developments in human genetics is the ability to diagnose some genetic diseases prior to birth. This ability depends on the capacity of cells in culture to express some defective functions characteristic of the whole animal with a genetic defect. Not only can such cells show an enzyme deficiency, but many or all of the disorders due to an abnormal chromosome structure or number can be demonstrated in cells in culture.

The usefulness of this development becomes apparent when these techniques are applied to cells derived from a fetus still in the mother's uterus. A fetus is bathed in a fluid that has many living cells floating about in it. These cells are derived largely from the skin and respiratory tract of the fetus. By means of a technique called *amniocentesis*, fluid from a pregnant woman can be sampled by needle puncture of the uterus usually 12 to 16 weeks after

Figure 8.8 The photomicrograph above shows a chromosome smear preparation taken from a dividing white blood cell of a normal male. The twenty-three chromosome sets have been repositioned by size to facilitate examination. Note the size disparity between the X and Y chromosomes. The photomicrograph below shows the chromosomes found in the white blood cells of a Down's syndrome male. Note the additional chromosome to set number 21. This chromosome abnormality is phenotypically expressed as a form of Mongolian idiocy.

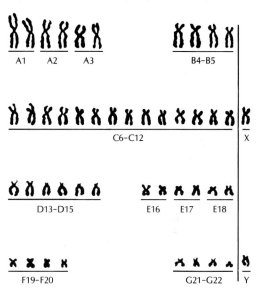

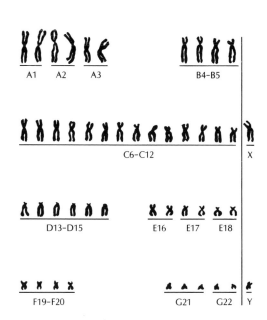

Figure 8.9 (above) In amniocentesis, a sterile needle is inserted into the amniotic cavity and a small sample of the fluid surrounding the fetus is drawn out. The fluid, which is derived mostly from fetal urine and secretions, contains cells of the fetus. These cells, when studied, can relate a great deal of important information, and potential genetic disease may be detected.

Figure 8.10 (below) Cell culture of fetal cells taken in amniocentesis. After fluid has been withdrawn, the sample is centrifuged to separate cells and fluid. These cells are then cultured.

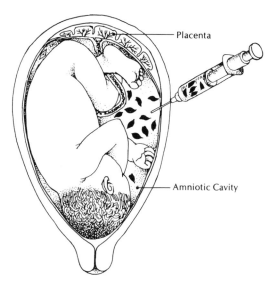

Placenta

Amniotic Cavity

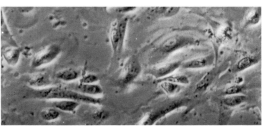

conception. A small amount of fluid can be removed and its cells cultured and examined either for biochemical defects or for abnormalities in the number or structure of chromosomes. In this way, diagnoses can be made for a growing number of human genetic diseases. All disorders of chromosome number usually can be detected by this technique early in pregnancy, and approximately forty different diseases involving enzyme deficiencies usually can be detected as well.

Tay-Sachs disease is one example of an enzyme deficiency disease that can be detected by amniocentesis. Because a discussion of this disease would be especially useful in demonstrating some of the principles of genetics, it will be examined in detail here.

Tay-Sachs disease was first described by the British opthamologist Warren Tay and the American neurologist Bernard Sachs in the 1890s. Of the people afflicted by this disease, 99 percent are people of the Jewish Ashkenazic (East European) ancestry. Among these Ashkenazy Jews, approximately one out of every thirty persons is a carrier of the disease.

Clinically, the disease is a severe neurological disorder that strikes what seem to be normal infants usually during the first year of life. An affected child becomes increasingly out of touch with

Figure 8.11 Structural formula of the ganglioside GM$_2$. This complex molecule is composed of the lipid fraction ceramide, which is linked to a polysaccharide containing glucose, galactose, N-acetyl galactosamine, and N-acetyl neuraminic acid. In Tay-Sachs disease, there is an absence of the enzyme hexosaminidase A. Because this enzyme is required to cleave the terminal sugar N-acetyl galactosamine from the ganglioside GM$_2$, an absence of the enzyme results in an accumulation of this ganglioside in nerve cells throughout the body. The arrow shows the cleavage point where the enzyme is active.

182

reality, and develops progressive muscle weakness, generalized paralysis, deafness, blindness, and convulsions. The child who is affected by this disease faces the prospect of an early death, usually in the third or fourth year of life.

Until recently, only limited information was available on possible causes of the disease. Pathological studies of tissues from patients with this disease showed that nerve cells in many parts of the body (including the brain) seemed to be full of some material not present in normal nerve cells.

In 1969 John O'Brien, together with his colleagues, provided the breakthrough for understanding this complex disease when they discovered that a patient's cells are absent in an enzyme called *hexosaminidase A*, which is required to carry out the degradation (breakdown) of the normal cell constituent GM_2 ganglioside. This enzyme is required to cleave the bond between galactose and N-acetyl galactosamine, as diagrammed in Figure 8.11. An absence of the enzyme results in the massive accumulation in many cells of the undegraded ganglioside, which interferes considerably with the normal function of the cells.

O'Brien also discovered that hexosaminidase A, which is present in the blood of normal humans, is absent in patients with the disease. Parents of such patients, who in most cases are carriers of the mutant gene, have a specific enzyme level in their blood and cells of approximately half that of normal humans. Most carriers, or heterozygotes, of an autosomal recessive disease involving an enzyme deficiency have one abnormal gene providing little or no active enzyme. The resulting average enzyme activity is 50 percent, and in most cases, this amount of enzyme is sufficient to prevent the disease.

It has been clear for a long time that Tay-Sachs disease results from a mutation in one of the autosomes rather than a mutant X

Figure 8.12   A diagram of fetal and maternal circulation, showing which substances can cross the placental barrier. A selectively porous membrane separates the two circulatory systems.

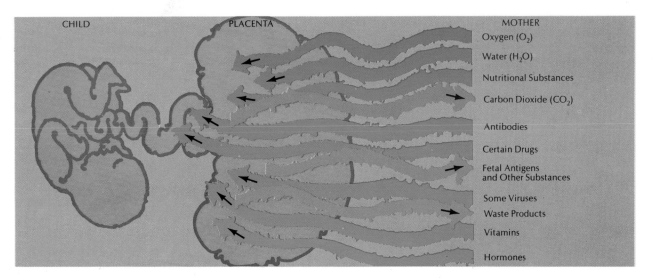

CHILD	PLACENTA	MOTHER
		Oxygen (O_2)
		Water (H_2O)
		Nutritional Substances
		Carbon Dioxide (CO_2)
		Antibodies
		Certain Drugs
		Fetal Antigens and Other Substances
		Some Viruses
		Waste Products
		Vitamins
		Hormones

Figure 8.13 Diagram showing how dominant normal and recessive Tay-Sachs alleles are transmitted from two carrier parents to possible offspring. In matings of this kind, one expects that approximately 25 percent of all offspring will be homozygous dominant (with two normal alleles), 50 percent will be heterozygous carriers (with one normal and one Tay-Sachs allele), and 25 percent will be homozygous recessive (with two Tay-Sachs alleles).

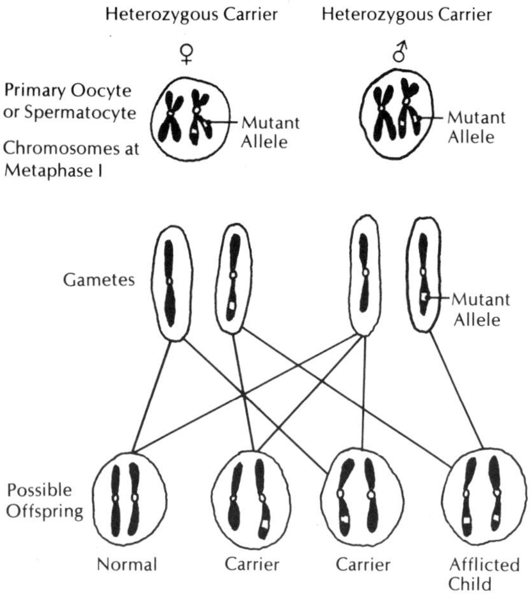

chromosome. That is, studies of the families of affected infants show that:

1. There is no significant difference between the number of affected males and affected females.
2. Except in rare cases, previous or subsequent generations are not affected with the disease.
3. There are likely to be a number of affected children in large families.

During spermatogenesis, a carrier (or heterozygous) male will produce some sperm carrying the normal copy of the gene for hexosaminidase A and other sperm carrying only the mutant form. Likewise, in a carrier (or heterozygous) female, oogenesis will result in either normal or mutant ova, with equal probability. Therefore, a homozygous (and affected) infant would necessarily acquire two copies of the mutant gene, one from each carrier parent, according to the diagram in Figure 8.13. Because all combinations of normal and mutant sperm and ova are equally likely, the average result of fertilization for every four gametes will be one fully normal embryo, two heterozygous carriers, and one homozygous embryo for the mutant gene (affected with Tay-Sachs disease). In other words, an average of 25 percent of all offspring produced from such a match would be normal, 50 percent would be heterozygous carriers, and 25 percent would be diseased homozygous mutant offspring.

Recently, it has been found that in the fetal cells suspended in the amniotic fluid of pregnant women, hexosaminidase A activity can be found in normal fetuses, but it is totally absent in the cells of fetuses that are homozygous for Tay-Sachs disease. This discovery has led to the current interest in being able to screen the Ashkenazy Jewish population in order to locate heterozygous couples who risk having a child with Tay-Sachs disease. If such carrier females are tested by amniocentesis, the following information can be determined. If amniotic cells contain full enzyme activity, the fetus is normal. If enzyme activity is 50 percent, the fetus is a carrier, having received the mutant gene from only one of his parents. However, if enzyme activity is absent, the child to be born is destined to have the disease. In these cases, it is becoming more acceptable to terminate the pregnancy to prevent the incalculable suffering that surrounds the short life of a child afflicted with Tay-Sachs disease.

Our society has unfortunately not yet resolved all of its conflicts regarding abortion as a goal for the control of genetic diseases. Some genetic diseases can be treated effectively by a variety of diets and drugs, but many cannot. And so geneticists, parents, and society are sometimes forced to use methods such as abortion as a means of preventing some kinds of severe genetic disorders. Cures for most genetic disorders are not possible now and are not likely

Table 8.1
Genetic Diseases Associated with Mental Retardation

Disorder	Defective Enzyme or Metabolic Arrangement

Chromosomal Abnormalities
  Down's Syndrome (extra #21)
  Klinefelter's Syndrome (XXY)
  Turner's Syndrome (X)
  "Tall Criminal" (XYY)
	Excess or Deficiency of Total
	  Genetic Material and Information

Arginosuccinic Aciduria	Arginosuccinase
Citrullinemia	Arginosuccinate Synthetase
Fucosidosis	Alpha-Fucosidase
Galactosemia	Galactose-1-Phosphate Uridyl Transferase
Gaucher's Disease	
Infantile Type	Absent Cerebrosidase
Adult Type	Deficient Cerebrosidase
Generalized Gangliosidosis	Absent Beta-Galactosidase
Juvenile GM$_1$ Gangliosidosis	Deficient Beta-Galactosidase
Juvenile GM$_2$ Gangliosidosis	Deficient Hexosaminidase A
Glycogen Storage Disease	
Type 2	Alpha-1,4 Glucosidase
Hunter's Disease	Increased Amniotic Fluid Heparitin Sulfate
Hurler's Disease	Increased Amniotic Fluid Heparitin Sulfate
I-Cell Disease	Multiple Lysosomal Hydrolases
Isovaleric Acidemia	Isovaleryl CoA Dehydrogenase
Lesch-Nyhan Syndrome	Hypoxanthine-Guanine Phosphoribose Transferase
Maple Syrup Urine Disease	Alpha-Keto Isocaproate Decarboxylase
Metachromatic Leucodystrophy	
Late Juvenile Type	Absent Arylsulfatase A
Juvenile and Adult Types	Deficient Arylsulfatase A
Methylmalonic Acidemia	Methylmalonyl CoA Carbonyl Mutase
Niemann-Pick Disease	Sphingomyelinase
Phenylketonuria	Phenylalanine Hydroxylase
Refsum's Disease	Phytanic Acid Alpha-Oxidase
Sandhoff's Disease	Hexosaminidase A and B
Sanfilippo Disease	Increased Amniotic Fluid Heparitin Sulfate
Tay-Sachs Disease	Hexosaminidase A
Wolman's Disease	Acid Lipase

Source: Adapted from T. Friedmann, "Prenatal Diagnosis of Genetic Disease," *Scientific American* (San Francisco: Freeman, 1971).

to become available soon. And until such a time arrives, prenatal detection and abortion will probably continue to be useful and effective means for the control of selected kinds of human genetic disease.

THE FUTURE

Genetic diseases are beginning to pose increasingly important problems in medicine. Today, approximately 5 to 10 percent of all admissions to a hospital pediatric ward are of patients whose illness has at least some genetic components. The percentage apparently will grow in the near future. What future developments might improve the medical care for these patients?

Today there are various treatments of the major genetic diseases. One treatment involves replacement of the protein *gene product* of the defective gene; injections of insulin for diabetes are an example of this kind of treatment. Another treatment involves the elimination of dietary constituents which, in a Garrod scheme, would accumulate to toxic levels. One example of this treatment involves the disease *phenylketonuria* (PKU), which is due to absence of the enzyme phenylalanine hydroxylase. Dietary restriction in infants of the amino acid phenylalanine has proven useful in

Figure 8.14   One phenotypic trait such as mental retardation may express the instructions of many genes (left). Conversely, the instructions from one gene, or genotype, may influence a number of phenotypic features (right).

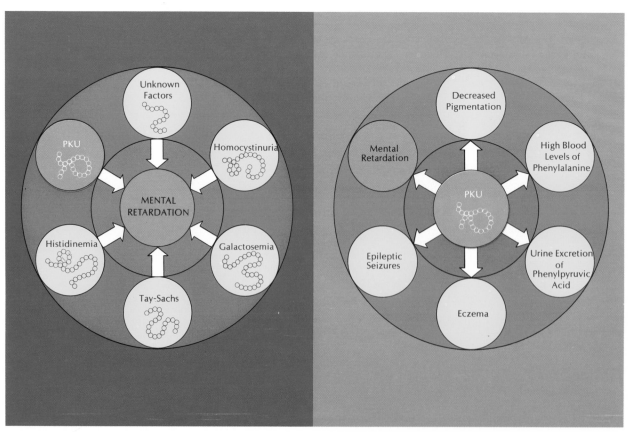

preventing the severe mental retardation that accompanies untreated PKU. Other treatments simply provide general support, as in institutionalization and care of patients with Down's syndrome. However, a great deal of research is being carried out to make treatment more effective and less cumbersome.

In some cases, an absent enzyme might be restored by providing whole new tissues or organs to an affected individual. For instance, transfusions of bone marrow have already been used to a limited extent to replace cells carrying the genetic information for some kinds of antibodies. Similarly, transplants of such organs as the kidney have been used in attempts to "cure" patients with rare disorders of amino acid and mucopolysaccharide metabolism.

In the future, replacement of defective gene products such as mutant enzymes might also be accomplished by putting normal enzymes into permeable capsules or onto insoluble granules and injecting or implanting them into enzyme-deficient individuals. In this way, researchers hope to make replacement enzymes more stable and to protect them from the body defenses, which recognize foreign proteins as "non-self" and cause them to be rejected (Chapter 11).

Further into the future is the possibility of replacing normal copies of the genes for a given product. This possibility is suggested by the recent development of methods to carry out the full, total synthesis of a gene, to fractionate genes, or to synthesize them from their transcribed RNA by use of an enzyme that can carry out this "reversed flow" of genetic information. For instance, scientists envision being able to insert into the cells of individuals suffering from Tay-Sachs disease the genes that code for normal copies of the missing enzyme hexosaminidase A. Such "transformations" of animal cells have not yet been accomplished. But there are indications that they will be carried out in the future, either with pure DNA or through infection with some kinds of viruses. If a virus is used to transfer the genes, the process would be called "transduction." These manipulations are still far off—at least clinically—but they will come eventually and benefit sick human beings.

When does life begin? Some people think life begins at fertilization, others at implantation, or when the newborn takes its first breath independent of its mother. Actually, life began only once, billions of years ago, and every living thing today has a direct connection to that beginning and to all of life.

Consider the fact that each human being is the direct descendant of two parents, four grandparents, eight great-grandparents, and so on, and that if you count back only thirty-two generations you will find that you have more ancestors than have ever lived at one time on earth. Try counting back to the first creatures to appear in the organic soup and the number soon becomes incalculable.

Or consider the knowledge that each person is made up of twenty-three pairs of chromosomes—half of which come from the mother and half from the father—and that each chromosome must carry at least a thousand genes. That person of necessity is a new and unique combination of those units, probably never before assembled. And yet every gene, or precursor of every gene, was handed down through all the generations, coming together now into a composite individual who is different from all others. But this "life" cannot be considered really "new."

Almost a hundred years ago August Weismann advanced a concept that there is an unbroken line of what he called "germ plasm" (reproductive cells) from the very beginning of life, and that with each generation the "somatoplasm" (body tissues), aside from the reproductive cells, must die, while the germ plasm continues on in another and uniquely different individual (Figure 9.1). It is somewhat like a relay race in which the baton is handed from one to another; the continuing element in the race is the baton—not the individuals carrying it. In this manner, each of us carries in our reproductive cells the essence of life, which will survive in another generation if the opportunity is provided. But the rest of our body will perish, its constituent elements returning once more to the cycles of nature.

TYPES OF ASEXUAL REPRODUCTION

For unicellular organisms, reproduction occurs when cellular structures are duplicated and the cell divides into a pair of daughter organisms. The parent organism does not die but contributes all of its materials to its offspring. Cell division seems to rejuvenate the unicellular organism, which otherwise would age and die. Even if a cell were immortal, however, it would still be susceptible to accidental death. The continued survival of a species therefore depends on its ability to produce new individuals at least as rapidly as old ones die.

For most procaryotic organisms, reproduction occurs through the simple process of *binary fission*. During a brief period of growth, all cellular components are duplicated. The cell then splits into two cells, each receiving exactly half of the chromosomal

9

Reproduction and Development

Figure 9.1   In the germ plasm continuity model, only the reproductive cells (black) carrying the hereditary information are sustained from generation to generation. The somatoplasm, or body tissues (yellow), are destined for temporary existence, then death.

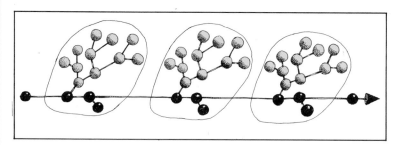

materials and about half of the other materials from the parent cell (Figure 9.2). The newly formed cells then begin to grow in preparation for the next division. In these cases, the offspring cell receives a single copy of the parental chromosome, which is duplicated again just before each division. Variations in genetic information can arise only through mutations—through spontaneous changes in the nucleotide sequence or through errors in the process of chromosome replication.

In single eucaryotic cells, chromosomes are distributed during mitosis, but little is known about the distribution of the other parts of the cell. For more complex unicellular organisms, a great number of organelles either must be duplicated before division and then distributed properly, or they must be produced after division according to instructions carried by the chromosomes.

For many kinds of single eucaryotic cells, reproduction occurs through mitotic cell division. In many organisms, however, both asexual and sexual reproduction take place (Figure 9.3). Meiotic division of the diploid adult cell produces haploid cells called *gametes*. A pair of these gametes (usually derived from different parental cells) then fuse to form a new diploid individual cell—a process called *cellular fusion*. In many cases, two kinds of gametes are produced by different parental cells: a flagellated gamete containing little cytoplasmic material and a nonmotile gamete containing a large amount of cytoplasm. Fusion occurs only when a flagellated gamete happens to run into a nonmotile gamete.

Most multicellular organisms produce single cells or small multicellular fragments that develop into new individuals. The parent, after completing a period of reproduction, usually enters an aging period called "senescence" and eventually dies. Asexual reproduction may involve either the formation of multicellular buds or fragments, or the production of cellular spores. Sexual reproduction usually produces unicellular gametes.

In the process of *budding*, one of the cells of the parent organism begins to grow and divide, much as if it were a newly formed zygote (Figure 9.4). For a time this bud continues to grow into a new organism while remaining attached to the parent and drawing nourishment from it.

At some stage in its growth, the developing bud may become separated from the parent and take up an independent life. In

Cell Wall   Nuclear Material

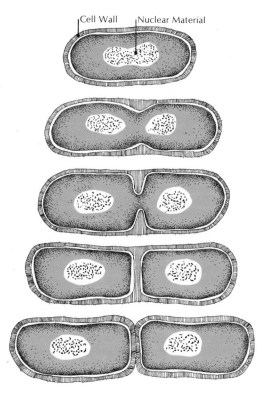

Figure 9.2   In this series of diagrams, a procaryotic cell is divided by means of binary fission into two cells. Note the lack of an elaborate mitotic apparatus, as is observed in eucaryotic cells.

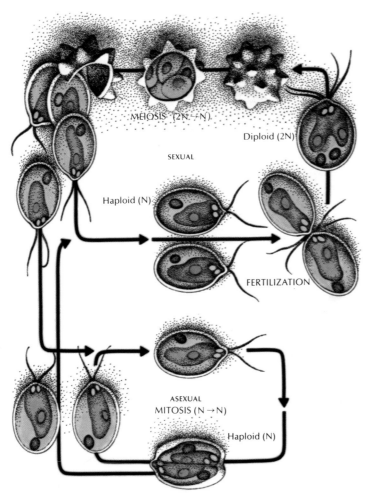

MEIOSIS (2N → N)

Diploid (2N)

SEXUAL

Haploid (N)

FERTILIZATION

ASEXUAL
MITOSIS (N → N)

Haploid (N)

Figure 9.3   The alternation of haploid and diploid generations. In many species of algae, large numbers of spores are formed by mitotic or meiotic divisions of one or more of the body cells of the parent organism. Each of the spores formed by mitotic division (mitospores) can develop into an individual similar to the parent plant. A spore formed by meiotic division (meiospore) will develop into a haploid organism that may be different from the diploid parent in appearance. In such species, the haploid individuals produce gametes, whose sexual fusion creates diploid offspring. Thus, haploid and diploid generations alternate.

Figure 9.4   *Volvox* (upper left), a green algal protistan, may be regarded as a multicellular plant or as a colony of unicellular organisms. Asexual reproduction occurs when one of the cells begins to divide, eventually forming a new sphere of cells within the hollow center of the parent colony. Only cells in the posterior half of the colony produce daughter colonies. The fresh-water coelenterate *Hydra* (upper right) is reproducing by means of budding. Each bud attached to the adult's stalk will form a miniature adult, and each will eventually pinch off to assume an independent existence. Newly formed cells continuously move outward as old cells die and drop off at the base and at the tips of the tentacles. Every few weeks the entire individual is completely replaced. Thus, the individual *Hydra* might be regarded as being in a continuous process of asexual reproduction. The yeast cells shown below are also reproducing by means of budding.

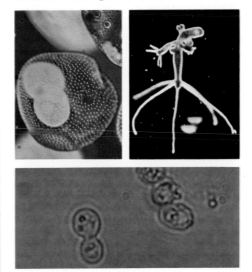

many cases, however, the offspring remain attached to the parent, forming a colony of potential individuals. Each bud is capable of survival if it should be cut off from the colony, but most retain bodily connections and an interchange of nutrients and other materials with the others.

In some kinds of algae and liverworts, older parts of the plant body may die, leaving the tips to develop into new individuals. Among many kinds of plants and fungi as well as some animals, fragments of the body can develop into new individuals if some exterior force breaks up the body. Flatworms such as *Planaria* are examples of multicellular organisms that spontaneously divide themselves into more-or-less equal fragments, each of which develops into a complete new individual.

Asexual reproduction through unicellular spores is found among protists, fungi, and plants, but not among animals. The spores of some species are flagellated; others are nonmotile and in most cases are transported by wind or water. The process of repro-

Figure 9.5 Generalized angiosperm life cycle. Like the cones of cycadophytes and conifers, the flowers of anthophytes are highly specialized reproductive organs produced at the tips of stems. The male stamens (the pollen-producing organs) and the female pistils (the seed-producing organs) can be seen in the lower portion of the diagram. As seen by the arrows, the cyclical sequence of events comprising pollination and fertilization begins at this point and can be followed around in a clockwise pattern.

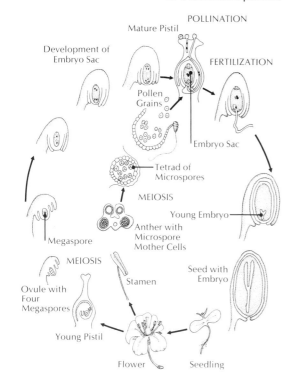

duction usually involves dispersion of the new individuals. Figure 9.5 shows a diagram of a typical plant reproductive cycle.

Another form of asexual reproduction is *parthenogenesis*, the spontaneous development of an unfertilized egg into an adult organism. Among honeybees, development of an unfertilized, haploid egg gives rise to a drone, a male adult with haploid cells. Fertilized honeybee eggs form diploid, female adults—either queens or workers, depending on the nutrients supplied during development (Figure 9.7).

In some cases of parthenogenesis, a diploid egg is formed either through fusion of the egg cell and one of the polar bodies or through a replication of the egg chromosomes without cell division. In other species where parthenogenesis is common (among the aphids, for example), eggs are formed without meiotic division. These diploid eggs can develop into diploid individuals with the same genetic information as the parents.

SEXUAL REPRODUCTION

Almost every kind of multicellular organism is capable of sexual reproduction, although this process may be supplemented by various forms of asexual reproduction. The mutations that occur in sexual reproduction are a source of genetic variation in a population. For example, suppose that two independent mutations, A and B, each appear on the average in about 1 of each 1,000 individuals. In a population that reproduces only asexually, about 1/10 percent of the population will show the mutant character A and another 1/10 percent will show the mutant character B. But only about 1 individual in each 1 million will undergo both A and B mutations simultaneously.

On the other hand, in a sexually reproducing population, there is a possibility that an individual with the A mutation will mate with an individual with the B mutation, and the A-B combination can be produced within a generation by *sexual recombination*.

Figure 9.6 In the event of amputation, the starfish is an example of an organism that is able to regenerate new limbs instead of merely healing over a resulting wound.

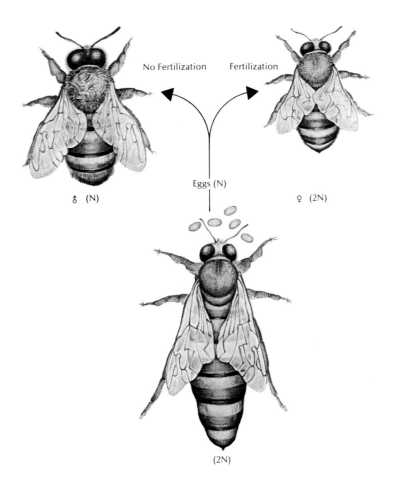

No Fertilization   Fertilization

Eggs (N)

♂ (N)   ♀ (2N)

(2N)

Figure 9.7   An example of parthenogenesis—the development of an unfertilized egg into an adult organism and a form of asexual reproduction. In the honeybees shown here, the queen produces eggs that if fertilized become female workers and if not fertilized become male drones.

Sexual recombinations can be made of all the different characteristics that happen to arise by mutation, so a population that reproduces sexually is likely to possess a much wider variety of genotypes than one that reproduces only asexually.

In all known cases, sexual recombination leads to the formation of a single cell, a *zygote*, which then develops through repeated divisions into a new adult. The result of this joint process of sexual recombination and reproduction is that *every cell of a multicellular organism probably contains the same genetic information*. This process has been favored in the evolution of all multicellular organisms. It also underlies natural selection, for it ensures that the reproductive cells of an organism will contain precisely the same type of genetic information as the body cells that initially determine the phenotype.

In its simplest form, then, sexual reproduction involves the formation of haploid gametes that fuse to form a diploid zygote. In some algae, such as *Ulothrix*, all gametes appear identical (Figure 9.8), and a gamete will fuse with any other gamete it happens to

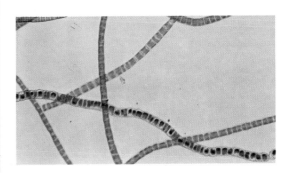

Figure 9.8   In the alga *Ulothrix* all gametes appear identical. The adult plant is haploid, and the diploid zygote undergoes meiosis early in its development. Only one of the haploid cells resulting from meiosis will survive to produce a new individual, thus ensuring that all cells of the adult plant will have the same genetic information. This condition, in which both joining gametes are of the same size and shape, is known as isogamy.

Figure 9.9 Each filament of *Ulothrix* is able to reproduce by sexual and asexual means. The adult plant is haploid and the diploid zygote divides meiotically early in development, but only a single resulting haploid cell survives to give rise to a new individual.

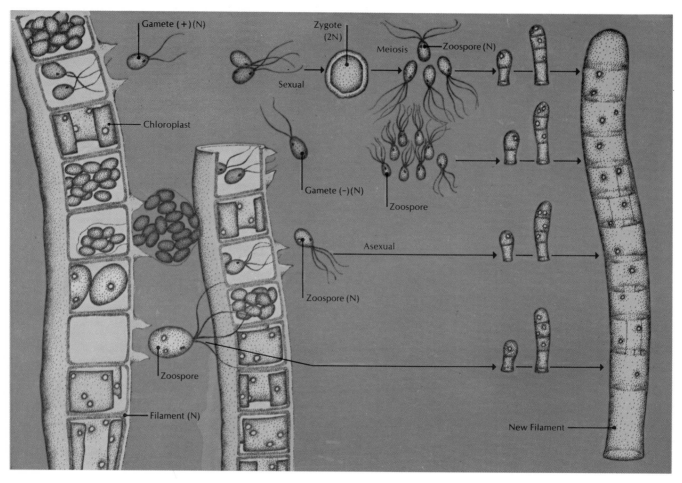

encounter. In contrast, fusion will occur in many species only if the gametes are from different parent individuals. The process of reproduction whereby gametes of identical size and shape join together is called *isogamy*.

Among most multicellular organisms, sexual reproduction involves two kinds of gametes. One kind tends to be small and highly motile; the other kind is less motile and carries a large supply of nutrients. In this case of *heterogamy*, or *anisogamy*, a particular adult usually produces only one kind of gamete. In the extreme development of heterogamy, one kind of gamete is the sperm—small, active, and equipped with flagella—and the other kind is the egg, or ovum—large, immotile, and packed with nutrients. This form of heterogamy is called *oogamy*; the individuals that produce sperm are males and those that produce eggs are females.

Although the pattern of oogamy is by far the most common form of sexual reproduction among multicellular organisms, it is not universal. In some cases of heterogamy the two kinds of gametes are quite similar in appearance, and the designation of one sex as female and the other as male may be arbitrary.

ANIMAL REPRODUCTIVE SYSTEMS

Although asexual reproduction through fission, budding, or fragmentation does occur in some of the invertebrate groups of animals, sexual reproduction is more common throughout the animal kingdom. In some of the simpler aquatic organisms, sperms and eggs are released into the water, where fertilization occurs. In most kinds of animals, some form of *copulation* occurs, with the male injecting sperms into the female's body. The fertilized zygote may develop for some time within the female's body or may be enclosed in an egg and later released from the female's body.

The gametes are produced by meiotic division of specialized primary reproductive cells in the organs called *gonads*. These organs are called *testes* in the male and *ovaries* in the female. The process of *gametogenesis*, or gamete production, varies somewhat in detail among the different groups of animals, but a generalized description of the process in mammals illustrates the major events.

The reproductive system of a mammal consists of the gonads, the reproductive tract through which the gametes move, and various associated glands. As in most higher animals, each mammalian species has male individuals that produce spermatozoa and female individuals that produce eggs. *Hermaphroditism*, in which a single individual produces both eggs and sperm, is found among some groups of animals. Fertilization occurs within the reproductive system of the female, and the embryo is retained within the female's body during the early stages of its development. In addition to the differences in the sexual organs of the two sexes (appropriate to their different roles in the formation of the zygote and nur-

Figure 9.10   The filamentous green alga *Spirogyra* reproduces sexually. The zygote undergoes meiosis upon germinating, but three of the four nuclei formed by meiotic division disintegrate, and the remaining nucleus undergoes mitotic division to form a new filament. This series of photographs shows the vegetative filament (far left); early conjugation (middle left), with protoplasmic bridge being formed between four sets of opposite cells in different filaments; later conjugation (middle right), with migration of cellular contents from one cell into the other cell; and complete conjugation (far right), with the formation of zygospores (zygotes), each capable of developing a new vegetative filament.
Courtesy Carolina Biological Supply Company

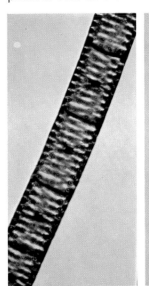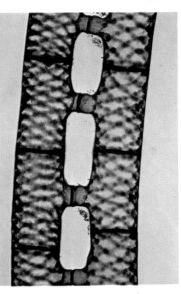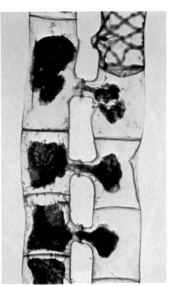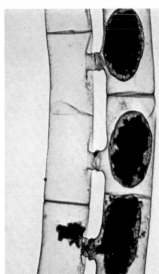

ture of the embryo), the sexes can be distinguished by various differences in the form of various other parts of the body.

The sex of an individual is determined primarily by genetic inheritance. As discussed in Chapter 8, a mammalian zygote acquiring an X chromosome from each of its parents will normally develop into a female (with an *XX* genotype), whereas a zygote obtaining one X and one Y chromosome will normally develop into a male (with an *XY* genotype). The determination of sex through sex chromosomes is widespread among animals but is not universal. Among bees, for example, males are haploid and females are diploid. Even among animals having sex chromosomes, the nature of sex determination varies widely. In birds, for example, males have a pair of similar sex chromosomes, whereas females have a pair of different sex chromosomes.

The determination of the primary reproductive cells, which will later form the eggs or sperm, occurs at a very early stage in the development of the animal embryo. Cells that form the gonads become determined at a much later stage of development. Thus, the primary reproductive cells are set aside for that fate early in development, long before the gonadal structures are visible. Even after the gonads have begun to develop, it is some time before they become determined as either male or female sex organs.

The male gonads, or testes, of a mammal contain a large number of tubules that are made up of primary reproductive cells and

Figure 9.11 As shown in this diagram, the mammalian male gonads, or testes, contain numerous seminiferous tubules (see A) that contain primary reproductive cells (also known as germ cells). These cells will eventually divide by meiosis and metamorphose into spermatozoa.

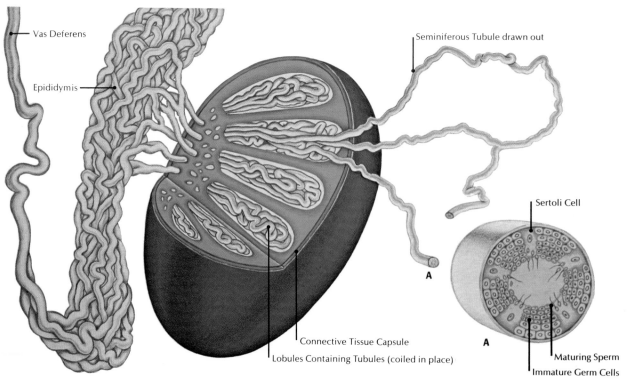

Vas Deferens

Epididymis

Seminiferous Tubule drawn out

Sertoli Cell

A

Connective Tissue Capsule

Lobules Containing Tubules (coiled in place)

A

Maturing Sperm

Immature Germ Cells

196

Sertoli cells. The tubules are complexly coiled within the testes and are surrounded by connective tissues and interstitial cells (Figure 9.11). In the mature mammal, the reproductive cells continue to divide mitotically, producing a continuous new supply. Some of these cells divide meiotically to produce haploid cells that develop into spermatozoa, a process called *spermatogenesis* (Figure 9.12). In some mammals, maturation of spermatozoa occurs only at certain seasons. In others, including man, mature spermatozoa are produced continuously during the reproductive portion of the life cycle. Mature spermatozoa leave the Sertoli cells, move through the tubule, and enter a duct leading to the exterior. The structure of the duct and the nature of the various glands that lubricate the passage of sperm or provide nourishment for sperm cells vary greatly among groups of animals.

The female gonads, or ovaries, of vertebrates consist of oval-shaped masses of cells, including vascular and connective tissues, as well as the primary reproductive cells, *follicle cells*, and *nurse cells*, which provide nutrients for the development of oocytes (Figure 9.13). The mammalian female primary reproductive cells cease mitotic division before birth, so that the newborn female possesses a complete lifetime supply of immature eggs. The primary reproductive cells grow larger than the other kinds of cells that surround them. The process by which the mature egg is produced, which is called *oogenesis*, begins with meiosis, but the meiotic division is

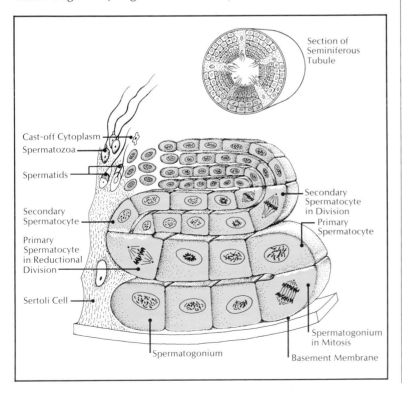

Figure 9.12 A schematic diagram showing the steps involved in spermatogenesis within the seminiferous tubules of an adult male. The sequence begins with the production of spermatogonia. Each spermatogonium divides twice mitotically to form four spermatocytes. Each spermatocyte yields four haploid spermatids through meiosis. The process is completed when the spermatids become sperm cells, or spermatozoa.

Section of Seminiferous Tubule

Cast-off Cytoplasm
Spermatozoa
Spermatids
Secondary Spermatocyte
Primary Spermatocyte in Reductional Division
Sertoli Cell
Secondary Spermatocyte in Division
Primary Spermatocyte
Spermatogonium
Spermatogonium in Mitosis
Basement Membrane

halted at prophase of the first division. The developing oocyte then grows larger as it absorbs nutrients from the follicle and nurse cells. The various substances that make up *yolk*—including protein, fat, glycogen, and RNA—are pumped into the oocyte.

Among some organisms, such as birds, the growth of the egg cell continues until it is many thousands of times as large as other cells of the body. The follicle cells form a thin layer over the growing egg cell. The mammalian egg is relatively small and contains only a small amount of yolk, but the layer of follicle cells becomes quite large around the egg cell. When accumulation of yolk has been completed, the oocyte resumes meiotic division. However, the division of cytoplasm during meiosis is unequal. The first division results in the pinching-off from the large egg cell of a very small cell containing little cytoplasm. This *first polar body* eventually disintegrates. In most vertebrates, meiosis is again halted after the first division and is not resumed until sometime after the egg has left the ovary. But whenever the second meiotic division occurs, it also results in the pinching-off of a small cell with very little cytoplasm. This *second polar body* also disintegrates. Thus, a single primary reproductive cell in the male gives rise to four spermatozoa, whereas a single primary reproductive cell in the female gives rise to only one mature egg.

In various vertebrates, large numbers of mature eggs may be produced at a single time in the life cycle of the female or at sea-

Figure 9.13   Diagram of an ovary, one of the mammalian female gonads, showing stages in the growth and maturation of the ovarian follicle (1–5). The body contained in the developing follicle is the egg cell, or oocyte, which will produce four ootids—three polar bodies and one egg. The mature egg, or ovum, will eventually emerge from the ruptured follicle (6).

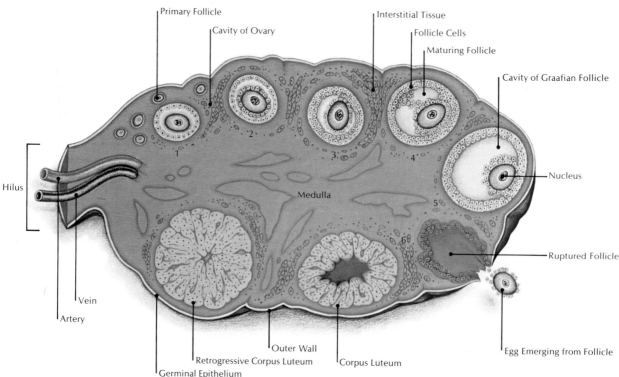

sonal intervals. Eggs of birds may be matured individually at about one-day intervals, but maturation of eggs is normally halted when a batch of eggs has been laid and must be incubated. In the human female, one egg is matured approximately each month during her reproductively active life, but egg maturation is suspended during pregnancy.

When the egg is matured, it moves out of the ovary as a result of the breakdown of some of the follicle cells and connective tissues. Moved by the action of cilia, the flow of fluids, or muscular contractions, the mammalian egg moves through the *oviduct* (the *fallopian tube* in human females) toward a chamber called the *uterus*. The uterus connects to a passage called the *vagina*, which opens to the exterior and is specialized to accept the male sex organ, or *penis*, permitting the introduction of sperm into the vagina and uterus during copulation (Figure 9.14). The sperm are carried into the uterus and the oviduct by their own swimming motions and by cilia or muscular contractions of the female reproductive tract. Among mammals, fertilization occurs in the oviduct; among some other groups of vertebrates, it occurs in the uterus. The fertilized mammalian egg becomes implanted in the wall of the uterus, where the early development of the embryo takes place.

The testes of most vertebrates lie inside the body, but in most mammals they are located in a sac called the *scrotum*, which hangs outside the body wall. This arrangement keeps the testes at the relatively low temperature needed for spermatogenesis and survival of the sperm. In birds and other vertebrates, various means of internal cooling are used to accomplish the same results. Only among primates and ungulates (hoofed mammals) do the testes remain permanently in the scrotum; in most other groups of mammals, the testes descend into the scrotum only during the mating season.

DEVELOPMENTAL PROCESSES

How does a single cell—the fertilized egg—differentiate into a multicellular organism? It is clear that the fertilized egg of both plants and animals carries in its chromosomes a complete set of genes for all cells in the mature organism. During development, each cell in the organism (excluding the gametes) contains a complete set of these genes. Although the genetic content remains constant, different sets of specialized cells arise during development, producing specific tissues and organs.

Morphogenesis

Morphogenesis is a general term used to describe processes by which tissues or embryonic tissue layers are shaped into organs and by which the organism acquires its overall adult shape and form. In plants, morphogenesis is accomplished chiefly through

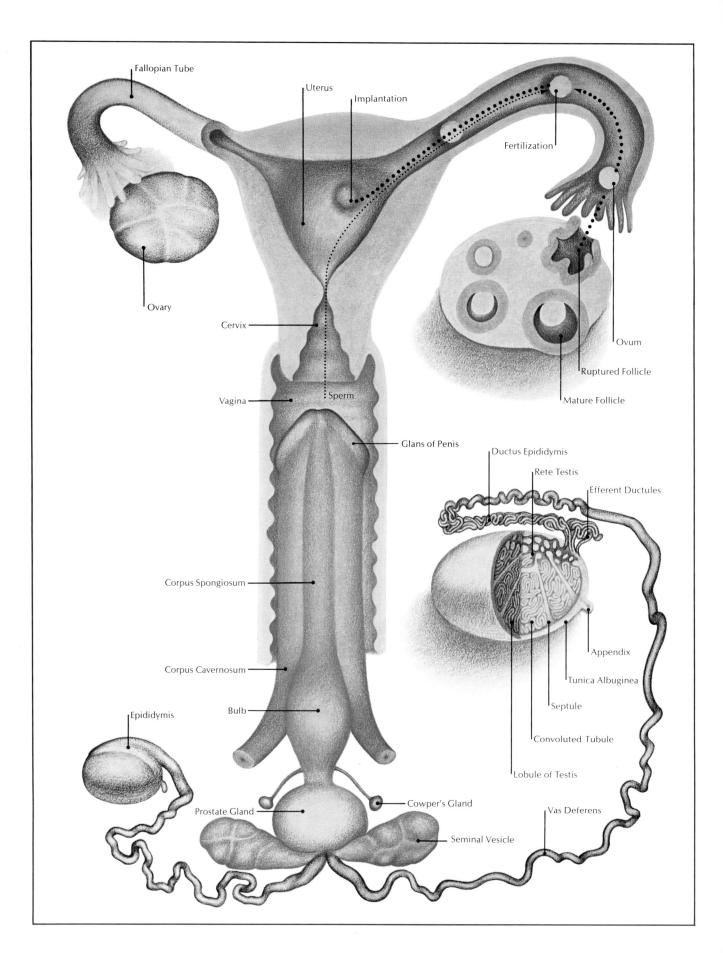

Fallopian Tube

Uterus

Implantation

Fertilization

Ovary

Cervix

Ovum

Ruptured Follicle

Mature Follicle

Vagina

Sperm

Glans of Penis

Ductus Epididymis

Rete Testis

Efferent Ductules

Corpus Spongiosum

Appendix

Tunica Albuginea

Septule

Corpus Cavernosum

Epididymis

Bulb

Convoluted Tubule

Lobule of Testis

Prostate Gland

Cowper's Gland

Vas Deferens

Seminal Vesicle

differential growth—that is, through tendencies for cells to elongate or to divide along particular planes and axes. In animal development, movements of cells—either by individual migration from one place to another or as sheets of cells—play a major role in morphogenesis.

One major difference between morphogenesis in plants and animals is that plants contain groups of relatively undifferentiated cells that continue to form new tissues and organs. These embryolike cells continue to divide throughout the life of the plant, retaining the capability of adding new tissues and organs continuously and indefinitely in response to environmental changes. On the other hand, morphogenetic movements in animals establish tissue and organ structures relatively early in development. Once the organ structures are fixed, the overall shape of the animal is determined and, except for regeneration or developmental abnormalities, no new organs are produced in the animal.

The ability of a plant to develop new organs throughout its life is called *indeterminate growth*. As a result of this morphogenetic potential, the shape and form of an adult plant varies greatly with the environmental conditions under which the plant grows. Although the individual organs and tissues show forms unique to the species, even the number of organs (such as leaves) may vary greatly from individual to individual. In contrast, because of the way morphogenesis occurs in animals, two individuals of the same species are apt to be quite similar in size and shape and certainly will have the same numbers of various organs.

Differentiation

Adult plants and animals are not merely enlarged copies of the fertilized egg. In a multicellular organism, the thousands or millions of cells produced by divisions of the zygote must become differentiated into many different kinds of specialized cells. Not only must the proper kinds of cells be produced but they must be produced in the proper numbers and assorted into the proper locations in the developing embryo. Each group of cells destined to produce a specific adult tissue passes through a series of biochemical and structural alterations that culminate in the formation of a tissue appropriately specialized for its function. This process is called *differentiation*.

During differentiation in a multicellular organism, cells acquire more and more specific determinations, and the paths open to each cell and its descendants become more and more restricted. In most cases, the process of determination cannot be detected by biochemical or structural changes in the cell. The cell in which determination has occurred appears identical to other nonspecialized cells, but observation shows that it now is committed to a particular course of development, which can be modified only

Figure 9.14   Diagram illustrating the major physiological events that comprise ovulation, fertilization, and implantation. The paths of the egg (large dotted line) and the sperm (small dotted line) converge at fertilization.

partially by outside influences. Specialized structures or chemicals within the cell may not become apparent until many division cycles after determination.

Growth

Growth occurs both by cell division and by enlargement of cells, and it is a universal feature of the development of an individual organism. In unicellular organisms, binary fission produces daughter cells that are similar in structure to the parent cells. After a period of growth in the daughter cells, during which time structures within the cell are duplicated or enlarged, the cell reaches a size at which it is prepared to divide again. In multicellular organisms, far more impressive feats of growth take place in the development from a single cell to the large body of the mature individual. Growth can be measured in terms of length, weight, number of cells, or amounts of various substances. The growth rate varies with time, in most cases reaching a maximum at some point during early life and becoming nearly zero (or even negative) in the mature organism.

Variations in growth patterns among various organisms are numerous. Many organisms have a simple S-shaped growth curve (Figure 9.16). Organisms passing through various abrupt changes in the life cycle—for example, the moltings and metamorphosis of an insect—may have several periods of rapid growth separated by intervals of zero or negative growth. Similarly, a woody plant

Figure 9.15 Stages in the development of carrot embryos. Single cells taken from carrot embryos can be cultured, and under proper conditions these cells will form embryolike structures, or embryoids, which pass through development stages resembling those of a normal embryo. A single cell from an embryo, isolated from its neighbors and supplied with a medium containing coconut milk, can develop into an apparently normal carrot plant that will flower.

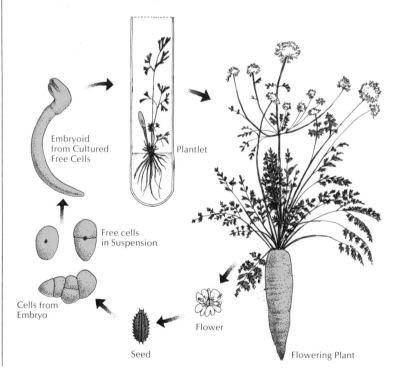

such as a pine tree grows much faster each year during certain favorable seasons.

In most multicellular organisms, the early stages of growth are marked by repeated mitoses of all the cells. As the number of cells increases, the amount of growth accomplished with each set of cell divisions increases. As development progresses, however, more and more cells become differentiated to serve specialized functions, and they cease to grow and divide. More and more of the machinery of the organism is diverted from growth to other specialized tasks, and the growth rate decreases. Even in the mature organism, some cells continue to grow and divide, but the rate at which new materials are added is barely enough to compensate for the materials lost through processes of aging.

The source of nutrients providing the energy necessary for growth during embryogenesis varies with different species. In most higher plants, insects, fish, amphibians, and birds, the energy reserves are stored materials such as plant endosperm and animal yolk. In mammals, nutrients from the mother nourish the embryo during intrauterine growth. Once the organism begins its development into a mature individual, nutrients for growth are provided from photosynthesis in plants and by active feeding by most animals.

Events that occur during growth are closely interrelated. Changes occurring in one part of the organism often trigger changes in other parts. This complex network of correlative effects results in the organized growth of an adult organism. In most cases, these interactions involve chemical messengers, or *hormones*, that can move from cells in one region of the organism to cells in another (as described in Chapter 11).

HUMAN DEVELOPMENT

One of the marvels of development is that purely reproductive cells arise at the right time, move in the right direction, and at the correct pace, so that they arrive at their place of abode just as that place is prepared to receive them. In fact, one of the earliest cellular commitments that a human embryo makes is to ensure the formation, preservation, and development of its reproductive cells, thus assuring the future of human life.

This event occurs 21 days after a human ovum is fertilized by a single sperm and 100 or so progenitors of these reproductive cells arise in the yolk sac. These cells, which will someday have the responsibility of continuing life, multiply and soon number 1,000 or more as they begin to migrate where the ovaries (or testes) will form in the developing embryo.

In a female embryo, 6 weeks after conception about 600,000 of these reproductive cells (oocytes) are present; 14 weeks after that they number 7 million. They will number about 400,000 at puberty, and at the height of sexual maturity (22 years), they may num-

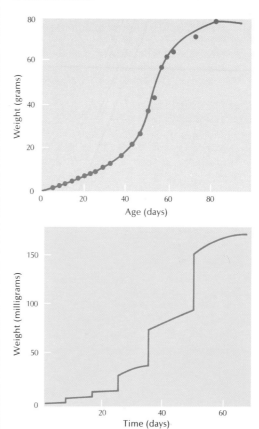

Figure 9.16 A typical S-shaped growth curve. The curve at top shows the increase in weight of a young corn plant. The growth rate of an insect is shown below.

Figure 9.17 (left) Living human ovum enlarged about 200 times. The human egg is approximately the size of one of the periods on this page. Unlike many other vertebrate ova, the human egg lacks large amounts of yolk and instead is dependent on nourishment from the mother's blood via the placenta.
From Rugh and Shettles, From Conception to Birth: The Drama of Life's Beginnings, *Harper & Row, 1971*

Figure 9.18 (right) Living, active human sperm highly magnified. Although the sperm has 85,000 times less volume than the human egg, it carries one-half the genetic material required for the complete development of the human infant.
From Rugh and Shettles, 1971

 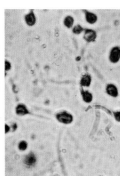

ber about 300,000. Of these 300,000 oocytes only about 400 can mature to ova and be fertilized, however—which is about one every 28 days for about 40 years. Theoretically no more than forty could be fertilized; the actual reported maximum is twenty-five.

In a human embryo destined to become a male, the same general sequence of events occurs. The reproductive cells (spermatocytes) undergo a maturing process that leads to motile and functional spermatozoa. This process becomes active at puberty and continues throughout the life of a human male. It goes on at such a phenomenal rate that millions of mature sperm are produced every day.

Now if each female has the potentiality of producing and having fertilized about forty ova during her lifetime, and if each male produces about 360 million mature sperm at any time, the two numbers multiplied together exceed 14 trillion. In other words, there is one chance in 14 trillion that a particular sperm and a particular ovum will come together to form an individual. We are, each of us, extremely fortunate to be alive today.

The First Trimester

If active sperm cells are in the female genital tract within the period from 48 hours before ovulation to about 12 hours after ovulation, fertilization is very likely to occur. Within seconds after the head of the sperm touches the egg, the fertilized ovum reacts with violent undulating movements that bring the sperm and ovum nuclei close together, thus restoring the full diploid set of twenty-three chromosome pairs. Fertilization occurs without the knowledge of the woman who will house the embryo and fetus, but this moment sets in action the sequence of processes that will lead 266 days later (plus or minus a week) to the delivery of a child composed of some trillions of cells that have arisen from the single fertilized ovum, about 1/175 of an inch in diameter.

The 266 days, or 9 months, of pregnancy can be divided into trimesters, each consisting of 3 months. The first trimester involves basic organization from an apparently unorganized zygote into a fetus with recognizable human features.

During the first month, the embryo reaches a length of about 1/8 of an inch. It is composed of millions of cells, all derived from the single-celled zygote. The embryo increases its weight by about 500 times, a change greater than at any later month. Some 30 hours after fertilization, the zygote divides into two equal cells (Figure 9.20), and 10 hours later into four cells. By 72 hours, there are sixteen cells—each identical in size and appearance to the original zygote (Figure 9.21). With each cleavage the cells become smaller and smaller. They all are enclosed and held together in a translucent membrane. This stage is called the *morula*.

During the next 3 days, the morula is free within the uterine cavity but then it finds a place to implant, usually in the posterior

Figure 9.19 A fertilized human egg, 12 hours after fertilization. Note the two discarded polar bodies, which have been produced during meiotic division of the egg nucleus. After fertilization, it generally takes at least 30 hours before the first cleavage of the egg.
Courtesy Dr. L. B. Shettles, Columbia Presbyterian Medical Center

Figure 9.20 (above) Human egg at the two-cell stage of division. This stage typically occurs between the thirtieth and fortieth hours after fertilization.
Courtesy Dr. L. B. Shettles

Figure 9.21 (below) Human embryo after four divisions, which result in sixteen cells. These sixteen cells are similar because differentiation has not begun.
Courtesy Dr. L. B. Shettles

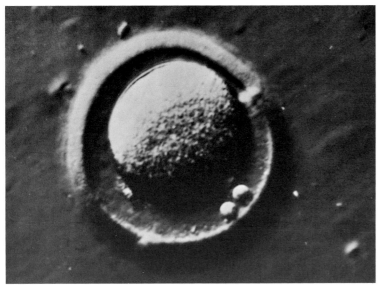

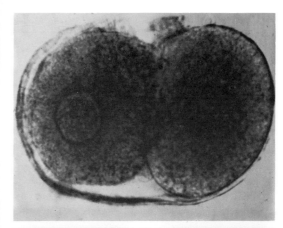

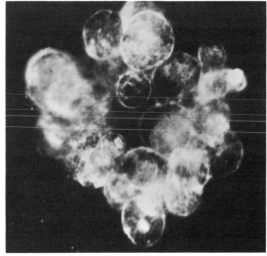

uterine wall. At the time of implantation the embryo is a hollow sphere called the *blastocyst*, which is filled with a fluid. The translucent membrane disappears, and then the embryo secretes an enzyme that helps it invade and penetrate the lining of the uterus. By day 8, the embryo has bored its way beneath the surface lining and is surrounded by bloody fluid rich in glycogen, a food needed by the embryo at this time. By day 12, implantation is completed.

By day 14, outgrowths from the embryo penetrate the maternal tissues and begin formation of the *placenta*, an organ formed from both the embryo and the mother. The placenta serves as a transfer area for wastes and carbon dioxide from the fetus and for nutrients and oxygen from the mother to the fetus. The placenta at birth is discarded, along with the umbilical cord. Together they make up what is known as the *afterbirth*.

Before the end of the first month, the human embryo forms a *primitive streak*, which is its main body axis. Membranes begin to form that will enclose the embryo and physically protect it. Of the various systems, the nervous system is the first to start development. It appears as a neural plate, neural fold, and then neural tube (day 24), which will become the brain and spinal cord and will give rise to many of the cranial and spinal nerves. Only the primary reproductive cells take precedence over brain and nervous system development. Associated with the brain are the sense organs (eyes, nose, and ears), which begin to appear at this time. On each side of this neural axis the embryo acquires thirty-two pairs of muscle blocks, or *somites*, which will give rise to most of the voluntary muscles of the body, much of the skeleton, and the skin cells.

Blood cells appear, and simultaneously (day 21) the tubular heart appears. By day 24, the heart begins to pulsate, even before

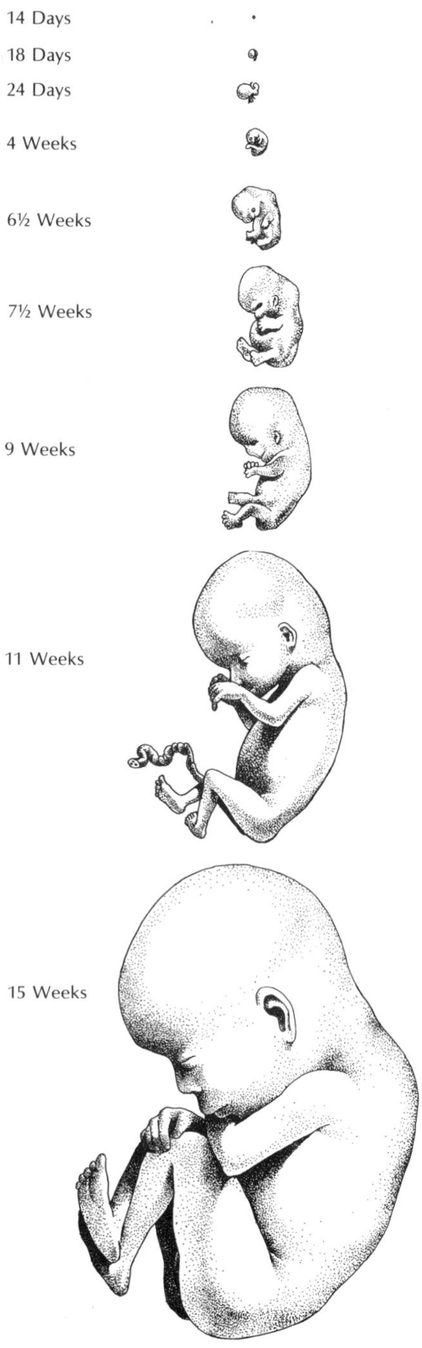

Figure 9.22   This size chart compares embryos of varying ages. It is clear that by the end of the first trimester, features of the developing fetus are distinctly human in appearance.

Figure 9.23 (lower left)   Note the paddlelike feet, the spinal cord, and the brain in this human embryo of 40 days.
From Rugh and Shettles, 1971

Figure 9.24 (lower right)   Human embryo at 40 days, lateral view. Note the deep cleft, or isthmus, in the brain, separating the midbrain from the hindbrain. By the sixth week, finger buds have formed.
Courtesy Dr. Roberts Rugh

14 Days

18 Days

24 Days

4 Weeks

6½ Weeks

7½ Weeks

9 Weeks

11 Weeks

15 Weeks

it is completely developed. Its first beats are irregular and slow, but they increase over a period of weeks until they may occur as rapidly as 180 per minute. Once the heart starts beating (day 24), it continues without any interruption for as long as the individual lives—perhaps 80 or 90 years—averaging about 100,000 beats per day.

Elements of the lower jaw appear, followed shortly by those of the upper jaw. The lung structure (day 27), and the beginnings of trachea and bronchi (day 31) are formed. By day 27, the embryo forms a pair of primitive kidneys. Budlike thickenings develop (day 28) that will become arms. The structures of the important endocrine organs such as the thyroid make their appearance just before the end of the first month.

Before the first month is over, all major systems have begun to form, although none is functional. The woman may not even know that she is pregnant, yet her embryo has become a complex and intricately associated group of organ structures, although the whole structure is smaller than the mother's smallest fingernail.

The events of the second month are vastly more complex. By the end of the second month, the growing organism measures about 1-1/4 inches and has small but definite limbs, a body with a distinct head, recognizable ears, a nose, open but lidless eyes, and an open mouth.

During this period, it is enclosed in a watery *amniotic sac*, which affords some physical protection against contusions, adhesions, and temperature changes. The amniotic fluid is kept clear by constant exchange with the mother's circulatory system. The embryo swallows the amniotic fluid, which is transferred through the

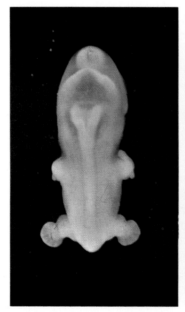
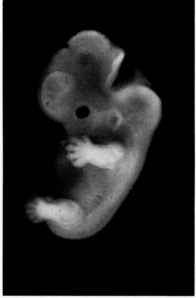

206

Figure 9.25   Photograph of an x-ray of a 2-month-
old fetus. Note the extensive development of the
skeleton, with centers of ossification appearing
as darkened regions. The limb bones and digits are
almost completely ossified, although the rib cage
remains mostly cartilage.
From Rugh and Shettles, 1971

digestive tract into the embryonic bloodstream. The waste material
in the amniotic fluid is then excreted via the body stalk, or *umbilical cord*. This structure joins the circulatory system of the embryo
to the placenta where nutritious and waste exchanges are made
with the mother's circulatory system.

Arm and leg buds are distinct by day 35. The eyes are forming
with lens and retina, the ear canals arise, and jaws develop. By
day 46, the reproductive organs begin to form ovaries or testes,
although secondary sex characteristics are not yet developed. Before the end of the second month, fingers and toes are distinguish-
able on the webbed paddlelike appendages. Ossification centers
appear in the jaws and clavicle regions, indicating the beginning
of skeletal formation.

The first 2 months of early development generally are considered to be the period of the embryo. By the end of the second
month, however, the growing organism resembles a miniature
human and is called a *fetus*.

By the end of the third month, the fetus is about 3 inches in
body length and weighs about 1/2 ounce. The muscles and their
nervous connections are much better developed. The fetus now
has numerous taste buds. Probably the most pronounced changes
occur in the reproductive and excretory systems, which arise in
proximity to each other and retain some mutual functions throughout life. The fetus excretes urine into the amniotic fluid, and its
waste products diffuse to the circulatory system of the mother.
X-rays show that many of the cartilage centers are changing into
bone; the skeleton is forming extensively. The skin has numerous
sweat glands, some of which now become the mammary glands.
At the end of the third month, the human fetus has acquired all
of its major organ systems. During the second trimester, there
will be refinements in all newly developing systems.

The Second Trimester

By the end of the fourth month, the fetus weighs 4 to 6 ounces and
is 7 to 8 inches long. Its back becomes straight as the internal viscera enlarge and are enclosed by the abdominal wall. The skin is
so well formed that the palms of the hands and soles of the feet
have unique patterns of lines. The fetus first reacts to stimuli with a
total body response and then gradually with typical reflexes. It
spontaneously stretches arms and legs, movements that the mother
may be able to feel. The heart now pumps regularly.

The brain is beginning to form convolutions to add to its surface
area and complexity. The eyes are beginning to be light sensitive,
but the ears are not yet sensitive to sound. The placenta now covers
a large portion of the inner surface of the uterus.

By the end of the fifth month, the fetus is 9 to 11 inches long
and weighs about 1/2 pound. It is now freely mobile within the
amniotic sac. Although the 5-month fetus would appear to be fully

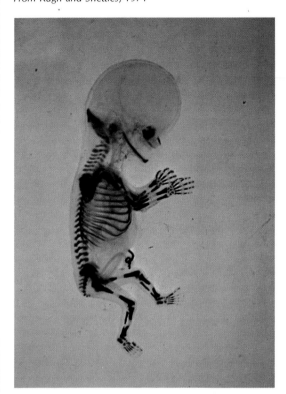

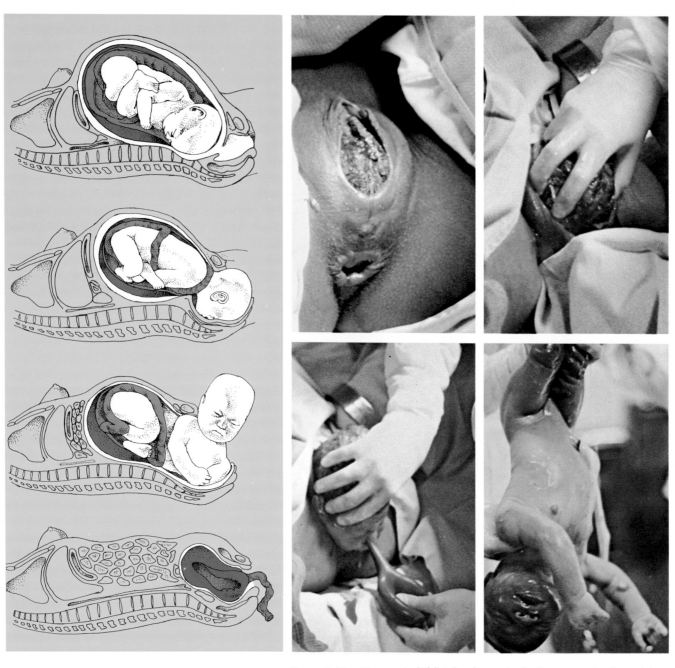

Figure 9.26   Stages of childbirth, shown both diagrammatically and in a sequence of photographs. The baby's head lies close to the cervix, which becomes fully dilated. Strong uterine contractions begin to force the head into the birth canal (vagina). After the head is born, the shoulders rotate in the birth canal and the rest of the body is expelled. In the last stage of delivery, the placenta and umbilical cord (sometimes called ''afterbirth'') are expelled.

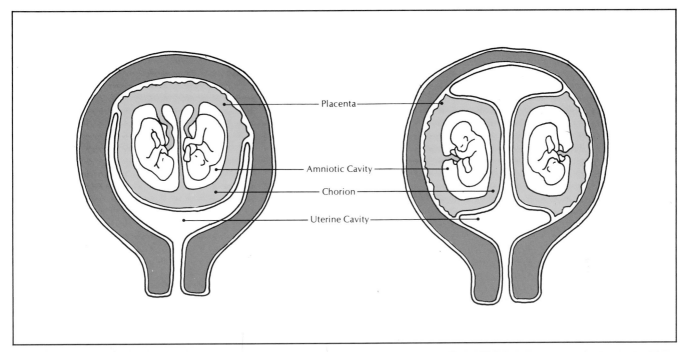

Figure 9.27   Identical twins (left) develop from the splitting of a single zygote; they share the same chorion and placenta. Fraternal twins (right) develop when two egg cells are released from the ovaries at the same time and are fertilized by different sperm; they have separate chorions and placentas. Monozygotic, or identical, twins are genetically identical and are also phenotypically alike. Because dizygotic, or fraternal, twins have separate chorions and placentas, they do not resemble each other in genotype or in phenotype.

Figure 9.28 Human fetus at 3 months with placenta attached. The placenta provides for physiological exchange between the developing fetus and the mother. By the end of the third month, the fetus is about 3 inches long; the head continues to be relatively large, with external ears level with the lower jaw, and eyelids that are sealed shut.

From Rugh and Shettles, 1971

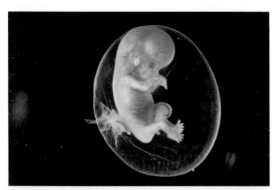

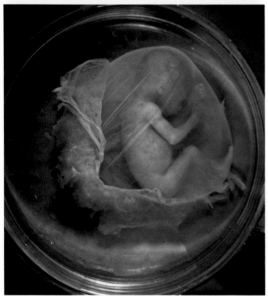

formed, if severed from the placenta it would be unable to survive outside the uterus.

All embryos of "higher" animals go through some embryonic stages having structures that resemble embryonic structures of "lower" animals. The human embryo and fetus are no exception. Every human embryo develops structures similar to those associated with the development of gills, a tail, and primitive kidneys on embryos of other animals. Vestiges of some 127 structures without apparent functions persist in the adult human, presumably remnants of distant animal ancestors. Some are retained because other associated organs are absolutely necessary to survival. Nevertheless, reduction or loss of ancestral and vestigial structures occurs extensively during the fifth month.

By the sixth month, the fetus measures about 12 to 14 inches and weighs about 1-1/2 pounds. Now the fetus is a well-proportioned miniature human being and may be able to survive if delivered prematurely by Caesarean section. Its skin is red and wrinkled and it looks rather like an old person. The skin of a 6-month-old fetus begins to form hair follicles. In addition, each square inch of skin is developing 700 sweat glands, 100 sebaceous (oil) glands, and 21,000 cells sensitive to heat, pressure, and pain. The digestive tract (gut) becomes filled with dead cells from its lining, plus a greenish secretion from the gall bladder (bile). This material remains in the gut until shortly after birth. Skeletal formation is active during the sixth month, and the mother must take in more calcium than she needs for herself in order to provide adequately for the mineralization of the fetal skeleton.

The Third Trimester

Antibodies of the immune systems can be transmitted between the maternal and fetal blood systems during the last trimester. Almost any infection or disease contracted by the mother will cause her to develop antibodies against the toxins of the disease. These antibodies will pass to the fetal bloodstream, giving the child the same immunity to those diseases for about 6 months after birth.

A third-trimester fetus may survive outside the uterus if it is removed by Caesarean section. Although only 10 percent of those delivered during the seventh month survive, the survival probability increases to 70 percent in the eighth month and 95 percent in the ninth month. In the last trimester, the fetus must obtain large amounts of calcium, iron, and nitrogen through the foods that the mother eats. During this period, 84 percent of the calcium that the mother consumes goes into the fetal skeleton. About 85 percent of the iron that she takes in goes into the hemoglobin of the fetal bloodstream. Nitrogen is needed as a major constituent of the many proteins being synthesized as the fetal nervous system and brain complete the final, rapid stages of their growth. The great importance of a proper maternal diet during the final trimester has

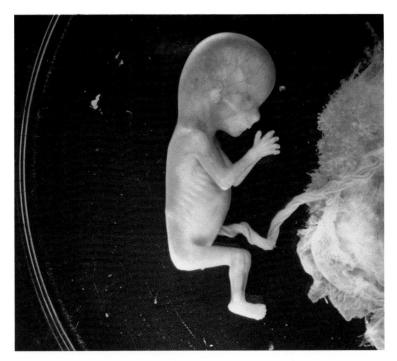

Figure 9.29 Human fetus at 4 months. At this time, the fetus is almost 6 inches in length and weighs about 4 ounces. Note the extensive number of superficial blood vessels in the head region. *From Rugh and Shettles, 1971*

been recognized only recently. It now appears that the tendency of persons from low socioeconomic classes to have lower than average scores on various kinds of intelligence tests may be more closely related to maternal diet during the final trimester of pregnancy than to either genetic inheritance or postnatal education.

During the ninth month, the rate of growth and development of the fetus slows. If it continued at the early growth rate after birth, it would weigh 200 pounds within a year. Fortunately there is some regulating mechanism that limits the growth of organisms to certain heights and weights.

The birth of the child is anticipated by certain events collectively known as labor and delivery, which is probably the single most critical period in the life of the child as it emerges from a warm and watery chamber into a cold, airy, and unlimited environment.

The moment of birth may be traumatic for the fetus. All of the systems so intricately developed over the preceding 9 months are shifted suddenly into full operational status. There is growing evidence that behavioral as well as physical development continues to progress steadily from the embryonic state through maturity. But as soon as a human child emerges into the world, the external environment becomes much more important. Without pause, he embarks on a lifelong stage of development and change—both physical and behavioral—that will continue, as before, with the reproduction of his kind.

Unit III
The Ascending Complexity of Life

For a number of the most strategic and salient structural elements of the mind there is already evidence of significant genetic determination. These genetic factors, and they may not be so many in number, define our intellectual and conceptual limits. I propose that through phylogenetic studies and through studies of the rare human genetic variants we can learn much concerning their basic cerebral components, in preparation for the day when we wish to begin to move back their limits.

And so perhaps, when we've mutated the genes and integrated the neurons and refined the biochemistry, our descendants will come to see us rather as we see Pooh: frail and slow in logic, weak in memory and pale in abstraction, but usually warm-hearted, generally compassionate, and on occasion possessed of innate common sense and uncommon perception—as when Pooh and Piglet walked home thoughtfully together in the golden evening, and for a long time they were silent.

"When you wake in the morning, Pooh," said Piglet at last, "what's the first thing you say to yourself?"

"What's for breakfast?" said Pooh. "What do you say, Piglet?"

"I say, I wonder what's going to happen exciting today?" said Piglet.

Pooh nodded thoughtfully. "It's the same thing," he said.

—Robert L. Sinsheimer (1971)

Unicellular organisms are microscopically small, and yet they exist in a surprising variety of shapes and structures. There are thousands of species of unicellular algae, fungi, and protozoans, and each is unique—in structural details, perhaps, or in metabolic processes, or in life cycles. Unicellular autotrophs need only inorganic molecules and sunlight to survive, whereas unicellular heterotrophs survive in any one of a variety of environments as long as they can obtain organic molecules as food. Some are encased in hard shells or slimy coatings; some use flagella for moving about. Still others attach themselves to fixed surfaces. Yet *every* unicellular organism must have a full complement of specialized structures if it is to survive as an independent creature.

Most species of fungi, plants, and animals are multicellular. Thousands, millions, even millions upon millions of individual cells work together as a single organism to maintain vital processes. Perhaps they come together as a weed, or a giraffe, or a human being. No matter how large a multicellular organism becomes, however, the size of the individual cells is still microscopic in all but a few cases. The reason is that the size of a cell is fixed according to the surface-to-volume ratio that best facilitates the rate of exchange of materials between the cell and its environment (Figure 10.1).

How is it possible to gain some understanding of the ways in which cells come together? The body of an average human being, for example, contains approximately 100 trillion cells of many different types. How can they possibly be sorted out for analysis? The answer is found in the way individual cells are organized into *groups* of cells to carry out certain *specialized* functions. Some cells are specialized for the synthesis of various substances that are essential for life processes. Others are specialized for transport of materials from one part of the organism to another. And still others are specialized for ensuring the integration or the reproduction of the organism.

Specialization, of course, has its price. During the development of the more advanced multicellular organisms, some cells have lost certain abilities as they have become specialized in others. In general, the more a cell's structure is modified for better performance of its specialized function, the less it is involved in carrying on other functions. Extreme examples are the nerve cells and skeletal muscle cells of vertebrate animals. Their structures are so highly modified, or differentiated, that these cells are no longer able to carry out a function that is basic to most living cells—the ability to reproduce. During differentiation, various parts of a cell structure may be modified; they may be increased or decreased in number, size, or amount in order to perform specific functions in the organism. Individual cells may become so specialized in one area that they cannot survive independently (except in laboratory cultures, where these cells can be supplied with the nutrients and

10

Levels of Integration

Figure 10.1 (above) Surface-to-volume relationships. How fast materials diffuse into a cell is limited by the physical characteristics of the outer membrane. If cell membranes were all essentially alike, then the rates of diffusion would be alike under comparable conditions of temperature, solubility, concentration of materials, and so forth. Assuming a constant diffusion rate of one unit of metabolic fuel per unit surface area per unit of time, an organism with the largest total cell surface will absorb the most food in a given time.

The reason is that as the volume of an object increases, the amount of surface does not increase as fast. In fact, the rate of decrease is linear. For example, with each doubling of one side of the cube shown above, the surface-to-volume ratio decreases by half. The greatest amount of cell surface area, then, would be provided by a large number of small cells or in cells with highly-folded plasma membranes, such as those with microvilli.

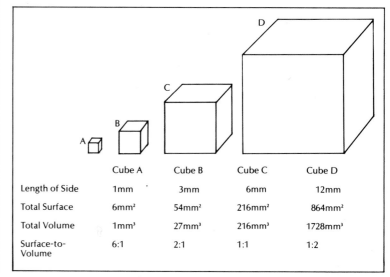

	Cube A	Cube B	Cube C	Cube D
Length of Side	1mm	3mm	6mm	12mm
Total Surface	6mm²	54mm²	216mm²	864mm²
Total Volume	1mm³	27mm³	216mm³	1728mm³
Surface-to-Volume	6:1	2:1	1:1	1:2

Figure 10.2 (below) Patterns of organization in biological systems, showing levels of specialization in various representative organisms.

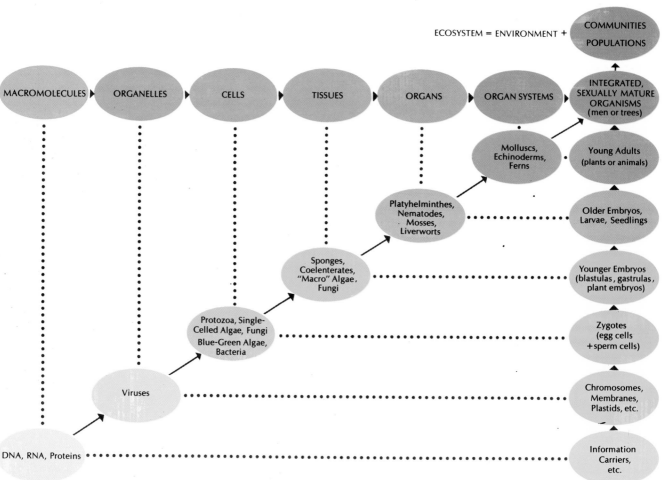

environment normally provided by other cells). They must depend on other cells in the organism for some essential life processes.

The degree of complexity achieved by groups of cells in a multicellular organism depends on how complex the organism is in terms of its relation to its environment. In fact, if animals and plants are grouped according to increasing levels of organizational complexity, a pattern emerges that seems to approximate the evolution of living things (Figure 10.2). That is why the forms of life at the lower levels of organizational complexity are often thought of as being more "primitive," even though this reasoning may not be entirely justified.

Figure 10.2 identifies different levels of biological organization together with representative developmental patterns and representative organisms. Organized groups of cells of similar type and function are called *tissues*. Various tissues may be coordinated to form an *organ*, which carries out a more complex yet unified function. Cell, tissue, organ—these are the levels of integration that mark how specialized an organism has become. The lines of demarcation are not always sharp, however, and there is considerable overlap. For example, some unicellular organisms join together to form filaments or bodies, but the cells in these groupings retain their individuality: any cell separated from the group can survive independently. Among some algae and fungi, two or more cells may fuse into a larger, multinucleate cell, thereby forming an organism called a *coenocyte*. The slime molds are particularly interesting because they can exist in various periods of their life cycle as coenocytic bodies, as multicellular structures with cellulose cell walls, and as independent cells (Figure 10.3).

Some protozoans form colonies. Strands of cytoplasm, for example, link individual cells of *Volvox* together into a colony. A *Volvox* colony is little more than flagellated, chloroplast-containing cells embedded in an extracellular material. Its cells are incapable of individual motility, and only a few reproductive cells in one part of the colony are able to reproduce by sexual *or* asexual means. *Volvox*, apparently, is a multicellular organism that is representative of the simplest kind of cell specialization.

Of greater complexity are the sponges. These creatures have food-gathering cells, food-transporting cells, skeleton-making cells, maintenance and repair cells, reproductive cells, and cells that attach the sponge onto a surface under the sea (Figure 10.4).

Interestingly enough, the various kinds of cells making up a protozoan or a sponge are not completely interdependent: any one kind of cell can become modified to perform the function of another. Because their cells retain some degree of independence, these organisms are regarded as primitive multicellular organisms.

The corals, jellyfish, sea anemones, and other coelenterates represent a more complex form of multicellularity. They exhibit a fairly high degree of cell specialization, including sensory cells of

Figure 10.3   A sequence in the development of the fruiting bodies of the cellular slime mold *Dictyostelium discoideum*. Single cells of this organism grow on decaying forest leaves and are almost identical to amoebae. They ingest bacteria and divide mitotically. When the individual cells run out of food, however, an amazing sequence of structural changes transforms a group of the cells into a multicellular organism. The cells gather together to form a fruiting body, within which some cells are transformed into encapsulated spores. The spores can survive extended periods of drought or cold without food. They later germinate to form new, individual, amoeboid cells.

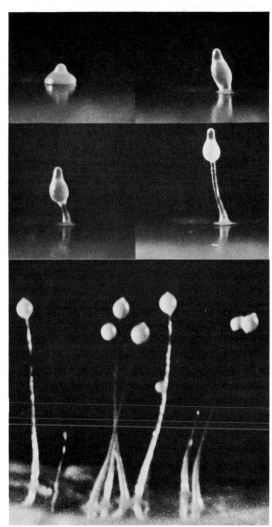

Figure 10.4   In this diagram of a sponge, there are hundreds of tiny porelike openings hidden in the body wall. Water is drawn into the internal cavity of the animal through these openings. Particulate matter in the water is trapped by the flagellated collar cells, and water passes on through the animal and out of the sponge via the excurrent opening.

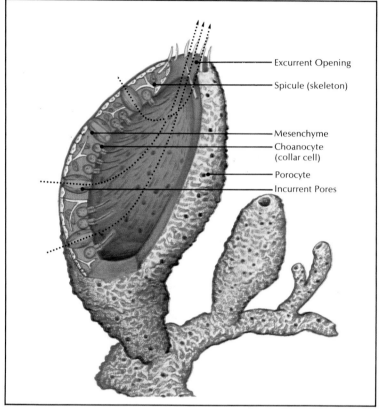

Excurrent Opening
Spicule (skeleton)
Mesenchyme
Choanocyte (collar cell)
Porocyte
Incurrent Pores

various types, a primitive sort of nerve cell, muscle cells, and the highly specialized stinging cells called "nematocysts." In addition, the coelenterates show a simple form of tissue organization. The outer surface of the organism is covered with a layer of specialized cells that form an *ectodermal* (skin) tissue. The inner surfaces are lined with an *endodermal* tissue composed of different specialized cells (Figure 10.5).

In the plant kingdom, organisms also can be arranged in levels of increasing cell specialization and tissue development. Between the groups of unicellular algae and the simple multicellular plants are intermediate forms that can be regarded either as colonies of unicellular organisms or as simple multicellular organisms.

The intermediate kinds of organisms are interesting because they offer clues about the ways that cell specialization, tissues, and organs might have evolved from simpler organisms. But how does cell specialization in itself make multicellularity possible? The kinds of specialized cells that exist are myriad, but a survey of a few representative types will help to answer that question.

SURFACE TISSUE

Consider, first, the aggregation of cells that cover and "contain" an animal. These cells adhere tightly together, forming a continuous

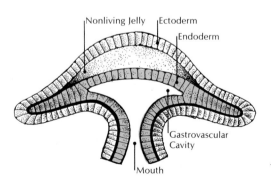

Nonliving Jelly   Ectoderm
Endoderm
Gastrovascular Cavity
Mouth

Figure 10.5   As seen in this diagram, the coelenterates show a simple form of tissue organization. The outer surface is covered with ectodermal cells, and endodermal cells line the inner surface.

sheet that is known as *epithelial tissue*. The cells work together to maintain the integrity of the surface, and if they are damaged in any way they work rapidly to replace themselves.

The way in which epithelial cells are attached to one another exemplifies the kinds of cell-to-cell attachments found in multicellular organisms. Between adjacent cells is a thin layer of *intercellular cement*. It is made primarily of a "mucopolysaccharide" that is secreted by the epithelial cells. The layer of epithelial cells rests on a base of strength-giving collagen fibers embedded in a matrix. The protein collagen and the matrix are also secreted by the epithelial cells. Some cells may also join together by intertwining their plasma membranes, much like the interlocking pieces of a jigsaw puzzle.

Epithelial cells may have their plasma membranes modified in various ways. For example, epithelial cells that line the respiratory cavities may have motile cilia that are covered with folded plasma membranes. Microvilli, which are examples of such folded membranes, are abundant in absorptive cells of the body, such as intestinal cells and cells lining parts of nephrons of the kidneys.

Many epithelial cells are specialized to secrete substances onto the surface of the organism. One common kind of secretion is *mucus*, a mucopolysaccharide that forms a protective covering over the outer cell surface. Others are poisonous secretions from certain cells of the skin that apparently have evolved from mucus-

Figure 10.6   Portuguese man-of-war. This coelenterate colony is made up of hundreds of individuals in four distinct classes based on function: feeding polyps, protective or defensive polyps, float polyps, and reproductive polyps.

Figure 10.7   Sea anemone (above) and star coral (below) on the Cayman Islands, British West Indies. Although the body organization of all coelenterates is simple in comparison with that of most other animals, it shows a more definite tissue development than that of the sponges.

Figure 10.8 (above) Diagrams showing the digestive tract of the cow. The cow stomach is divided into four compartments: rumen, reticulum, omasum, and abomasum. In the first two compartments, bacterial and protozoan action break down cellulose to simple sugars. (This population of bacteria and protozoa also synthesize many of the vitamins needed by the cow.) The omasum functions as a water conservation device, whereas the conventional digestive activity of the stomach takes place in the abomasum. After food is thoroughly mixed with mucous fluid, acid, pepsin, and other enzymes, the posterior sphincter opens, the stomach contracts, and the food is forced into the small intestine.

Figure 10.9 (below) Photograph of the gastric mucosa in cross section. The gastric mucosa includes several kinds of cells. The outer parts of folds in the lining are covered mostly with mucous cells, which secrete a viscous fluid that coats the stomach lining. This fluid protects the stomach lining from the actions of digestive enzymes and dilutes the food mixture. Parietal cells, which secrete hydrochloric acid, and chief cells, which secrete the protein-hydrolyzing enzyme pepsin, are abundant between folds of the lining.

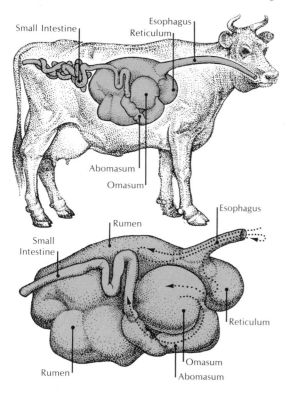

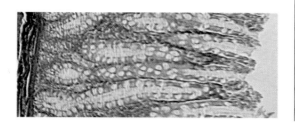

secreting cells. Epithelial cells lining the interior of digestive tracts secrete enzymes that break down molecules of ingested food so that the material can be absorbed by other cells of the intestinal epithelium.

Some epithelial cells are specialized as pigment-bearing cells. Some kinds work to protect an organism by absorbing harmful radiation; others work to conceal it through protective coloration that hopefully will thwart predators. Bioluminescent cells, which are capable of converting chemical energy into light, are another example of epithelial specialization involving pigments.

Like the epithelial tissue of an animal, *epidermal tissue* covers and protects the surfaces of roots, stems, and leaves in most plants (Figure 10.11). Epidermal tissue also is made up of a thin layer of flattened, interconnected cells. Most epidermal plant cells have thickened outer walls, relatively large central vacuoles, and a relatively small amount of cytoplasm.

Many of the epidermal cells on the aerial parts of the plant secrete *cuticle*, a water-resistant, waxy substance that forms a surface layer, protecting the plant from dehydration and from invasion by parasites. The epidermal cells of the roots have no cuticle covering but are specialized to absorb water; some form long, hairlike extensions into the soil. Some of the epidermal cells on the aerial part of the plant may form spines, hairs, or glands, all of which play roles in the protection or functioning of the plant.

SUPPORTIVE AND CONNECTIVE TISSUE

Still other forms of tissue work to support the structure of organisms. In animals, *skeletal tissues* such as chitin, cartilage, and bone provide a more-or-less rigid framework that gives structural support and provides a system of levers by which the organism can move parts of its body. *Connective tissues* are made up of generalized cells called "fibroblasts," which synthesize extracellular fibers such as collagen and elastin (Figure 10.13). These fibers bind tissues together, provide support for the tissues and for the organs formed from them, and join skeletal members to one another and to the muscles that move them. Connective fibers are an important part of the structure of an animal. Collagen, in fact, is the most abundant single protein in the human body and the most common protein in the entire animal kingdom.

In plants, certain kinds of tissues continue to produce new cells—and thereby new tissues and organs—throughout the life of a plant. In most higher animals, in contrast, almost all tissues and organs form early in life and most cells of the body lose the ability to divide. The undifferentiated plant tissues that are sites of active cell division are called *meristems*. Although there is a great deal of variation in characteristics among meristematic cells, most tend to be thin-walled, small, closely packed, and filled largely with cytoplasm. Meristematic tissues form the growing tips of roots and

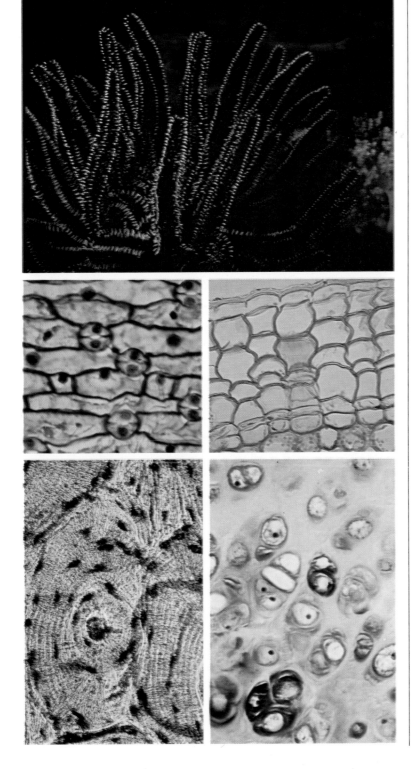

Figure 10.10 (above)   Photograph of the bioluminescent feathered sea star. Bioluminescence in this organism produces the golden glow effect and is due to cells that are able to convert chemical energy into light. The emission of light is dependent upon nervous stimulation of specialized cells in light-producing organs.

Figure 10.11 (middle)   The photograph at left shows allium leaf epidermis. Note the thick, dark cell walls and the relatively large central vacuoles of these cells. Epidermal cells such as these typically secrete a waxy cuticle to prevent dehydration and to protect the plant from mechanical injury. Also visible in this photograph are numerous stomata-guard cell complexes, which serve to regulate plant transpiration. The photograph at right shows a partial section through a woody stem. In adult trees, the epidermis is replaced by another tissue, the periderm, which is composed of cork cells. Cork cells secrete a waterproof coating of suberin and then die, so that the surface of the periderm is a thick layer of hollow, water- and injury-resistant cork cells. The periderm forms the familiar bark of a tree.

Figure 10.12 (below)   At the left is a photograph of human bone in cross section. The dark spaces contain bone cells that communicate with one another through small canals by fine pseudopods. Hyaline cartilage is shown at right. Note the clear, almost transparent matrix in which are embedded lacunae, or cell spaces. Each lacuna contains one or more chondryocytes, or cartilage cells.
Courtesy Carolina Biological Supply Company

Figure 10.13 In this photograph of fibroelastic connective tissue, the most conspicuous components are the large number of threadlike fibers, some of them tough and strong (collagen) and other elastic and flexible (elastin). Various connective tissue cells and background substances are made up of fluids, gels, or firm materials.
Courtesy Carolina Biological Supply Company

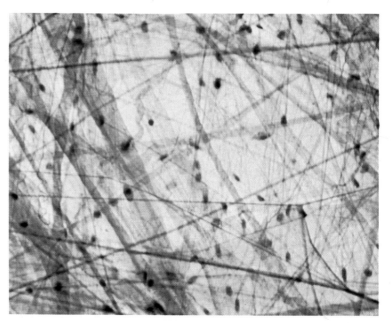

shoots. In many plants, there are layers of meristematic tissue near the surface of branches and stems. These lateral meristems enable the plant body to continue to grow thicker throughout its life.

Most of the plant body is made up of *fundamental tissues*, each of which is composed largely of a single kind of specialized cell. *Sclerenchyma* is a supportive tissue of the plant body composed of cells that secrete material rich in lignin. This material forms into thick cell walls and later dies, leaving the mature tissue as a network of lignified walls with minute pores that were once the cell interiors. Some sclerenchymatic cells become elongated into fibers, such as those of flax and hemp. *Collenchyma*, which is another supportive

Figure 10.14 In the lefthand photograph, a typical dicot stem is shown in cross section. Note the concentric ringlike arrangement of vascular tissue. In the righthand photograph, a typical monocot stem is shown in cross section, with well developed vascular bundles scattered throughout the ground tissue.
Courtesy Carolina Biological Supply Company

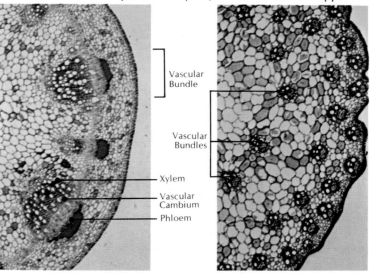

Vascular Bundle

Vascular Bundles

Xylem

Vascular Cambium

Phloem

tissue with thickened cell walls, remains alive throughout most of the plant's life span.

Much of the body of a plant—particularly of the lower plants—is made up of living tissue composed of nonspecialized, thin-walled, large-vacuole cells that may occasionally begin meristematic activity or cell specialization to form other new tissues. This *parenchymatic tissue* gives support and rigidity to the plant body and serves as a site for storing nutrients and water.

ANIMAL NERVOUS TISSUE

Some epithelial cells in animals specialize to form nervous tissue, which integrates the organism's activities at all levels. This tissue is made up of two types of cells, *neurons* and *interstitial cells*, both of which have extensive fiberlike projections. Neurons create the electrochemical signal called the *nerve impulse* and transmit this impulse to other cells—neurons, muscle cells, or gland cells. Interstitial cells play a supportive role—binding neurons together, forming insulating coverings for the nerve fibers, or providing the neurons with certain necessary nutrients.

A typical neuron consists of a cell body (*soma*) and one or more cytoplasmic extensions (Figure 10.15). These extensions are of two types: the *dendrites*, which carry impulses toward the cell body, and the *axons*, which carry nerve impulses away from the cell

Figure 10.15 Diagram illustrating different types of neurons (left). Neurons may differ in shape and size of the cell body (soma), in dendrite processes, in length of the axon, and in the presence or absence of a myelin sheath covering the axon. Also shown is a photograph depicting a slice of cerebellar cortex stained to outline neurons (right).

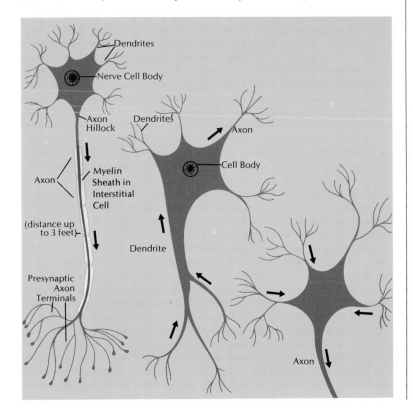

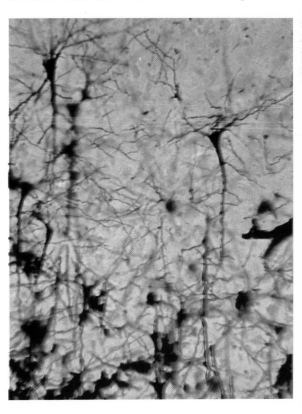

Figure 10.16 At left is shown skeletal muscle (voluntary striated muscle). The lower portion of this light micrograph shows a longitudinal section of parts of several muscle fibers with cross striations. The upper portion of the photograph shows cross sections of muscle fibers with many individual fibrils barely visible. A few of the many peripheral nuclei are visible. The blue area is connective tissue fibers and cells. This tissue allows for passage of blood vessels and nerves, and holds muscle cells in alignment so that an individual muscle can act as an integrated organ. In the middle photograph, several fibers of human cardiac muscle (involuntary striated muscle) are shown in longitudinal section. There are several nuclei per cell and they are more centrally located. Cardiac muscle cells are shorter than skeletal muscle cells and may be branched. The wider, nonbanded light areas are called intercalated discs and appear as folded, intertwined cell membranes when viewed with the aid of an electron microscope. The mechanism of muscle contraction, using interacting actin and myosin protein filaments, is similar to that of skeletal muscle. Smooth muscle (involuntary nonstriated muscle) is shown at right. Each of the spindle-shaped cells have one large, centrally located nucleus. In contrast to striated muscle cells, they have a very slow contraction time. Individual cells are surrounded by closely-woven connective tissue, which allows for passage of fine nerve processes and some small blood vessels. Smooth muscle cells lack the characteristic striations of skeletal and cardiac muscle but do have both fine actin filaments and larger myosin filaments. Smooth muscle is found in numerous places in the vertebrate body including the walls of the intestinal tract, the blood vessels, and the urinary and reproductive tracts.

body. Most neurons have multiple dendrites and a single axon. The axons of many neurons are wrapped in a *myelin sheath*, another example of cell specialization. The myelin sheath is composed of the membranes of interstitial cells and is wrapped around the axons to form several concentric layers. The axon, with its surrounding sheath, is called a nerve fiber.

Some axons are exceedingly long. For example, a nerve terminating in a blood vessel in the foot of a giraffe has its cell body and dendrites in the spinal cord. Its axon is a single, continuous fiber, perhaps 8 or 9 feet in length, extending from the spinal cord to the end of the animal's foot. The role of the neuron and its corresponding parts will be discussed in Chapter 12.

ANIMAL MUSCLE TISSUE

Animal muscle tissue represents one of the most complex levels of cell organization. The cells in this tissue have two extensions from the base cell that run parallel to the surface of the organism. Within these extensions are fiberlike cells that respond to nerve impulses by contracting. Three different types of muscle tissue are found in higher animals: skeletal, cardiac, and smooth muscle.

Under a light microscope, the cells of *skeletal muscle tissue* look like fibers. Each fiber is a single cell with several nuclei surrounded by a thin but tough cell membrane. And each displays a banded pattern of alternating light and dark transverse lines (Figure 10.16). The thick and thin filaments making up these banded regions are linked by *cross-bridges*, which probably play an important role in contraction.

Each of these striated cells, or fibers, is made up of many smaller threadlike structures of protein called *myofibrils*, which are also striated. Between the myofibrils in the cell are mitochondria

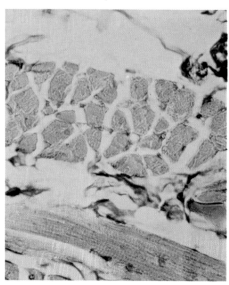
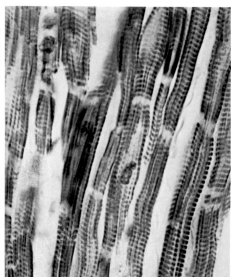
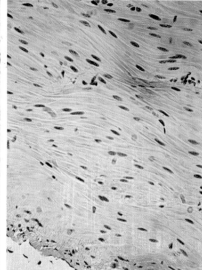

Figure 10.17   These schematic diagrams are progressively enlarged to show the internal structure of frog skeletal muscle tissue. This fast-contracting muscle is made up of many interacting muscle fibers (cells), which appear striated under magnification. Individual fibers include myofibrils within which repeated patterns of light and dark bands can be distinguished.

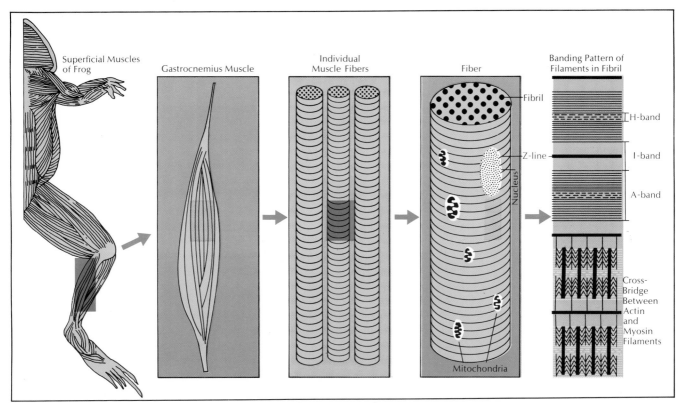

Superficial Muscles of Frog

Gastrocnemius Muscle

Individual Muscle Fibers

Fiber

Banding Pattern of Filaments in Fibril

Fibril

H-band

Z-line

I-band

Nucleus

A-band

Mitochondria

Cross-Bridge Between Actin and Myosin Filaments

and a complex membrane structure that is a modified endoplasmic reticulum. The nuclei are found adjacent to the cell membrane.

Remarkably, electron micrographs reveal that myofibrils are made up of still smaller *myofilaments*. There are two kinds of myofilaments in each myofibril; thicker filaments of about 100 angstroms in diameter and thinner filaments of about one-half that diameter. The dark "A-band" in muscle cells is made of both thick and thin filaments; the light "I-band" is made of thin filaments alone, whereas the light "H-band" is made up only of thick filaments. The "Z-line" is a narrow zone of very dense material not arranged in filaments (Figure 10.17).

The proteins in the myofilaments are composed chiefly of *myosin* and *actin*, with smaller amounts of *tropomyosin*. Myosin makes up about one-half the protein in the myofibril. When myosin is extracted from the myofibril, this protein forms filaments having numerous side projections that appear to be similar in spacing to the cross-bridges of the myofibrils. The hypothesis that the thick filaments are composed of myosin is confirmed by the observation that the dark bands disappear from a myofibril when myosin is extracted from it. The other major protein of myofibrils, actin, forms a globular molecule in pure solution. When placed in a solution with salt and ATP concentrations similar to those of the muscle cell, actin forms long fibers. The assumption that actin forms the

Figure 10.18 This model of muscle contraction was derived by H. E. Huxley and J. Hanson in their electron microscope studies of striated muscle. They observed muscles fixed in a stretched condition, at rest length, and contracted. Each of these teams drew the same conclusion from their studies: the lengths of the filaments in striated muscle do not change during contraction; rather, the two kinds of filaments slide past one another. This mechanism of muscle shortening is called the sliding-filament model of muscle contraction. The schematic drawings show myofibril filaments in both a stretched muscle (A) and in a resting muscle (B).

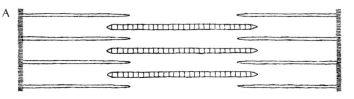

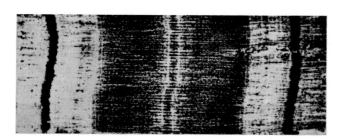

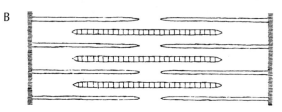

Figure 10.19 Electron micrograph of a cross section through several myofibrils in skeletal muscle cell. The fine dots in the micrograph represent cross sections of myosin and actin filaments.

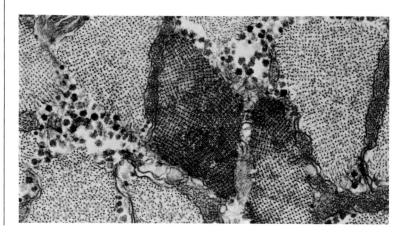

thin filaments is confirmed by the observation that the structures of the light I-band disappear when actin is extracted from the skeletal muscle cell.

Cardiac muscle tissue, found only in vertebrate animals, forms the bulk of the organ called the heart. It also may extend a short distance along the walls of the large arteries that emerge from the heart. A modified type of cardiac muscle constitutes the heart's so-called neuromuscular tissue, which functions as an internal impulse-conducting system within the organ. Cardiac muscle is intermediate in some respects between skeletal muscle and smooth muscle. It is striated like skeletal muscle but is not under voluntary control.

Smooth muscle tissue contracts or relaxes slowly. In vertebrates, smooth muscle is found in internal organs such as the stomach, intestines, and blood vessels, and its contraction is involuntary. This muscle tissue lacks the striated pattern formed by the orderly array of thick and thin filaments in skeletal and cardiac muscle cells. Smooth muscle cells are about 10 microns in diameter, tapering with a single, centrally located nucleus. Smooth muscle cells prepared for electron microscopy appear to contain only thin (actin) filaments. Recent results with different techniques of fixation, however, show that smooth muscles may also have a thick type of filament. The "true" ultrastructure of smooth muscle remains to be discovered.

227

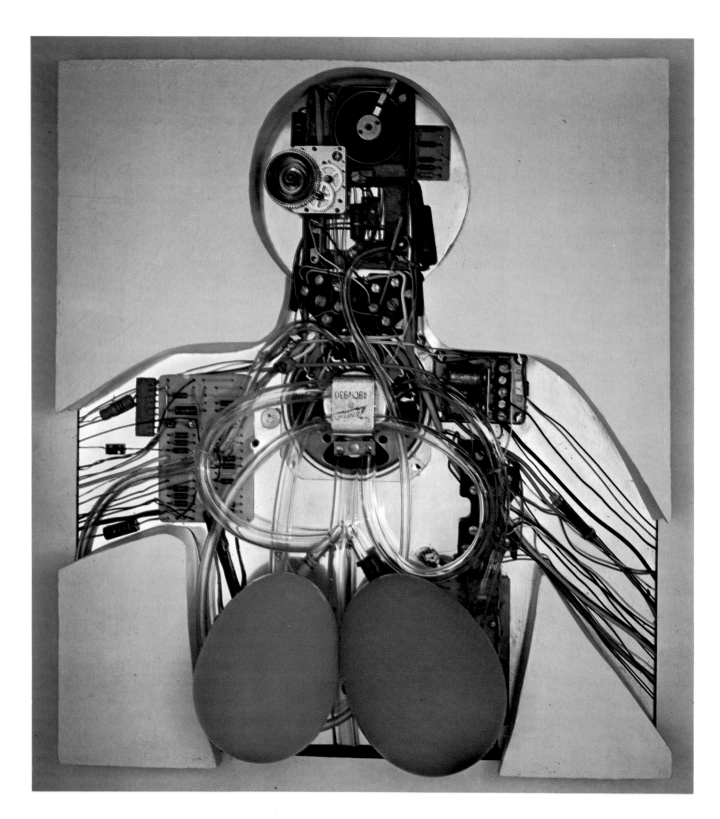

What enables countless millions of single cells to coordinate their activities as effectively as they do in an individual organism? Think of yourself for a moment, not as you see yourself in a mirror but as an enormous population of individual cells. Look at your hand and visualize the millions of cells that make up the skin alone. How does each cell get its food, and from where? How does it get its oxygen, and how does it get rid of its wastes? How does it regulate its own metabolism, while at the same time making its contribution to the structure and function of the hand? A skin cell can reproduce by mitosis; why does it do so only when your hand needs more skin cells in that region?

A living cell in your hand must cope with essentially the same challenges that all living systems must face. Each living creature must obtain from the environment the materials needed to build and maintain the structures of its body, to repair those structures, and to reproduce. Each must capture energy to build and maintain an ordered structure and to carry out the physical and chemical processes of life. Materials must be transported internally from place to place and made available where needed. Waste products of metabolism must be eliminated selectively from the body. Heat must be selectively retained or dissipated to keep the internal environment within the rather narrow temperature limits that permit efficient biochemical reactions.

Every living system, then, continually adjusts its physiological functions in order to maintain itself. This state of "dynamic equilibrium" is called *homeostasis*. The concept of homeostasis is one of the most important in the study of biological systems.

NUTRITION AND THE DIGESTIVE SYSTEM

In Chapter 3 you learned that nutrition refers to the intake by an organism of the materials, and therefore the energy, needed to sustain life. Autotrophs take in simple inorganic substances and synthesize all the organic molecules they require; heterotrophs, which are unable to perform all the required syntheses, obtain organic molecules by ingesting other organisms or their products.

Organic substances are scarce in most environments. As a consequence, heterotrophs often must extract food particles from large volumes of the surrounding fluid, if that is what comprises their environment, or they must actively pursue and capture their food. Large food particles are taken into some sort of internal digestive system, where they are physically and chemically broken down to small molecules that can be absorbed.

In all higher animals beginning with the roundworm, the digestive tract is a continuous tube with a mouth at one end and an anus at the other. Food is progressively digested as it moves through the tube, or gut, and processing is not interrupted by each new ingestion. Movement of food through the tube is accomplished by rhythmic contractions, or *peristalsis*, of muscles in the

11

Integration of Function

gut walls. In more complex organisms, the length of the digestive system is increased by coiling or folding of the tube; the surface area available for absorption of small organic molecules is increased by foldings or projections of the tube lining; and the tube is differentiated into specialized passageways and cavities within which various parts of the digestive process occur.

The basic chemical process of digestion is one of *hydrolysis*, whereby macromolecules (proteins, fats, and carbohydrates) are broken down to simpler organic molecules such as amino acids, glucose, glycerol, and fatty acids. These smaller molecules are then absorbed across the lining of the digestive tract. Hydrolytic

Figure 11.1   The human digestive system functions in the ingestion, digestion, and absorption of food. Ingested food frequently must be broken down, or digested, before the nutrients can be absorbed into the blood by a capillary network. Certain materials are neither digested nor absorbed but are eliminated as part of the feces.

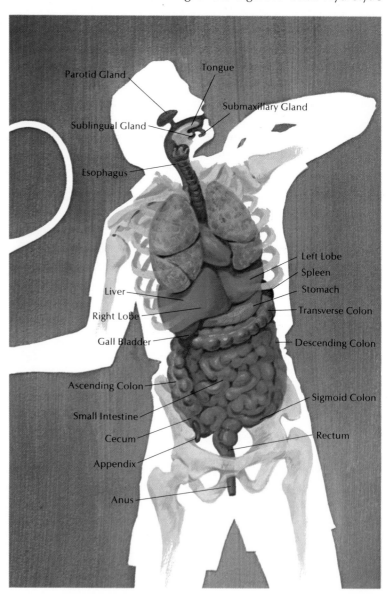

enzymes are released into the digestive tract from specialized secretory cells and organs. For these chemical reactions to occur, large chunks of food must first be physically broken down into smaller pieces, allowing access to the digestive enzymes.

The *mouth* region is specialized for taking food into the digestive tract. In many animals, specialized structures such as teeth assist in the physical disintegration of food. For animals that feed on green plants, the preliminary cutting and crushing of food is particularly important, because the tough cellulose walls of plant cells are resistant to the action of digestive enzymes. In mammals, the food mixture is lubricated with *saliva*, which is secreted within the mouth and mixed with the food during chewing. This secretion contains an enzyme that hydrolyzes starches into disaccharides. From the mouth, the food moves into the gullet, or *esophagus*, which is a tube serving primarily to transport food to the stomach. Muscle contractions move food along the tube, and lubricating fluids are secreted from the glands in the tube walls.

The *stomach* of a mammal is a muscular, saclike organ that can be closed at both ends by *sphincters*, or rings of muscle. Muscle contractions in the stomach wall churn and squeeze the food, thoroughly mixing it with substances secreted by cells and glands in the stomach lining. After the food mixture is mixed thoroughly with mucous fluid, acid, and enzymes such as pepsin, the posterior sphincter opens, the stomach contracts, and the food is forced into the *small intestine*.

Within the long, coiled tube of the small intestine, digestion is completed and the resulting small organic molecules are absorbed into cells that line the tube. The surface of the small intestine is covered with minute, fingerlike projections called *villi*. The cell surfaces lining the villi are themselves covered with submicroscopic projections, or *microvilli*. These structures produce a great surface area—about 2,000 square feet in the human small intestine—which ensures absorption of nutrients.

The short segment of small intestine closest to the stomach is known as the *duodenum*. Here secretions from the *pancreas* and *liver* enter the gut. Pancreatic juice contains enzymes that break down the most common types of biological macromolecules. These protein-digesting enzymes are synthesized by the pancreas in an inactive form so that they do not destroy themselves or other pancreatic proteins; only when they arrive in the duodenum do these enzymes become activated. In addition to enzymes, pancreatic juice contains sodium bicarbonate, which neutralizes the acid passed down from the stomach and makes the contents of the intestine slightly alkaline.

Another secretion aiding in digestion is produced by the liver in the form of *bile*, which is stored in the gall bladder until it is needed in the duodenum. Bile contains sodium salts of certain complex organic acids that function as detergents, breaking up insoluble

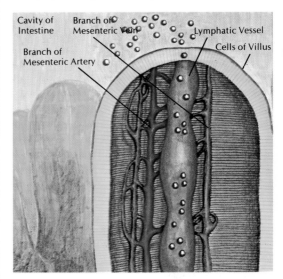

Figure 11.2 Diagram of a single intestinal villus. Many of these fingerlike projections line the small intestine, and large molecules such as fats enter the lymphatic system through these villi.

Cavity of Intestine

Branch of Mesenteric Vein

Lymphatic Vessel

Cells of Villus

Branch of Mesenteric Artery

droplets of fats and making the fat molecules accessible to enzymes in the pancreas.

Nutrients are absorbed from the intestine and then diffused into the circulatory system. The absorbed nutrients are carried away from the intestine by a capillary network that feeds into the *hepatic portal vein*, which transports dissolved nutrients to another capillary network in the liver. The liver removes excess nutrients for storage and releases nutrients from storage if the blood level of some substance falls too low. In these ways, the liver helps maintain optimum levels of glucose and amino acids in the blood throughout periods of fasting and gluttony.

The major part of the large intestine, the *colon*, is a corrugated tube with a smooth lining. The colon serves primarily to remove excess water from undigested material. It houses large numbers of bacteria, principally the species *Escherichia coli*, which are also found in lesser numbers elsewhere in the gut. These bacteria synthesize significant amounts of some vitamins, which are subsequently absorbed by the colon and used by the body. At the end of the large intestine is a short section, the *rectum*, where undigested waste material is stored until eliminated from the body through the orifice called the anus.

GAS EXCHANGE AND THE RESPIRATORY SYSTEM

Although all cells—and therefore all tissues and organs—engage in respiration, one system of organs is specialized for procuring oxygen, delivering it to the blood, and for removing carbon dioxide. This interrelated group of organs is called the respiratory system.

Land animals possess internal respiratory systems that are braced against collapse. Small land organisms such as snails and some crustaceans have simple lungs. Insects breathe through a system of tracheae, which are thin tubes that open on the body surface and ramify extensively throughout every tissue, bringing the air supply close to the cells and minimizing the role of the body fluid in the transport of gases.

Most large animals possess lungs, which are alternately emptied and filled with fresh air by active movements known as ventilation. All oxygen enters the respiratory system of a land animal in the gaseous state. It is then dissolved in a thin film of water on the lining of the respiratory tract and diffuses across cell membranes in solution.

Air, which consists of approximately 20 percent oxygen, is drawn during normal breathing through the nostrils into the chamber that lies above the soft palate in the mouth and opens directly into the throat. Opening into the throat are the esophagus and the windpipe, or *trachea*. The mammalian trachea extends as a single, open tube from the throat to the chest, where it immediately branches into left and right *bronchi*. The trachea and bronchi are lined with ciliated mucus epithelium. Mucus secretions keep the

Figure 11.3 A diagrammatic representation of the human respiratory system. In the smaller diagram at lower left, grapelike clusters of pulmonary alveoli are shown at the end of each terminal bronchiole. Alveoli are surrounded by a capillary network in the lower right diagram.

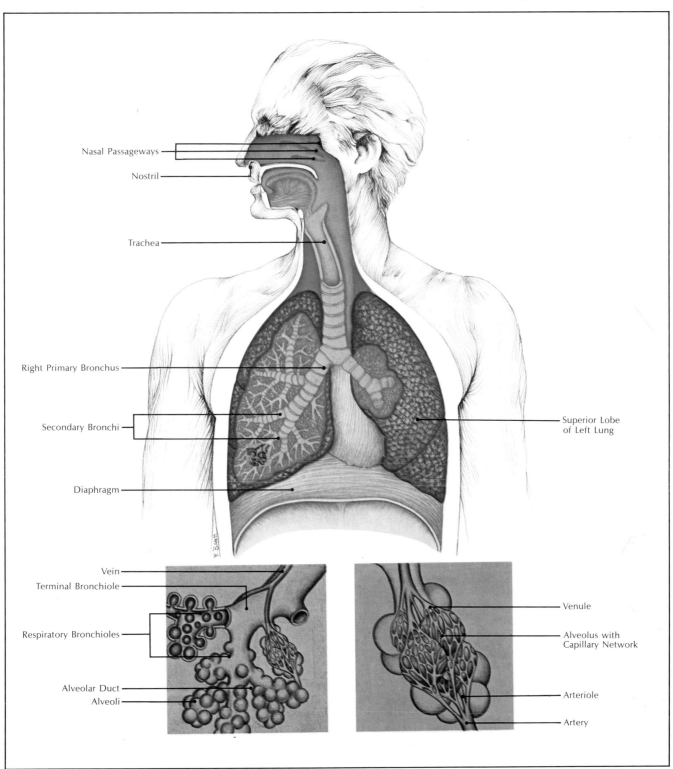

Figure 11.4   These diagrams show position changes in the diaphragm and rib cage during exhalation and inhalation.

Exhalation

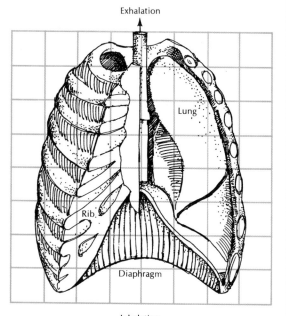

Inhalation

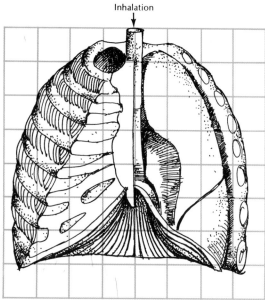

epithelial surface moist, and the cilia, whose direction of beating is invariably toward the throat, maintain a constant flow of the mucus film away from the lungs and into the throat. In this way, the respiratory passages are constantly cleansed of dust, microorganisms, and other inhaled particles.

The bronchi lead into the lungs, where they branch into a number of secondary bronchi and then subdivide into smaller muscular tubes, the *bronchioles*. The lungs are a pair of large, spongy sacs that lie within the chest cavity. These sacs are made up of millions of microscopic pouches called *pulmonary alveoli* that are arranged in clusters at the end of each terminal bronchiole. The alveoli consist of thin epithelium that is in direct contact with an equally thin capillary wall. The tissue separation between air and blood at this site is about 1/5 micron; or about one-third the diameter of a red blood cell. The average total amount of area represented by the alveolar epithelium in one human being is 750 square feet.

The lungs occupy most of the chest cavity and are protected by the somewhat flexible rib cage. The floor of the chest cavity is a sheet of muscle called the *diaphragm*. Inhalation is caused by expansion of the rib cage and contraction of the diaphragm. When the diaphragm contracts, it flattens and thus enlarges the volume of the chest cavity. Atmospheric pressure forces air into the lungs so that they expand to fill the enlarged cavity. Exhalation occurs when the diaphragm relaxes and bulges upward, reducing the volume of the chest cavity and forcing air out of the lungs.

The rate of breathing and the extent to which the lungs are refilled on each breath are under partial voluntary control. But the rate and volume are normally controlled involuntarily by a respiratory center at the base of the brain. This respiratory center is affected, among other things, by the concentration of carbon dioxide in the blood.

INTERNAL TRANSPORT AND THE CIRCULATORY SYSTEM

The simplest means of distributing molecules is by diffusion. Organisms that are only a few cells in thickness usually require diffusion and active transport; here the slow rate of diffusion is not a problem because the distances to be crossed are short. Even in larger organisms, diffusion is the major mechanism for transporting substances into cells. But once nutrient materials have entered the organism, they still must be transported to sites where they can be used in chemical reactions. Cellular coordination of a large organism requires a rapid transport mechanism for quickly moving materials to distant places where diffusion can then take over. The human circulatory system is such a mechanism.

Human beings have a circulatory system in which a heart pumps into large vessels called *arteries*, which then carry the blood to various parts of the body. The arteries branch repeatedly,

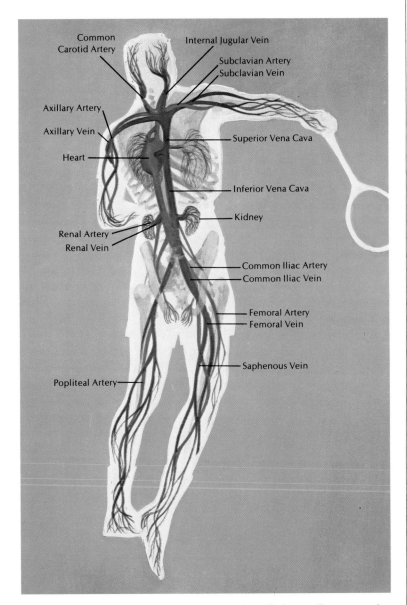

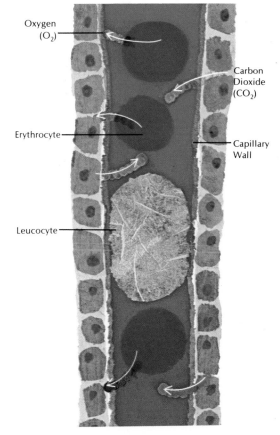

Figure 11.5   The circulatory system is vital to all tissues in the human body. Capillaries are the small, thin-walled circulatory structures that form networks through which blood is brought in contact with all body elements. Nutrients and oxygen pass through capillary walls to reach nearby cells, and cellular waste products flow the other way into the blood to be transported elsewhere in the body.

Figure 11.6   As shown in this diagram, there is an exchange of oxygen and carbon dioxide in the capillaries. Red blood cells, which carry oxygen and carbon dioxide between the tissues and the lungs (via hemoglobin), pass through one at a time, whereas the larger leucocytes must elongate in order to squeeze through the capillaries.

finally becoming tiny, thin-walled vessels called *capillaries*. As the blood moves through the capillaries, materials such as oxygen and carbon dioxide are exchanged with neighboring cells. The blood then moves into larger, thicker-walled *veins*, which transport it back to the heart.

The Circulatory Fluids

The blood of human beings, like all vertebrates, is composed of *plasma*, a fluid in which many proteins, ions, nutrients, and waste materials are dissolved. As shown in Figure 11.7, floating in the plasma are *red blood cells* (or "erythrocytes"), several types of

Figure 11.7   In the red blood cells at left, which function in the transportation and distribution of oxygen in mammals, note the absence of nuclei. The structures at right, called platelets, usually are not whole cells but are membrane-covered cell fragments, often without nuclei. Platelets are involved in the blood-clotting mechanism.
Walter Dawn

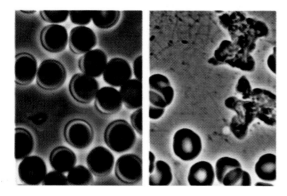

white blood cells (or "leucocytes"), and cell fragments called *platelets* (or "thrombocytes").

Materials move between blood and cells by diffusion. Some of these substances are simply dissolved in the blood plasma. However, the red blood cells play a special role in oxygen transport. These cells contain the protein hemoglobin, which gives blood its red color. Hemoglobin combines reversibly with oxygen so that blood is capable of transporting about 60 times more oxygen than would dissolve in plasma. When the oxygen concentration in cells around blood is high (as it is in the capillaries of the lungs), hemoglobin combines readily with oxygen. In body parts where the oxygen concentration is low, hemoglobin readily gives up oxygen and allows it to diffuse out of the blood into the body tissues.

Hemoglobin contains four subunits, each of which consists of a protein chain, globin, and *heme*—a complex polycyclic, porphyrin ring structure containing iron (Figure 11.8). Each subunit can bind one oxygen molecule. The four subunits interact in a cooperative way so that after one subunit binds an oxygen molecule, the other subunits bind additional oxygen molecules more readily. In this way, hemoglobin performs efficiently in its role of binding oxygen in the lungs and releasing it to the body tissues.

Carbon dioxide, which is produced as a waste product of metabolic respiration, is carried by the blood from the body cells to the

Figure 11.8   Hemoglobin molecule. The structure of the molecule has been deduced primarily from x-ray diffraction studies. The molecule consists of four closely associated polypeptide chains. Each molecule contains two alpha and two beta chains, each of which binds an oxygen-carrying heme group. (See also Figure 8.4.)
Reproduced by permission from The Structure and Action of Proteins *by Richard E. Dickerson and Irving Geis, Harper & Row, Publishers. Copyright 1969 by Richard E. Dickerson and Irving Geis.*

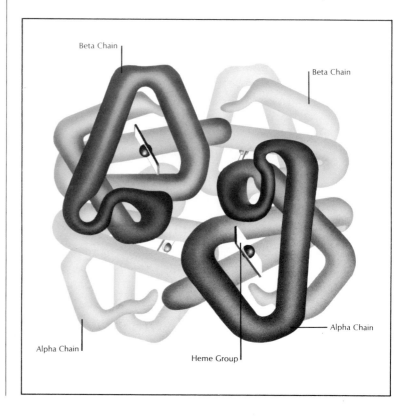

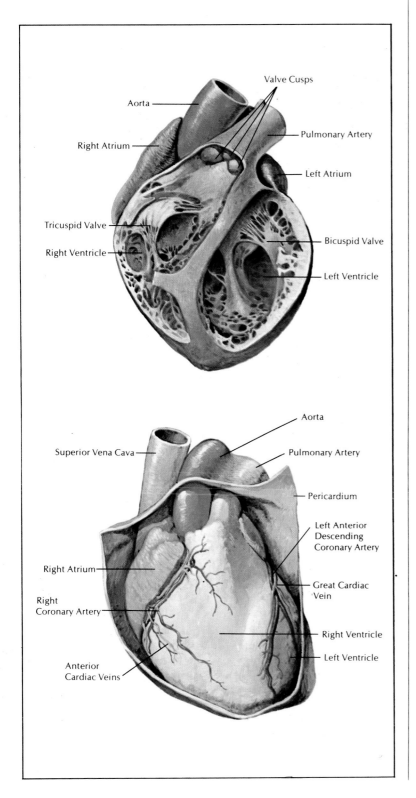

Figure 11.9   Diagrammatic representation of the heart. In the illustration at top, note the fibrous interior of the heart. It is made up of chambers, which contract with regular rhythmic motion, and of valves, which, as in other muscle systems, work in coordinated pairs—as one valve opens to admit blood, its antagonistic counterpart snaps closed. As seen below, the pericardium is the membranous sac that encloses the heart. The exterior muscle wall is fed by an intricate system of veins and arteries.

Valve Cusps

Aorta

Pulmonary Artery

Right Atrium

Left Atrium

Tricuspid Valve

Bicuspid Valve

Right Ventricle

Left Ventricle

Aorta

Superior Vena Cava

Pulmonary Artery

Pericardium

Left Anterior Descending Coronary Artery

Right Atrium

Great Cardiac Vein

Right Coronary Artery

Anterior Cardiac Veins

Right Ventricle

Left Ventricle

Figure 11.10 The processes involved in blood clotting. When normal blood is drawn from the body and allowed to sit at room temperature for several minutes, the fibrin-associated clotting phenomenon occurs. The plasma, which separates from the cells when centrifuged, makes up 55 percent of the blood

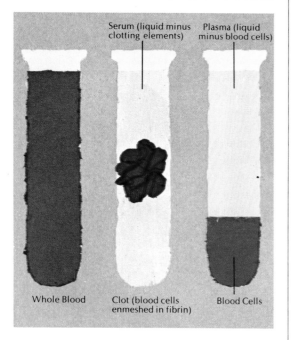

Serum (liquid minus clotting elements)

Plasma (liquid minus blood cells)

Whole Blood

Clot (blood cells enmeshed in fibrin)

Blood Cells

lungs, where it is excreted. In the blood, dissolved carbon dioxide gas is in equilibrium with carbonic acid and a bicarbonate ion. Because the hydration of carbon dioxide (and the dehydration of carbonic acid) is a relatively slow reaction, red blood cells contain the enzyme carbonic anhydrase to catalyze this reaction. This catalysis speeds up the reaction so that bicarbonate can be converted to gaseous carbon dioxide for exchange during the rapid flow of blood through the capillaries of the lungs.

Dissolved proteins and the platelets are involved in *blood clotting*, a complex series of reactions that occur in case of injury to the circulatory system. The end result of these reactions is the formation of a clot, which temporarily seals off the injured area until the damage is repaired. An outline of the processes involved in blood clotting is presented in Figure 11.10.

The Circulatory Pump

Like all vertebrates, human beings have hearts made up of a specialized striated muscle that contracts spontaneously. The mammalian heartbeat is initiated and coordinated by a patch of muscle cells that are specialized for electric conduction rather than contraction. This *pacemaker* initiates electric activity that spreads through the cardiac muscle fibers of the *atria* to another specialized region, the *atrioventricular node*, located in the wall between the two atria just above the ventricles. The atrioventricular node passes the wave of excitation to a bundle of similar conducting tissue that extends throughout the ventricles.

Because the heart muscle is constantly working, it requires a plentiful supply of oxygen to carry out its energy-releasing metabolism. The heart muscle contains its own system of blood vessels. If these vessels become blocked and a portion of the heart muscle becomes short of oxygen, the heart fails to function.

The Transport Cycle

Blood arriving at the heart from the body tissues is high in carbon dioxide and lower in oxygen. It enters the right atrium, which pumps blood into the right ventricle. The right ventricle sends blood to the lungs via the pulmonary artery. After the blood exchanges its carbon dioxide for oxygen in the capillaries of the lungs, it returns to the left atrium of the heart by way of the pulmonary vein. The left atrium pumps blood into the left ventricle, which pumps blood out to all parts of the body through a large artery, the *aorta*. The heart, like all pumps, must have valves in order to prevent backward flow.

The arteries have relatively thick walls of muscular tissue, which hold blood under pressure as it moves away from the heart. The contraction of the ventricle pushes blood into the aorta under a pressure equivalent to about 120 millimeters of mercury. As blood moves away from the heart, the arteries branch to form

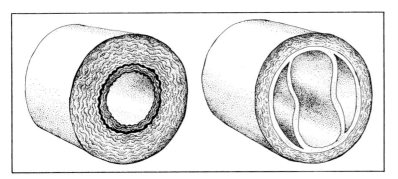

Figure 11.11   When blood returns to the heart through the veins, valves prevent the blood from flowing in the wrong direction, while muscular action serves to push the blood along. In these cutaway views of an artery (left) and a vein (right), note that arteries have thicker walls than do veins because they must expand and contract with greater force to direct blood flow to tissues and organs.

numerous smaller arteries and eventually lead into the capillary system. Capillary walls consist of a single layer of epithelial cells, permitting ready diffusion of substances from blood to tissue cells and vice versa. The number of capillaries in the body is enormous—there are, for example, a few thousand in a single cubic millimeter of skeletal muscle tissue. And every cell of the body is within a cell or two of a capillary.

Veins have thinner walls and are less elastic than arteries. A layer of fibrous tissue and a thin layer of muscle surround the epithelial cells that line veins. Veins lack sufficient elasticity to keep blood under pressure. Blood arrives in veins from capillaries under low pressure, and its movement back through veins to the heart is largely dependent on contraction of skeletal muscles. As the muscles contract, they squeeze blood through the veins that pass through muscle tissue. Valves along the veins keep blood from moving back toward the capillaries. The flow of blood through the circulatory system is controlled both by the rate at which the heart beats and by the relaxation or constriction of the muscles inside artery walls.

Nerves leading from special centers in the brain can speed up or slow down the heartbeat. These centers are activated by various sensory inputs, including cells that detect unusual stretching of arteries (leading to a slowing of heartbeat) or of veins (leading to a speeding of heartbeat).

The *lymphatic system* is an independent transport system found in vertebrates. It is involved in movement of tissue fluids outside blood vessels. Some of the fluids that leave arterial capillaries, together with fluids produced by tissue cells themselves, diffuse into small tubes called lymph capillaries. The tissue fluid, or *lymph*, that enters these vessels moves through the lymph capillaries into larger lymph vessels and eventually into compact, ovoid organs called *lymph nodes*. Foreign particles or microorganisms that wander into tissues are likely to be carried into lymph nodes, for walls of blood capillaries are quite resistant to the passage of such intruders. The small white blood cells known as "lymphocytes" are found in lymph as well as in blood. Lymph nodes are drained by other lymph vessels that come together to form two large ducts

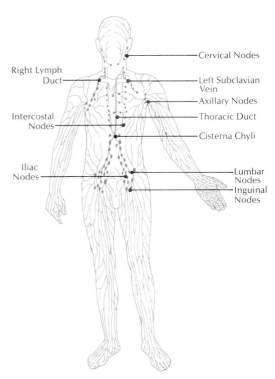

Figure 11.12   The human lymphatic system. Lymphatic vessels function as an auxiliary branch of the main cardiovascular system by returning tissue fluid back to the main bloodstream. Lymph nodes, located along the vessels, trap foreign matter, including invading bacteria.
From "The Lymphatic System" by H. S. Mayerson.
© 1963 by Scientific American, Inc.

Figure 11.13   The human excretory system.
Note the detailed diagram of
a nephron (yellow), one of the functional
units of the kidney.

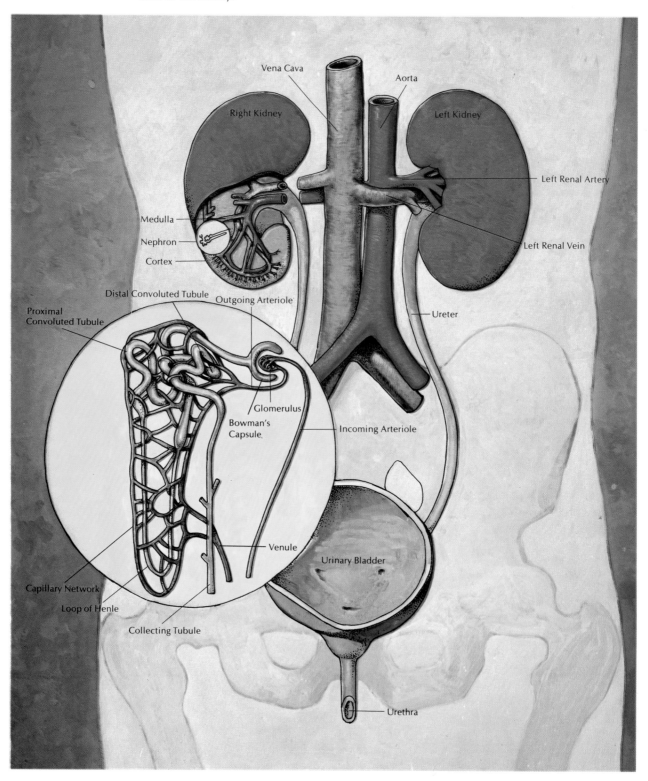

that empty the filtered lymph into large veins. The lymphatic system is important to the body's immune responses.

WASTE ELIMINATION AND THE EXCRETORY SYSTEM

To maintain proper internal conditions, an organism must not only obtain nutrients, it must regulate its internal composition and get rid of waste products. The principal wastes are carbon dioxide (the major waste product of cellular respiration), salts, and nitrogenous wastes of protein metabolism. A proper water balance must also be maintained. The functions of maintaining salt and water balance and disposing of nitrogenous wastes are performed by excretory systems.

Land animals, including man, must obtain water by drinking and eating. Because excess salts are frequently included in their food, their excretory systems are specialized for elimination of salts and retention of water. In many land animals, salts are also eliminated through the body surface by sweating.

Much of the food that a human takes in consists of proteins. In the digestive process, proteins are hydrolyzed into amino acids, which are absorbed by the cells. Some of these amino acids are reused in synthesizing needed proteins and other molecules, but the greater part of this supply is further broken down by the liver as a source of chemical energy. The first step in the metabolism of an amino acid is its conversion into an organic acid by the removal of amino groups. The amino groups are converted into *ammonia* during this process of "deamination."

Ammonia is toxic to humans if it accumulates in high concentrations. Ammonia does not diffuse readily into the air, as does the waste product carbon dioxide. A water solution of ammonia must be excreted. Because only dilute ammonia solutions can be tolerated, ammonia is combined with carbon dioxide to form *urea*, a substance that can be tolerated by humans in concentrations substantially higher than that of ammonia alone. Even with urea, a substantial amount of water must be used in excretion. In certain egg-laying animals, ammonia is converted to uric acid, a substance that precipitates out of solution to form a solid; this solid can be expelled from the body with relatively little water loss.

The human excretory system consists of paired kidneys, which concentrate waste products. Each *kidney* is an organ that adjusts the concentrations of various ions. In addition, the kidney helps regulate the blood concentration of glucose, and it excretes nitrogenous wastes such as urea, products of hemoglobin breakdown, and creatinine formed as a waste product of muscular activity. Useless materials that find their way into the organism across the intestinal or respiratory epithelia are also excreted by the kidney.

Each human kidney contains about one million functional units called *nephrons*. Each nephron consists of a network of blood capillaries, called the *glomerulus*; a cup-shaped structure called the

Bowman's capsule (which surrounds the glomerulus); and a series of connecting *renal tubules*.

Blood filtration takes place from the glomerulus into the Bowman's capsule. Because the exit from the glomerular capillaries is smaller than the entrance, considerable blood pressure is built up in these capillaries. Under this high pressure, about one-fifth of the fluid portion of blood is forced through spaces in the capillary walls into Bowman's capsule, leaving only blood cells, plasma proteins, and fluid within the glomerulus. The ultrafiltrate in the Bowman's capsule at this stage then contains the same concentration of small molecules as blood, including nutrients and salts in addition to waste products.

This ultrafiltrate moves through the renal tubules, where it again comes into proximity with the surrounding capillary bed that contains the blood from which it was filtered. Here, cells specialized for carrying out active transport pump glucose, amino acids, and some ions out of the tubule fluid, and some of these materials then diffuse back into the blood.

The remaining fluid in the tubule continues into the *collecting tubule*, which empties into the *ureter*. The final fluid waste product is called *urine*. Each ureter is a bilateral, thick-walled, muscular tube that conveys fluid from the kidney to the urinary *bladder*. The bladder serves as a storage vessel for urine. Its epithelial lining is composed of five or six layers of bulging, pear-shaped cells. As urine accumulates and distends the bladder, these cells slide past one another and spread to form a thinner membrane of greater surface area. Urine passes from the bladder to the exterior of the organism through the *urethra*, which can be opened and closed by muscles that are under voluntary control.

CHEMICAL REGULATION AND THE ENDOCRINE SYSTEM

Hormones are chemical substances that are produced in one part of the body and transported by the blood to influence activities in another part of the body. Some hormones, such as those produced by the digestive tract, are produced in tissues that are involved primarily in other functions; others are secreted by tissues or organs that are specifically modified for hormone production. In either case, the hormones are secreted into the blood circulatory system. In contrast to *exocrine glands*, which discharge their secretions through ducts, the tissues that produce hormones secrete them directly without benefit of ducts. For this reason, glands that secrete hormones are known as the ductless glands, or the *endocrine glands*.

The major glands of the endocrine system are depicted in Figure 11.14. The *pituitary gland* is composed of two distinct divisions, which have different embryological origins. The posterior pituitary develops from neural tissue, whereas the anterior pituitary originates in tissues from the roof of the mouth. The posterior pituitary releases two hormones: one called *antidiuretic hormone* (ADH), which raises blood pressure and enhances retention of body water, and *oxytocin*, which induces uterine contractions in the female mammal at the end of pregnancy. Oxytocin is used clinically to induce labor when this procedure appears medically desirable. The anterior pituitary is especially important: It secretes a hormone that affects growth and a variety of hormones, most of which act on other endocrine and exocrine glands, including the thyroid, the ovaries, and the mammary glands.

Each of the glands to which the pituitary sends activating secretions in turn produces hormones. For example, under the influence of a hormone called *thyroid-stimulating hormone* (TSH) that is produced by the pituitary, the thyroid glands secrete *thyroxin*, which affects the rate of certain metabolic processes. A rise in the level of thyroxin in turn causes a reduction in the secretion of TSH, thus reducing the level of circulating thyroxin. This self-regulating circuit, when it works properly, ensures a level of circulating thyroxin that is optimal for normal bodily function. People with excessive levels of thyroxin (hyperthyroid) may be tense and have trouble sleeping; people with insufficient thyroxin (hypothyroid) may be listless and apathetic, or may suffer from a more serious and chronic condition known as *cretinism*.

The *adrenal glands*, located next to the kidneys, consist of two sections: an internal core known as the *adrenal medulla* and an external covering known as the *adrenal cortex*. The adrenal medulla secretes the hormones *norepinephrine* and *epinephrine*. These hormones have effects similar to the stimulative effects of strong emotions on the sympathetic nervous system, as you will read about in Chapter 12.

The similarity of these two effects is not very surprising when you consider the fact that these two hormones, epinephrine and norepinephrine, are transmitter substances in the sympathetic nervous system. Moreover, the adrenal medulla is stimulated to release epinephrine and norepinephrine by direct neural connections from the sympathetic system.

The preceding example represents what is known as a "positive feedback circuit." In other words, the sympathetic activity has the effect of facilitating the adrenal medullary activity which, in turn, aids further sympathetic activity.

The adrenal cortex is stimulated by the *adrenocorticotrophic hormone* (ACTH) rather than by direct neural input. Under the influence of ACTH, the adrenal cortex releases a variety of steroid hormones that regulate water balance and energy expenditure. One of the most important roles of the adrenal cortex is to help animals and human beings withstand stress.

The mechanism of hormone action is only now beginning to be understood, despite the fact that their effects and, in many cases, their chemical structure have been known for many years. A major

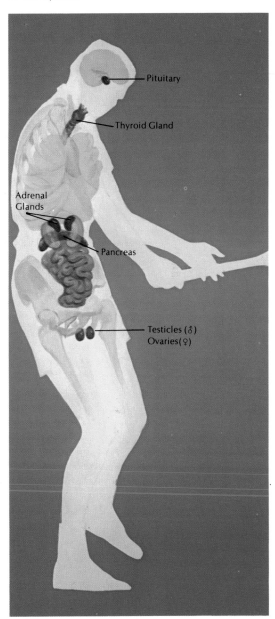

Figure 11.14 The control of the human endocrine system is carefully balanced to provide sufficient hormones to meet the body's needs. The pituitary gland, the thyroid gland, and the adrenal glands are the major hormone-producing glands of this system.

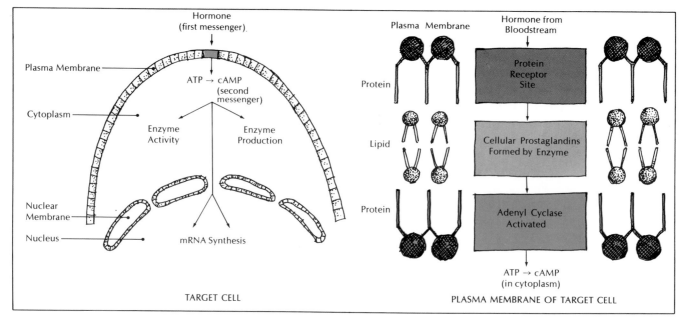

Figure 11.15 Hormones produced by endocrine glands travel via the bloodstream to target cells of the body. Plasma membranes of these target cells have protein receptor sites that are specific for a given hormone. Biochemical effects produced by changes in the levels of cyclic AMP (cAMP) will induce various physiological effects in cells and organisms.

area of research in physiology and medicine concerns how hormones work, and in 1971 Dr. Earl Sutherland of Vanderbilt University won the Nobel prize for his studies in this area.

One hypothesis is that when hormone molecules come in contact with plasma membranes of target cells they function as "first messengers." The hormone acts on *specific protein receptor* sites on the surface of plasma membranes. The specific protein receptors set off the synthesis of a group of substances called *prostaglandins* within the plasma membrane, which are active in the synthesis of the enzyme adenyl cyclase (Figure 11.15). This enzyme helps to break down ATP to adenosine 3'5' monophosphate, or *cyclic AMP* (cAMP), in the cytoplasm.

Cyclic AMP acts as a "second messenger" and may affect cells in the following ways: First, it activates or modifies enzyme activity. Second, it stimulates the production of specific proteins including enzymes. And third, it may prevent messenger RNA production. These biochemical effects can trigger physiological effects such as muscle contraction and blood pressure control, glandular secretions, growth and differentiation of cells and tissues, as well as feedback activities stimulating new hormone production.

It is too early to speculate whether the above hypothesis for hormone action will be substantiated in all cells. Nevertheless, a more complete understanding may help lead to the control of basic life processes.

Table 11.1 lists the major hormones that have been described in man, and indicates their tissue or organ of origin, their target tissue or organ, and their effect on the body. The effect of certain hormones, such as the *thyroid-stimulating hormone* (TSH), is very

Table 11.1
Important Mammalian Hormones

Source	Hormone	Principal Effects
Pyloric mucosa of stomach	Gastrin	Stimulates secretion of gastric juice
Mucosa of duodenum	Secretin	Stimulates secretion of pancreatic juice
	Cholecystokinin	Stimulates release of bile by gallbladder
	Enterogastrone	Inhibits secretion of gastric juice
Damaged tissues	Histamine	Increases capillary permeability
Pancreas	Insulin	Stimulates glycogen formation and storage; stimulates carbohydrate oxidation; inhibits formation of glucose
	Glucagon	Stimulates coversion of glycogen into glucose
Kidney plus blood	Hypertensin	Stimulates vasoconstriction, causing rise in blood pressure
Thymus	Thymic hormone	Stimulates immunologic competence in lymphoid tissues
Testes	Testosterone	Stimulates male secondary sexual characteristics
Ovaries	Estrogens	Stimulates female secondary sexual characteristics
	Progesterone	Stimulates female secondary sexual characteristics and maintains pregnancy
Thyroid	Thyroxin	Stimulates oxidative metabolism
Parathyroids	Parathormone	Regulates calcium-phosphate metabolism
Adrenal medulla	Epinephrine	Stimulates syndrome of reactions commonly termed "fight or flight"
	Norepinephrine	Stimulates reactions similar to those produced by adrenalin
Adrenal cortex	Glucocorticoids (corticosterone, cortisone, hydrocortisone, and so on)	Stimulate formation and storage of glycogen; help maintain normal blood-sugar level
	Mineralocorticoids (aldosterone, deoxycorticosterone, and so on)	Regulate sodium-potassium metabolism
	Cortical sex hormones (adrenosterone, and so on)	Stimulate secondary sexual characteristics, particularly those of the male
Anterior pituitary	Growth hormone	Stimulates growth
	Thyroid-stimulating hormone (TSH)	Stimulates the thyroid
	Adrenocorticotrophic hormone (ACTH)	Stimulates the adrenal cortex
	Follicle-stimulating hormone (FSH)	Stimulates growth of ovarian follicles and of seminiferous tubules of the testes
	Interstitial Cell-stimulating hormone (ICSH)	Stimulates production of testosterone in testes
	Luteinizing hormone (LH)	Stimulates conversion of follicles into corpora lutea; stimulates secretion of sex hormones by ovaries and testes
	Prolactin	Stimulates milk secretion by mammary glands
Intermediate lobe of pituitary	Melanocyte-stimulating hormone	Controls cutaneous pigmentation
Posterior pituitary	Oxytocin	Stimulates contraction of uterine muscles; stimulates release of milk by mammary glands
	Vasopressin	Stimulates increased water reabsorption by kidneys; stimulates constriction of blood vessels (and other smooth muscle)

Source: Adapted from W. T. Keeton, *Biological Science* (New York: Norton, 1970).

Figure 11.16 (opposite page, above) Primary immune response of inflammatory cells to antigen injected into a mouse (not previously exposed to this antigen). First (upper left), large granular mast cells in the body tissues come in contact with the antigen and break down. This cellular disintegration releases large numbers of lysosomes, whose enzymes and other substances destroy other cells in the vicinity, initiating inflammation. The first moving defensive cells to arrive at the site are neutrophils, which swallow up some of the foreign particles but soon disintegrate themselves. Next (upper middle), lymphocytes and monocytes reach the area and feed upon the foreign particles and cellular debris. This process causes some of the lymphocytes to enlarge and become macrophages (upper right). Eventually, these cells ingest all the foreign matter and the inflammation subsides. Most of the antigen is broken down into amino acids and sugars by enzymes, but some is maintained in the macrophages by combining with RNA.
After Scientific American, *1964*

Figure 11.17 (opposite page, below) Secondary immune response of inflammatory cells to antigen injected into a mouse (previously exposed to this antigen). Neutrophils arrive at the site but in fewer numbers, while macrophages arrive in larger numbers (lower left). Some of the macrophages contain antigen in combination with RNA, and these cells interact with eosinophils, which cause them to be broken open. More macrophages then move in and engulf pieces of broken cells (lower middle). Some antigen escapes destruction in combination with RNA in the macrophages (lower right). The rapid arrival of macrophages and the ease with which they ingest foreign particles result in a shorter period of secondary inflammation and one that is less severe.
After Scientific American, *1964*

circumscribed, for it acts only upon secretory cells of the thyroid gland. Other hormones, such as the sex hormones, affect a variety of target tissues and organs and thus have manifold effects.

A reciprocal relation is frequently encountered between endocrine glands that is known as "feedback" control. This relation is especially pronounced between the pituitary gland and its target endocrine glands. For example, the anterior lobe of the pituitary gland secretes TSH, which stimulates the thyroid gland to synthesize and to secrete thyroxin. As thyroxin levels rise in the blood, a feedback effect is observed, for the secretion of TSH is depressed. The pituitary is involved in several similar relationships with other endocrine glands. It is often referred to as the "master gland."

Hormonal regulation of bodily functions differs from nervous regulation in an important way. Whereas the action of the nervous system is rapid and usually of very short duration, the effect of the endocrine system is somewhat delayed in time, but prolonged. Both mechanisms are chemical, in final analysis, and in many cases their effects duplicate each other; in other cases, their effects complement each other; and in still others they are quite separate. Between them, all of the functions of the body that reside in separate tissues and organs are coordinated and integrated for the reproduction and maintenance of the individual as a whole.

BODY DEFENSES AND THE IMMUNOLOGICAL SYSTEM

The human body is fortunate in having a coordinated system of defenses against potentially damaging invaders of its cells and tissues. The nervous system and endocrine system enable the individual to cope in one way or another with endangering situations that can be perceived by way of its sense organs—to run away from a predator that can be seen, heard, or smelled; to dodge a hurled stick or stone; to drop a hot ember; to fend off a threatening blow. But how does it protect itself from organisms too small to be seen, felt, heard, or smelled—microorganisms that enter the body through a scratch, through a mosquito sting, or hidden in a morsel of food? These situations call for another integrated line of defense, the *immune response.*

In a sense, immunity—as it exists in man and other warm-blooded vertebrates—is a communications system. If foreign material enters the body, the first step in the chain of communication is recognition that the invading material *is* foreign—*not-self* rather than self. The second step involves multiplication and activation of the cells appropriate to combat the particular alien material present. Finally, these cells or their specialized chemical products must find and destroy the foreign cell, microorganism, or protein. This three-stage process represents the simplest outline of a *primary immune response* to any sort of foreign material.

Immunity is usually thought of in terms of the resistance to further infections that follows an attack of infectious disease. For cen-

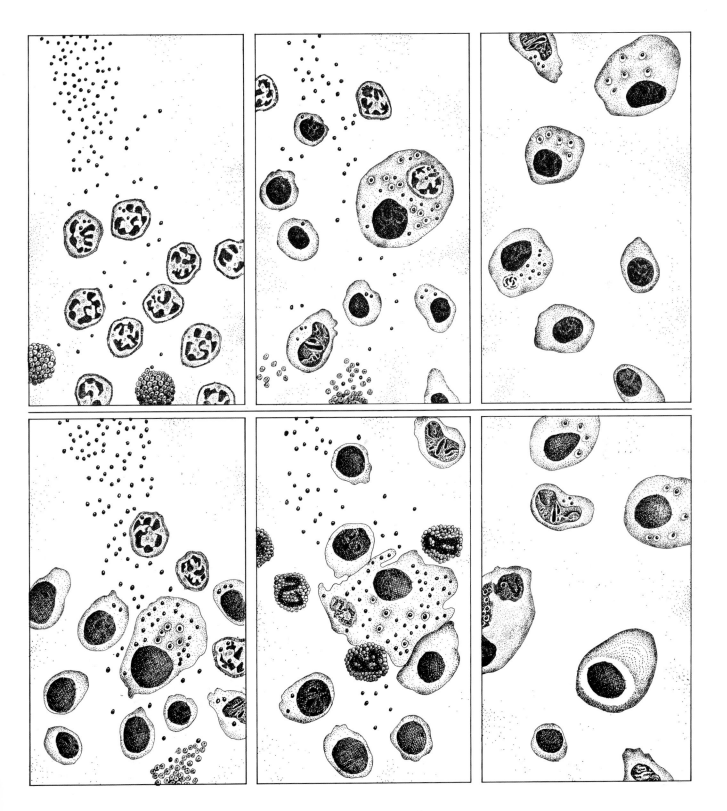

turies it has been known that once a man has been affected by smallpox, he never contracts the disease again. Allowing for an immense variety of details, immunity to smallpox is a prototype of all *secondary immune responses*. Once foreign material has been successfully dealt with, there is an enlarged population of cells able to deal with that particular foreign substance and able to be called into action more rapidly and more effectively upon subsequent exposures.

The system primarily concerned with the immune response is the *lymphatic system*, which consists of a large number of different tissues and organs as well as circulating cells that emanate from them. The most prominent tissues and organs are the spleen, thymus, tonsils, and lymph nodes. The lymph nodes are compact bodies about the size of a pea. They are distributed in clusters in strategic areas—primarily the throat and neck, armpits, groin, and the mesenteries of the intestinal tract and lungs. Somewhat less conspicuous is lymphoid tissue embedded in the wall of the intestinal tract and in the bone marrow.

The presence of a foreign substance in the body triggers a dramatic sequence of events known as the *inflammatory response*. The major participants in this response are the white blood cells, or *leucocytes*, which are amoeboid cells capable of moving through intercellular fluids and tissues of the body as well as the bloodstream. Lymphocytes, granulocytes—neutrophils, eosinophils, and basophils—and monocytes are all types of leucocytes. Most of the leucocytes in the body under normal conditions are *lymphocytes* and *neutrophilic granulocytes* (*neutrophils*). Leucocytes are manufactured in bone marrow, the lymphatic system, and the thymus gland. In general, leucocytes combat foreign materials by engulfing and digesting them, but the details of the inflammatory response are relatively complex.

Any substance introduced into the tissues of the body that can provoke an immune response is called an *antigen*. Most antigens have one thing in common: *foreignness* to the individual. Antigens are chemical substances that are not in the chemical make-up of the individual. Thus the protein of type A blood is antigenic to a person who has type B blood; hen's egg albumin is antigenic to any human being, as are most of the proteins of any nonhuman organism, including those of bacteria and viruses; and the polysaccharides of bacterial cell walls are antigenic to all vertebrates. They are not antigenic if ingested because, before they can be absorbed into the body's tissues, the digestive enzymes break them down into their smaller nonantigenic components.

The nature of the reaction that is triggered by an antigen is not completely understood. Many immunologists believe that if circulating or wandering lymphocytes in the body come in contact with material that is not-self, the lymphocytes will pick up some of this antigen and transport it to the lymphoid tissue. Here the antigen is

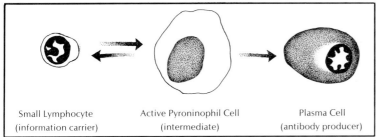

| Small Lymphocyte | Active Pyroninophil Cell | Plasma Cell |
| (information carrier) | (intermediate) | (antibody producer) |

Figure 11.18   The surface of an immunocyte contains antibody-type combining sites that can combine with a specific antigen. Binding of an antigen to the cell surface may cause the cell to enlarge and proliferate to produce a clone of similar cells. These cells then become capable of producing large amounts of antibodies as well as multiplying as plasma cells.

"recognized" by lymphocytes that, from about the time of birth of the individual, have been "preprogrammed" for this kind of antigen. The recognition is based on the shape of the antigen molecule and is manifested by rapid proliferation of the recognizing lymphocytes. Each of them gives rise by mitosis to a *clone*, or a large number of identical cells, which then stop dividing and become transformed into *plasma cells*. Plasma cells synthesize and secrete large numbers of protein molecules called *antibodies*.

Antibodies synthesized by plasma cells are secreted into the blood. They are the circulating antibodies and include the well-known antibodies against the variety of infections that beset mankind. Another class of antibodies is associated with lymphocytes that originate in the thymus gland during early infancy. These antibodies are called *cell-mediated antibodies*. Their action appears to be directed against foreign molecules. Accordingly, it is the cell-mediated response that is important in the rejection reactions after tissue and organ grafts or transplants of skin, hearts, lungs, kidneys, and so on.

The thymus gland is unlike any other mammalian organ in that it is relatively large in infancy and gradually fades away to a few fibrous shreds in old age. The first clues to its importance came in 1960, when a method was devised for surgically removing the thymus gland from newborn mice. Such mice show serious disturbances in their immune responses, and detailed analyses of their various anomalies has proved highly fruitful to researchers. But for an aid to understanding the function of the thymus, it has been valuable to use what one scientist calls "experiments of nature"—genetic diseases that in one way or another involve the immune system. These observations have led to the same conclusions as a wide range of experiments in mice, chickens, rats, and other animals. It is clear that there are two major families of immunocytes. One set of cells is thymus-dependent; the ancestors of these cells have come from the thymus. The other immunocytes can be called thymus-independent for the present time, although eventually they will probably be found to be related to some other tissue that plays a role similar to that of the thymus.

Antibodies and antigens are thought to fit together as a lock and its key (Figure 11.20), which is reminiscent of the relationship thought to apply between enzymes and their substrates. This rela-

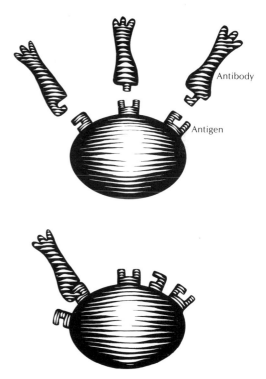

Antibody

Antigen

Figure 11.19   The "lock and key" hypothesis explaining the mechanism of immunity was proposed by the German bacteriologist Paul Ehrlich in the late 1800s. Blood cells with different "locks" will bind to specific antigen-antibody "keys," causing agglutination, or clumping, to occur. Thus, when the key fits the lock immunity is established.

Figure 11.20 Lock and key antigen-antibody reaction. Evidence suggests that an antibody molecule has two identical halves, each structured from one large and one small component. A particular antibody reacts with a particular antigen because the configuration of its combining site interlocks with that of the antigen. When the antigen and antibody make contact at the proper angle to bring the complementary patterns together, a union between antigen and antibody is formed, thus immobilizing the antigen.

Antigen   Antibody   Antigen

Antigen-Antibody Complex

Figure 11.21 One of the most important aspects of recent immunological research has been the recognition of the importance of the human thymus gland. In babies, it is a mass of actively multiplying lymphocytes lying just behind the top of the breastbone. Immunocytes produced directly by the thymus or the ancestors of cells produced by the thymus appear to be concerned with cell-mediated immune responses.

tionship is postulated as the basis of the specificity that characterizes both systems and explains why antibody X is ineffective against antigen Y, and so on. It also explains how an antibody that is specific for one antigen can react with another antigen that is structurally similar, yet slightly different. For example, the antibodies that are stimulated to form against the cowpox virus are effective against the closely related smallpox virus. Because cowpox is harmless to human beings, it is used to induce immunity against smallpox and is the basis of smallpox vaccination.

Despite the highly versatile and efficient immunity mechanisms that have evolved in the "higher" animals such as man, there are some situations in which they appear to be ineffective, others in which they get out of control, and still others in which they "backfire." Examples of situations in which they are apparently ineffective include many of the animal parasites that have evolved counter-antibody mechanisms of their own, which enable them to survive in the host's tissues. Such parasites include malarial and trypanosomal protozoans, filarian worms, liver flukes, and blood flukes, to name a few, which are endemic to certain areas of the

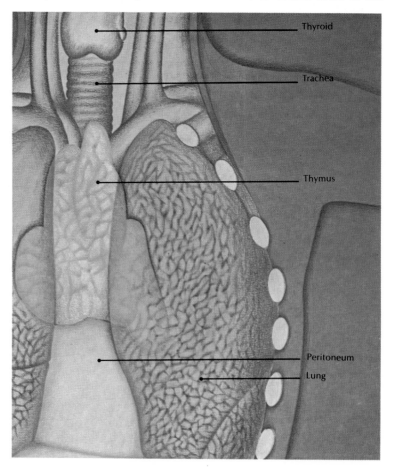

250

world and play prominent roles in the economy, politics, and social mores of entire nations.

Allergies are an example of an immune response that is somewhat out of control. One of the side effects of this immune response is the release of histamine and other chemicals from cells in the area adjacent to or surrounding the site of invasion by the foreign substance. These chemicals cause the fine arterioles to dilate, so that more blood is brought to the area. They also make the capillaries "leaky," so that the large serum proteins including the antibody proteins (serum globulins) can get into the tissues to combat the infection. In the case of an allergy, a person has become sensitized to some ordinarily innocuous substance in such a way that exposure to that substance elicits these secondary immune responses. The release of histamine in the skin in response to surface contact with such a substance causes localized edema, which we recognize as a rash or a welt or hives. If the offending substance is inhaled, the histamine is released in the tissues of the respiratory passages, causing their constriction and resultant interference with air flow — in other words, asthma. Although this allergy

Figure 11.22   Electron micrograph showing antigens that have become trapped on the branches of dendritic phagocytic cells within a lymph node. Also present are the nuclei of small lymphocytes.

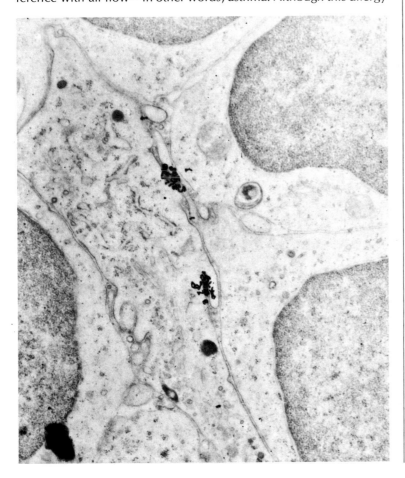

Figure 11.23   Both of these photographs show allergy skin tests in which numerous allergens have been injected under the skin in different areas. The area is "mapped" so that if a response appears, the doctor will know what substance the patient is allergic to. The most common allergens are pollens, animal dander, and dust, but there are many cases in which the allergen is milk, eggs, or a common chemical.

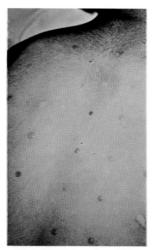

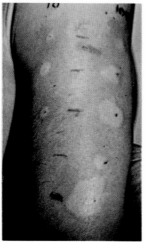

can be severe, the most severe reaction is one that brings about collapse of one of the major blood vessels, a drop in blood pressure, shock, and even death. This is called *anaphylactic shock.*

Examples of situations in which the immune response "backfires" are the so-called *autoimmune diseases,* such as hemolytic anemia. In hemolytic anemia, the individual's immunity system fails to recognize that its own blood cells are not foreign and produces antibodies against them.

Improved control of the immunity system is one of the major thrusts of applied biology. For example, attempts are being made to stimulate the production of antibodies against cancer. Another kind of control focuses on means by which antibody production can be suppressed or overridden. Some degree of success has been achieved here as well. Many kidney transplantations have been achieved, for example, and the rejection of transplants and grafts of other organs and tissues has been delayed. Surgeons and immunologists alike look forward to the day when they can control cancer as successfully as they can measles, smallpox, and polio, and make organ transplants not only between people but between different species of animals so that life may be prolonged.

TEMPERATURE REGULATION

Enzymes catalyze the biochemical reactions that make up the life processes of all organisms, but enzymes can function effectively only over a limited range of temperatures. In order to survive, organisms must somehow control their internal temperature, either by finding an environment with a suitable temperature range or making some physiological compromise with the environment. The body heat of living organisms comes from the oxidation of foods. Some of the energy obtained in this way is stored in the chemical bonds of ATP, and the rest is released as heat. In one sense, heat produced in metabolism is wasted energy because it cannot be used to synthesize new molecules for growth and maintenance. On the other hand, if used well, this heat can maintain the body temperature at a level above that of the environment.

All plants and animals, except birds and mammals, have little control over their body temperatures. Cold-blooded animals cannot regulate their body temperatures. Plants and cold-blooded animals that live in climates that are cold for a part of the year have two alternatives—they may live for only one season or they may become dormant in some fashion during the cold season. Bacteria and fungi can form spores, which are resistant to cold, and certain seed plants (called annuals) form seeds and die at the end of a growing season. When the environment becomes favorable again, the spores or seeds germinate, producing a new generation of organisms.

Birds and mammals, on the other hand, can regulate their body temperatures. In these warm-blooded animals, a portion of the brain, the hypothalamus, acts as a thermostat. Alterations in the external temperature are sensed in various ways. There are heat and cold receptors at various points on the body surface. The brain also responds to changes in the temperature of blood and sets various processes in motion to return the body temperature to normal.

If the body temperature rises, several mechanisms are used to dissipate heat. Blood vessels in the skin expand, increasing the amount of blood in a position to be cooled. Sweat glands secrete liquid, which spreads over the skin and evaporates, a physical process that removes heat. Animals that are covered with fur have sweat glands; they use evaporation of water from their tongues as a cooling mechanism. Panting moves air over the tongue and speeds evaporation.

If the body temperature falls, the thermoregulatory system tries to minimize heat loss by constricting the blood vessels in the skin. Extra heat is also produced by muscle movements such as shivering and increased activity. If a warm-blooded animal is unprotected in cold environmental temperatures for a prolonged period, it dies because of loss of body heat. Thus, there are many physiological, anatomical, and behavioral adaptations that allow plants and animals to live under all sorts of extreme conditions.

There is a surprising degree of similarity in the functions performed by all organisms, even though they represent a diversity of body structures. At the molecular or biochemical level, the basic structural components and activities are similar throughout most of the spectrum of living things, and in many ways these basic similarities are more striking than the differences.

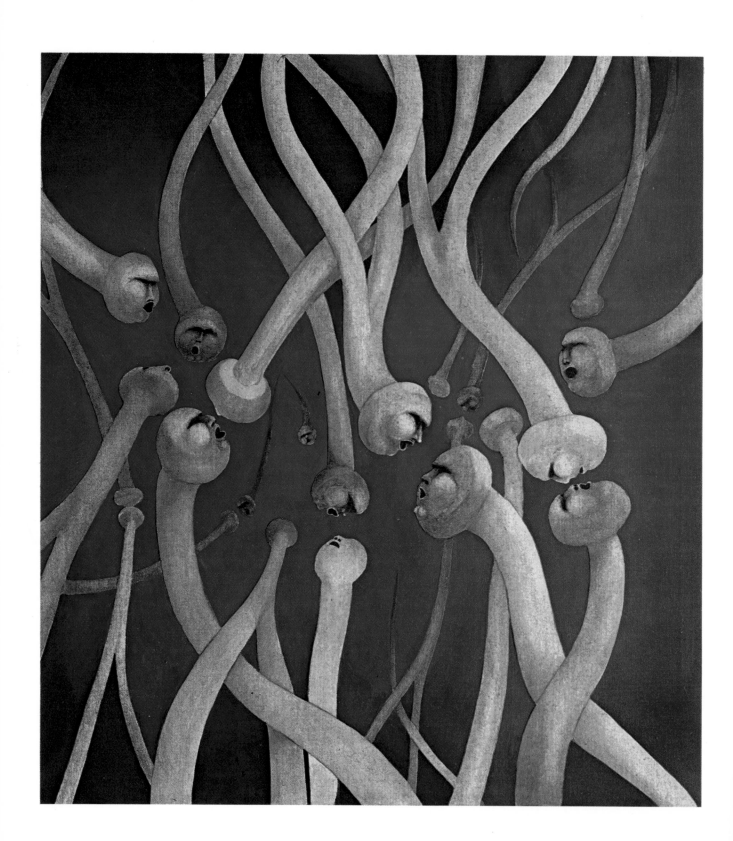

We, like all other animals, survive in part because we are able to gather information about our environment and about how we happen to fit into it. Perhaps the information we receive at any one moment has to do merely with an annoying change in temperature; perhaps it is of a more dynamic nature—gathering food, escaping from danger, bearing children. Whatever the situation, we carry within our body a complex nervous system that allows us to evaluate all such information and to coordinate responses appropriate to our well-being.

Our nervous system is organized into interconnecting groups and levels with almost unimaginable complexity. The gathering of information is performed by cells called *receptors* and *sensory neurons*. The coordinated activities are executed by organs and systems called *effectors*, which may be muscles or glands, and the effectors in their turn are controlled by *motor neurons*—cells again. The nervous system also contains an immense number of *interneurons*—neither sensory nor motor—whose role is to process the sensory input, evaluate it, and command the motor output. These cells perform an associative function, forming connective links between sensory and motor levels.

All this neuronal activity is required for the perception of even the simplest sensation; responses involving comparisons, value judgments, correlations with remembered experiences, aesthetic evaluations, and so on take much longer. The weaving of these patterns into a tapestry of response in space and time occurs in a complex way that C. Sherrington in 1906 likened to the workings of an "enchanted loom."

EVOLUTION OF THE NERVOUS SYSTEM

Even the simplest unicellular organism exhibits some form of *irritability*—the capacity to respond with action to certain kinds of stimuli (described in Chapter 2). An amoeba, for example, lacks apparent specialized structures for sensory reception or response, but it does show a regular behavior pattern. It responds to small food particles or certain chemicals with feeding behavior (phagocytosis) and responds to almost any other stimulus by withdrawing.

The simplest multicellular animals behave as if they are collections of relatively independent cells. Chemical signals spread slowly from one cell to another and, except where contact stimuli occur between touching cells, little communication or coordination takes place. In most multicellular animals, however, specialized cells of the nervous system speed communication among the cells of the body.

The coelenterates are the simplest multicellular animals that have any sort of collection of neurons to mediate the property of irritability. They have neither a head nor a brain, but they do have two "nets" of neurons to receive stimuli from equilibrium receptors and light receptors located around the periphery of their body

12

Nervous Systems

Figure 12.1   The nervous system of the fresh-water *Hydra* is best described as a collection of two-dimensional nerve nets; groups of neurons scattered over a surface whose processes (dendrites and axons) cross in a netlike fashion and intermingle. Note the absence of a centralized nerve tract and ganglia. The majority of *Hydra* behavior is limited to feeding and defensive contractile responses.

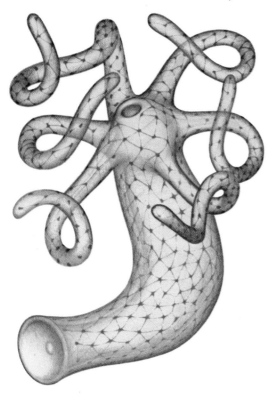

(Figure 12.1). These *nerve nets* direct the two kinds of muscular activity exhibited by a jellyfish: a rhythmic, rapid twitch that propels the animal, and localized, variable, and slow contractions associated with feeding and avoidance activities.

Most other animals that have highly differentiated, complex nervous systems are *bilaterally symmetrical*. Locomotion is usually in a direction such that the head leads, and most major sensory receptors are located on or about the head. Because most of these bilateral organisms are segmented, various elements of the nervous system tend to be repeated in each segment.

In bilaterally symmetrical organisms of increasing complexity, several trends in nervous system organization can be detected. First, there is a trend toward increased *cephalization*, or development of the size and complexity of the brain. Second, there is an increase in the number of interneurons, or neurons whose processes synapse only with other neurons and not with sensory or effector cells. Third, there is an increased variety of structurally different kinds of neurons and interstitial cells. Fourth, there is an increase in the variety and differentiation of synaptic regions within the brain.

In animals that are more complex, the brain has more neurons and a far more complex organization of subsystems than any other part of the nervous system.

There are two major anatomical patterns of organization. In the evolutionary branch containing the molluscs, annelids, and arthropods, the central nervous system is organized into *ganglia*, which are concentrated masses of nervous cell bodies. Each body segment often has one pair of ganglia. Each ganglion contains an outer ring of neuron cell bodies and an interior region where axons meet dendrites and nerve fiber tracts, which connect the various ganglia. In general, each ganglion controls half of one body segment, but in most species the ganglia of several anterior segments are fused to form the brain.

In the chordate branch, in which vertebrates (and man) are found, the primary organization consists of a continuous neural tube of nervous tissue extending the length of the organism. Neuron cell bodies, glia, and gray matter are mixed together in the wall of the tube; they also have ganglia in addition to the neural tube. Bundles of axons (or fiber tracts) are separate from the gray matter. They are called the white matter because of the glistening appearance of the myelin material formed by specialized cells.

In contrast to the somewhat limited behavioral powers of the earthworm, the octopus (a cephalopod mollusc) is better at learning than many vertebrates. Octopi have the greatest degree of cephalization of any mollusc, and perhaps of any invertebrate. Their brains are structurally similar to those of mammals in terms of organization and complexity. These creatures can be taught to discriminate between objects on the basis of touch or sight. Their

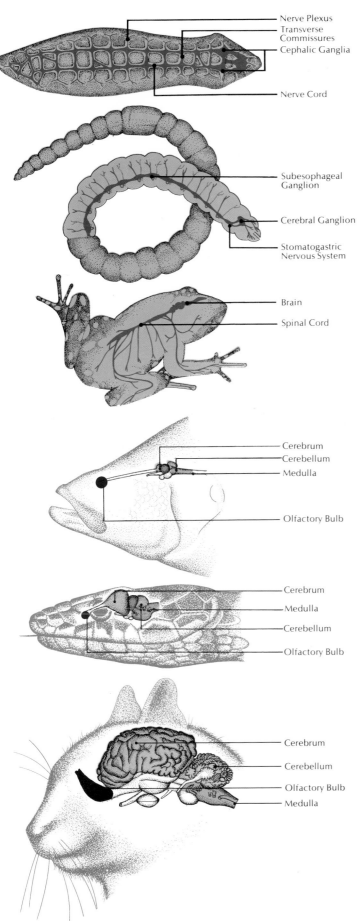

Nerve Plexus
Transverse Commissures
Cephalic Ganglia

Nerve Cord

Subesophageal Ganglion

Cerebral Ganglion

Stomatogastric Nervous System

Brain

Spinal Cord

Cerebrum
Cerebellum
Medulla

Olfactory Bulb

Cerebrum
Medulla
Cerebellum
Olfactory Bulb

Cerebrum
Cerebellum
Olfactory Bulb
Medulla

Figure 12.2   Diagrams of representative invertebrate and vertebrate animals selected to portray the phylogenetic progression in the evolution of nervous systems. The top diagram shows the bilateral ladderlike nervous system of a triclad flatworm. Note the beginning of cephalization, or brain development, as denoted by the cephalic ganglia. In the middle diagram, the ventral solid nerve tract of the earthworm is shown with its segmental arrangement of ganglia. In addition to the central nervous system, it has a subepidermal nerve net, but this net apparently functions only in localized and minor responses. Locomotion is effected by waves of contraction that pass along each segment of the body. In the lower diagram, the primitive vertebrate nervous system is represented by an amphibian. The central nervous system of all chordates (including amphibians) has a dorsal, hollow, fluid-filled nerve cord.

Figure 12.3   Diagrams showing comparative mapping of sensory and motor areas of the cerebral cortex in various representative vertebrate chordates. In mammals, each sensory system is connected to a specific region of the cortex, and each motor system arises from another specific cortical region. In addition to the sensory and motor regions of the cortex, there are association regions that have often been assumed to be involved in higher intellectual activities. From the fish diagram downward, note the progressive enlargement of association areas. In higher mammals, the size of the cortex is greater and the amount of association cortex much larger than in lower mammals. The relative positions of the various motor and sensory regions are much the same in all mammalian species.

Figure 12.4 (left)   In an effort to determine the process by which a single cell "learns," the gill-withdrawal reflex was studied in the aplysia, a snaillike marine animal. In this illustration, normal gill position is shown above (colored area). A gentle touching on the mantle shelf causes it to withdraw the gill (below). After repeated presentations of the stimulus, the animal becomes habituated to it, and responds less frequently.
From "Nerve Cell and Behavior," Erik Kandel. © 1970 by Scientific American, Inc. All rights reserved

Figure 12.5 (right)   Free-living planarians (flatworms) can be trained in both classical conditioning experiments and in simple maze problems. In the top two photographs, the planarian is in a normal extended position during locomotion. In the lower photographs, the planarian shows a response to an unconditioned stimulus, an electric shock. Worms trained using a variety of experimental regimens have also been used in the study of memory transfer and the chemical theory of learning.

Tactile
Stimulus

patterns of behavior include both stereotyped and plastic elements. It is in vertebrates, however, that the greatest plasticity of behavior occurs.

NERVE CELLS IN VERTEBRATES

The nervous system in vertebrate animals is composed of the brain (encased in a protective body skull), the spinal cord, and a variety of peripheral nerves. Its primary function is to process information, a task that it accomplishes when individual nerve cells respond to stimuli and transmit signals through adjacent nerve cells.

The human nervous system contains billions of individual cells. Between the inside and outside of each cell, there is normally a difference in electric potential of a fraction of a volt. This *resting potential* is maintained by cellular processes that pump charged ions through the cell membrane. When the cell is appropriately stimulated, localized changes in the membrane bring about a shift in potential: the nerve cell discharges, or *fires*, and the localized change in potential (called a "spike") sweeps along the entire

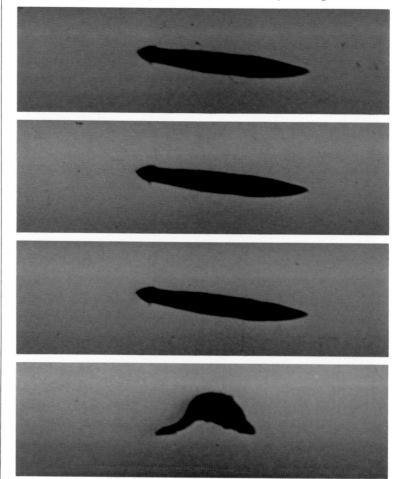

Figure 12.6   This photomicrograph of two neuroglia (or glial cells) among larger nerve cells shows how the neuroglia and their extensions are complexly interwoven with neurons. Note their attachment to the walls of the blood vessel (upper righthand corner). Although many researchers are studying them, the functions of the neuroglia are as yet undetermined.

length of the cell membrane. Under certain conditions, the firing of one nerve cell alters the resting potential of adjacent nerve cells and causes them to fire. In this fashion, patterns of electric excitation move from one part of the nervous system to another.

There are two kinds of cells in the nervous system of the vertebrate organism: *neurons* and *neuroglia*, or glial cells (one type of interstitial cell described in Chapter 10). The neurons are the specialized cells that transmit electric activity. The functions of the glial cells are not clear. They are usually thought to play a supportive or nourishing role, or they may store specialized molecules. In general, glial cells have extensions that weave around neurons and are often attached to the walls of blood vessels, as shown in the photomicrograph in Figure 12.6.

The structures of two typical neurons are depicted in Figure 12.7. As you learned in Chapter 10, the neuron has three basic parts: the *soma* (cell body); the *dendrites* (the structures that conduct signals toward the cell body); and the *axons* (the long conductive structures that carry signals away from the cell body). There are many different types of neurons and many ways of classifying them—according to the shape and size of the cell body, the presence or absence of specialized dendritic processes, the length of the axons, and the presence or absence of the *myelin sheath*. Despite these variations, however, some general points can be made about the electric activity of neurons and about the associated chemical events.

The Electric Activity of Neurons

If a small glass tube or pipette, drawn to a very fine tip, is filled with potassium chloride (a conducting fluid) and inserted through the cell membrane and into an axon, it is possible to record a dif-

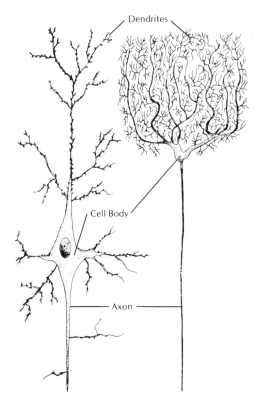

Figure 12.7   Diagram showing two types of neurons found in human beings. A typical neuron is composed of a cell body (or soma); dendrites (fibers that conduct impulses toward the soma); and axons (fibers that conduct impulses away from the soma).

Figure 12.8 Photomicrograph of a motor neuron with its radiating dendrites and axon. Superimposed on the photograph is a drawing of a microelectrode—a glass pipette filled with conducting solution. This glass microelectrode has been enlarged five times but is shown as it would be located for intracellular recording.

ference in the electric potentials between the outside of the cell and the fluid in the axon (Figure 12.8). In a typical inactive nerve cell, this potential difference might be on the order of -70 millivolts (about 5 percent as much energy as a flashlight battery generates), with the inside of the neuron carrying a negative charge relative to the outside of the neuron. The resting potential of such a cell, then, is -70 millivolts.

If the neuron is excited with appropriate chemical or electric stimuli, its potential changes. First, there is a positive shift in voltage, which produces a decrease in the potential difference recorded across the cell membrane. When voltage has decreased to a critical level (the *threshold value*), an *action potential* is generated. The action potential is called a *spike* because at a given point in a neuron it rises to a peak and then declines. It is a rapid change in voltage that moves along the neuron like the crest of a wave moving along the surface of a lake. The change results in a brief reversal of the resting-potential condition. In other words, the inside of the membrane momentarily becomes positively charged with respect to the outside. Then the resting potential is restored. The sequence of changes in electric potential as the cell fires is depicted in Figure 12.9.

Accompanying this sequence of changes in voltage is a parallel sequence of changes in the excitability of the neuron. Immediately following an action potential, there is a very brief *absolute refractory period* during which another action potential cannot be generated, no matter how strong the stimulus. Then there is a brief *relative refractory period* during which a stronger than usual stimulus is required to evoke a second spike.

There are two key characteristics of action potentials. First, they are "*all-or-none.*" If the threshold value is reached, the cell fires; the size of the action potential depends on the cell, not on the magnitude of the original stimulus. The second characteristic is that action potentials are *nondecremental*; it does not decrease in size as it travels from origin to destination. As an action potential generated near the cell body travels to the farthest ends of the axon, it neither fades nor dies out.

The speed with which the action potential travels varies from one neuron to the next, with a minimum velocity of about 10 feet per second and a maximum of 390 feet per second. Generally, the larger the diameter of the nerve fiber, the faster the conduction. Large fibers also tend to have lower thresholds and shorter refractory periods than small ones. In other words, large fibers fire more easily, conduct faster, and recover faster.

Although the magnitude of a stimulus does not affect the size of a cell's action potential, it does affect the rate at which the cell fires. Figure 12.10 shows the effect of an increasingly strong stimulus on the firing rate of a neuron. A stronger stimulus produces a more rapid rate of action potentials by reactivating the cell earlier

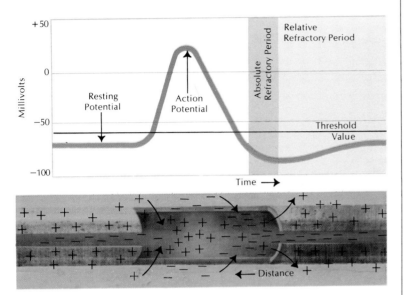

in the relative refractory period. In addition, because different neurons have different thresholds of excitation, an intense stimulus causes more nerve cells to fire than a weak one. These two factors—how many cells are firing and how rapidly—are central to the transmission of information from one part of the nervous system to another.

So far, we have been talking about action potentials within a single nerve cell. How is information transmitted from cell to cell?

The typical vertebrate neuron, when it fires, releases certain chemicals from the end of its axon into the tiny gap (only a few hundred angstroms wide) between its axon and the dendrites of a neighboring cell. These chemicals, called *transmitter substances*, diffuse across the gap and induce changes in the membrane of the receiving cell that tend to depolarize it and produce an action potential. There is a meeting point of the axonal ending of one cell and the membrane that surrounds the dendrites or the nerve cell body of the next neuron across the intervening gap. This meeting point is known as the *synapse*, or *synaptic junction* (Figure 12.12).

Figure 12.10 The occurrence of an action potential is immediately followed by a very brief period during which a spike cannot be generated regardless of the strength of the stimulus. This absolute refractory period can range from 4/10 to 1 millisecond, and it is followed by the longer relative refractory period during which a stronger-than-usual stimulus is required to evoke an action potential. To the right are sample recordings of the activity of a single neuron under varying levels of stimulation. The vertical lines represent action potentials.

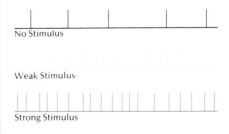

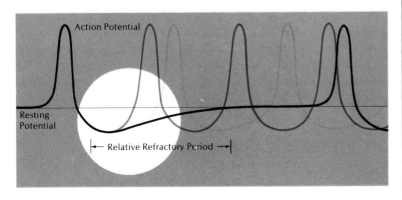

Figure 12.11   Scanning electron micrograph showing several axons converging upon a single cell. By means of such complex synaptic connections, one postsynaptic cell can integrate large amounts of information.

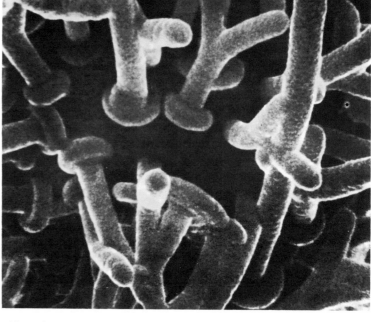

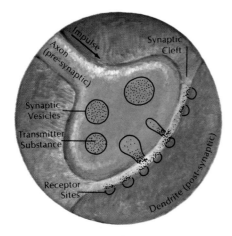

Figure 12.12   Schematic diagram of a synaptic junction. The slightly bulged axonal ending contains vesicles, or packets, that may serve to store chemical transmitters before they are released across the synapse. Specialization may also exist on the postsynaptic side of the synapse: note that there is a change in the small segment of the postsynaptic cell membrane immediately opposite the presynaptic axonal ending.

Although evidence indicating the existence of synapses has been accumulating since the turn of the century, it was only with the development of the electron microscope that synapses could actually be seen.

The release of some chemical transmitter substances excites a postsynaptic cell (the cell whose dendrites receive substances across the synapse) and leads to production of an action potential. Other substances (as yet largely unidentified) *inhibit* the discharge of the cell instead of exciting it. In other words, they reduce its rate of discharge. If enough of the inhibitory chemical is present, the cell may not fire at all for a period of time.

Throughout the central nervous system, inhibitory synapses counteract the generation of impulses by excitatory synapses. At every synapse of the central nervous system that has been thoroughly investigated, there is a conflict of excitatory and inhibitory action on a single neuron. Apparently, few excitatory synapses have the unchallenged power to excite a nerve impulse in the postsynaptic cell. If there were no inhibitory synapses, a single impulse might cause an explosive spread of excitation throughout the neuronal networks of the nervous system—in other words, convulsions such as those that occur in epilepsy.

A great many axon terminals converge on most individual neurons, and each terminal can release a quantity of transmitter substance that contributes to excitation or inhibition of the cell. In general, then, frequency of discharge for a particular cell at a particular time depends on (1) the relative quantities of excitatory and inhibitory transmitter substances acting on the membrane of the postsynaptic cell; and (2) the intrinsic excitability of the cell (how recently it has discharged, its metabolic status, and so on).

Neurochemical Considerations

Cellular chemistry must be considered in a discussion of information processing in the nervous system because it is basic both to the resting and action potentials of a cell and to the mechanisms of transmission at synaptic junctions.

The resting potential results from the fact that there are unequal amounts of certain chemicals on the two sides of the cell membrane. This distribution can occur because certain substances, when dissolved in water, interact with water molecules and dissociate into atoms with an excess or deficiency of electrons. When common table salt (NaCl), for example, is dissolved in water, it separates into its constituent atoms (Na^+ and Cl^-). The sodium (Na) loses an electron in the process and is left with a net positive charge; the chlorine (Cl) gains an electron and acquires a net negative charge. These charged particles are known as *ions*.

The main ions involved in the resting and action potentials of nerve cells are sodium, potassium, chlorine, and some large organic molecules (confined to the interior of the cell) that carry a negative charge. In the resting cell, there is a net excess of negatively charged ions on the inside of the cell membrane, which results in the recorded resting potential (such as that shown earlier, in Figure 12.9). During the generation of an action potential, there is a rapid influx of sodium ions. These positively charged ions momentarily cause the interior of the membrane to become positively charged with respect to the exterior of the cell. Electric currents created by the localized influx cause neighboring areas of the membrane to depolarize, so that the action potential moves along the axon.

When the action potential reaches the end of the axon, a transmitter substance is released into the synaptic junction, as was described earlier. A variety of transmitter substances function in the

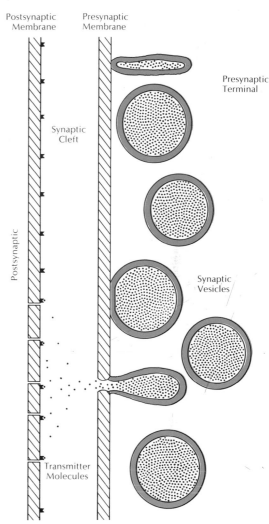

Figure 12.13   Schematic diagram representing a portion of a synaptic cleft with synaptic vesicles in close proximity in the presynaptic terminal. Note the vesicles discharging transmitter molecules into the synaptic cleft. Some transmitter molecules are shown combined with receptor sites on the postsynaptic membrane, thus facilitating membrane pore dilation.

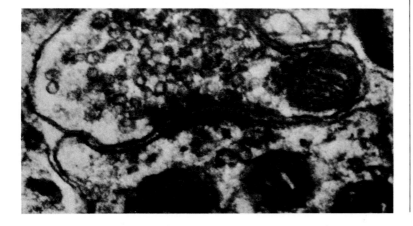

Figure 12.14.   Electron micrograph of a synaptic knob containing small round bodies called synaptic vesicles. The large round bodies are mitochondria.

Figure 12.15 The spinal cord is a relatively simple neural system that receives and processes information from sensory receptors and then delivers appropriate impulses to stimulate actions by effectors. It serves as an interface between the brain and the effectors and between sense organs and the brain. The spinal cord is organized into an inner region of gray matter, where the integration of sensory input and motor output occurs, and a marginal region of tracts that carry information between segments of the cord and to or from the brain. Input arrives over axons in the dorsal roots, and output leaves via the ventral roots. There is a pair of dorsal roots and a pair of ventral roots for each spinal segment. Several segments may carry the information to and from each limb.

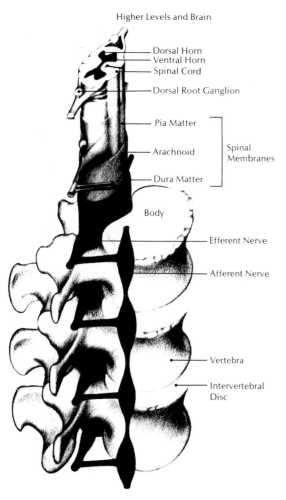

Higher Levels and Brain

Dorsal Horn
Ventral Horn
Spinal Cord

Dorsal Root Ganglion

Pia Matter
Arachnoid — Spinal Membranes
Dura Matter

Body

Efferent Nerve

Afferent Nerve

Vertebra

Intervertebral Disc

activity of the vertebrate nervous system, but it is believed that a particular neuron releases only one of them. A common and much studied transmitter substance is *acetylcholine* (ACh). This substance operates in the parasympathetic nervous system (discussed later in the chapter), probably within parts of the brain and spinal cord, and at junctions of neurons and muscles.

In addition to the transmitter substance released by a neuron, other chemical compounds are present near the axonal endings. They inactivate the transmitter by transforming it chemically. If inactivation did not occur, transmitter substances would accumulate indefinitely in the region of the postsynaptic membrane, resulting in the continuous stimulation of the cell and a blurring of message transmission. Together, the release of a small quantity of transmitter substance and the rapid inactivation of the transmitter by local chemical compounds permit the transmission of discrete messages across synaptic junctions.

Many drugs that are commonly used as stimulants or depressants act at the synapses in the central nervous system. The exact nature of their effect cannot be known because not all of the separate connections in the brain nor all of the transmitter substances are known. Most common nonprescription sleeping pills contain scopolamine, which acts to depress transmission at synaptic junctions where the release of ACh is the critical event. The drug acts on synapses in parts of the nervous system and may produce such side effects as blurred vision and increased heart rate.

THE VERTEBRATE NERVOUS SYSTEM

The vertebrate nervous system is composed of two basic subdivisions. The first, the *central nervous system*, consists of the brain and spinal cord. The second is the *peripheral nervous system*, whose neurons work to carry sensory information from outside the body to the central nervous system and convey messages that activate glands and muscles.

Another useful way of dividing the vertebrate nervous system involves consideration of two major parts: the *somatic nervous system*, which is involved in the control of *voluntary* responses to stimuli; and the *autonomic nervous system*, which regulates *involuntary* activities. Both of these systems include parts of the central and peripheral nervous systems, although these divisions are most frequently used in discussion of the peripheral nervous system. The autonomic nervous system will be explored in more detail later in the chapter.

The Spinal Cord

The spinal cord is of particular interest for two reasons. First, it controls many reflexes that are crucial to the survival of an organism. Secondly, it constitutes a simplified model (simplified, that is, in comparison to the brain) of a neurological system that receives information, processes it, and then delivers impulses to the muscles for the initiation and integration of motor sequences. A diagram of the spinal cord is shown in Figure 12.15.

The simplest spinal reflex arc involves just two neurons. An *afferent* sensory neuron conveys information about stimulation *from* a receptor at the periphery to the spinal cord. An *efferent* motor neuron carries an activating impulse from the spinal cord *to* the muscles. The knee jerk elicited by a tap below the kneecap is the result of the simultaneous activation of a number of such two-neuron arcs. Two-neuron arcs are relatively uncommon in our functioning spinal reflexes, however. In most cases, one or more interneurons intervenes between the afferent and efferent neurons. Their participation greatly increases the possibilities for complex reflex activity.

Inhibitory and excitatory fibers play a central role in reflex activity. In the case of the knee jerk, at the same time that the muscles on the front of the leg are being excited and are contracting, the motor nerves of the muscles on the back of the leg are being inhibited from discharging action potentials. That is one of the reasons that several two-neuron arcs make up the reflex: one for excitation of one set of muscles, and one for inhibition of the other. This inhibition prevents the simultaneous contraction of the two opposing muscle groups, ensuring a smooth, coordinated extension of the lower leg (the word "jerk" is something of a misnomer). This coordination of motor movements, involving the excitation of some motor neurons and the simultaneous inhibition of others, is the key characteristic of spinal reflex regulation of behavior.

In a reflex, muscle activation occurs before nerve impulses have traveled up the spinal cord to the brain. Conscious awareness of stimuli and responses comes only after the reflex action has been triggered. The spinal cord is responsible not only for reflexes but for translation of commands from the brain for movement into coordinated contraction and relaxation of different muscle groups. As a consequence of both anatomical and physiological studies of the spinal cord, surgeons can detect disease or damage in this organ.

The Brain

The vertebrate brain is divided into three regions: the forebrain, the midbrain, and the hindbrain. The forebrain consists of the two cerebral hemispheres, including such interior structures as the thalamus and hypothalamus. The hindbrain includes the medulla, pons, and cerebellum. The midbrain is the intermediate zone between the forebrain and hindbrain. Figure 12.17 shows the left and right cerebral hemispheres.

The surface of the cerebral hemispheres is called the *cortex*. The human cortex is intricately convoluted, and its indentations, known as *fissures*, divide the cortex into a complex set of folds.

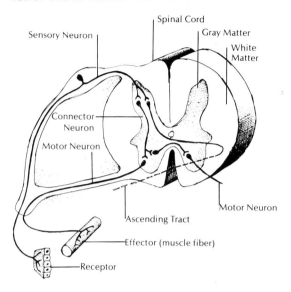

Figure 12.16   This diagram shows a simple type of neural pathway (involving three neurons) in human beings—the reflex arc. The simplest reflex arc involves only two neurons: an afferent sensory neuron and an efferent motor neuron.

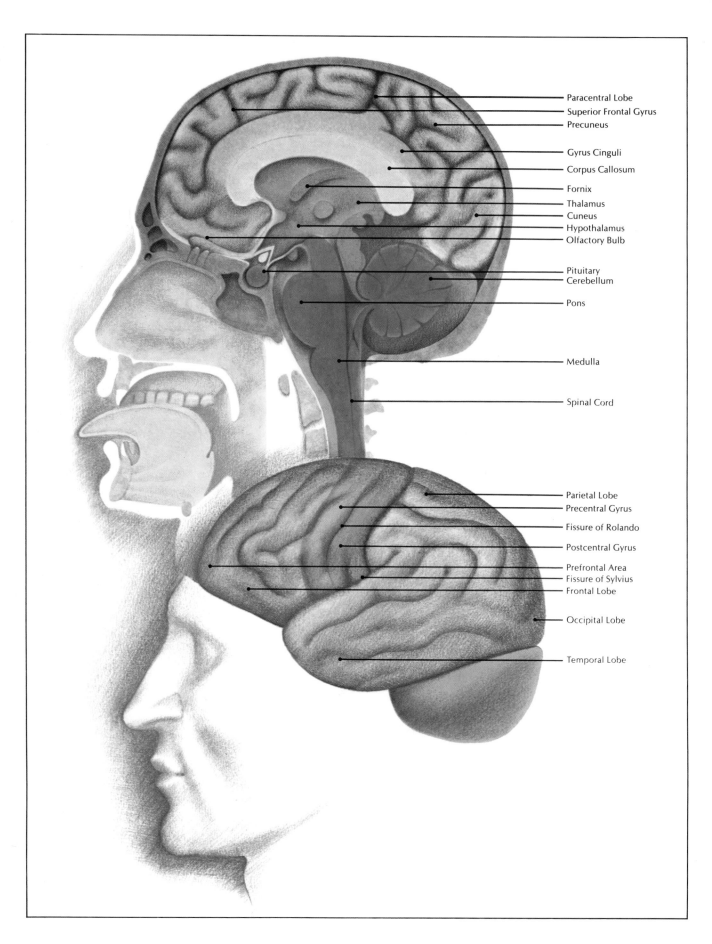

Paracentral Lobe
Superior Frontal Gyrus
Precuneus

Gyrus Cinguli
Corpus Callosum

Fornix
Thalamus
Cuneus
Hypothalamus
Olfactory Bulb

Pituitary
Cerebellum

Pons

Medulla

Spinal Cord

Parietal Lobe
Precentral Gyrus

Fissure of Rolando

Postcentral Gyrus

Prefrontal Area
Fissure of Sylvius
Frontal Lobe

Occipital Lobe

Temporal Lobe

Figure 12.17 (opposite page)   This view of the human brain in sagittal section illustrates the relationships between the forebrain—consisting of the cerebral hemispheres, the thalamus, and the hypothalamus; the hindbrain—consisting of the medulla, the pons, and the cerebellum; and the intermediately placed midbrain. The intricately convoluted outer surface of the human brain is composed of the cerebral hemispheres. The degree of convolution is directly proportional to the effective surface area of the cortex, which is thought to be related to information processing capacity. Rodents and primitive mammals have a relatively smooth cortical surface, whereas primates and marine mammals have the most convolutions.

Not all vertebrates have intensively convoluted cortexes. The surface of the rat brain, for example, is virtually smooth, except for a large fissure that separates the two hemispheres and another fissure that can be seen on the side of the hemispheres.

There are many different types of cortex, classified according to the kinds of cells they contain and the way these cells are distributed. A highly developed cortex is largely a mammalian characteristic. Fish and amphibia have only the most primitive cortical tissue; the reptiles represent a transitional group. Although simple learning can take place in animals with no cortical tissue, extensive studies in comparative learning ability show that cortical development and intellectual capacity tend to correlate.

The basic divisions of the human cortex are the four *lobes* of each hemisphere. The *occipital lobe*, located at the back of the brain, is particularly concerned with the reception and analysis of visual information. Injury to this portion of the cortex can produce blindness in part of the visual field.

The *temporal lobes* are located on the side of each hemisphere; their boundary is the lateral fissure of Sylvius. The auditory reception areas are located in the temporal lobes, as are certain areas for the processing of visual information. In a series of studies with epileptic patients undergoing surgery, the Canadian neurologist W. Penfield elicited some dramatic responses by applying electric stimulation to certain points on the temporal lobe. Stimulation at some points caused complex auditory or visual illusions; stimulation at other points seemed to reactivate sensations with such vividness that the patient felt as if he was reliving the experience rather than merely remembering it (Figure 12.18).

In the *frontal lobe* at the front of each hemisphere, the area next to the fissure of Rolando (Figure 12.17) is primarily concerned with the regulation of fine voluntary movements. Another part of the frontal lobe has to do with the use of language. In 1861, P. Broca, a French physician, discovered that damage in this part of the dominant hemisphere (the left, in most people) affected the ability to use speech. Broca's discovery was the first indication that different parts of the brain might control different types of behavior—that is, that the brain might show *localization of function*. Since Broca's time, many parts of the brain have been found to be associated with particular functions.

The prefrontal area of the frontal lobe (Figure 12.17), although not associated with particular sensory or motor activities, has at times been considered the repository of intellectual ability or emotional control. The belief that it controlled intellectual ability undoubtedly sprang from the fact that this region is much larger in animals exhibiting more advanced levels of organization than in lower species. Most modern researchers would not accept the view that the frontal lobe regulates intellectual performance; its involvement is more subtle than that. Some people do as well on

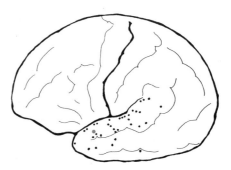

Figure 12.18   In a series of experiments with epileptic patients undergoing surgery, stimulation at the point on the temporal lobe indicated by the red dot caused a patient to say, "Dream is starting —there are a lot of people in the living room—one of them is my mother."
After Penfield and Roberts, 1946

intelligence tests even after massive removal of frontal lobe tissue as they did before. The defects in behavior resulting from frontal lobe lesions, it now appears, involve the ability to order stimuli and sort out information rather than general intellectual ability.

The fourth pair of lobes is the *parietal lobes*, which contain the primary receiving areas for the skin senses and our sense of bodily position. Damage to the somesthetic cortex produces deficits in the sense of touch. A person with extensive parietal lobe damage extending beyond the primary sensory area also has unusual difficulty with spatial organization of the environment and has a distorted perception of his own body image.

Somesthetic sensations from the various regions of the body are represented in an orderly fashion in the postcentral fold, as shown on the left side of Figure 12.19. The right side of that figure shows a similar arrangement of body regions in the precentral fold, which controls fine muscular movements. There are several supplementary areas for both skin reception and motor activity, but these two can be used to illustrate two important facts. First, both the motor

Figure 12.19   The relative proportions of body region representation in the motor cortex (anterior to the central fissure) and in the somesthetic cortex (the post central gyrus).

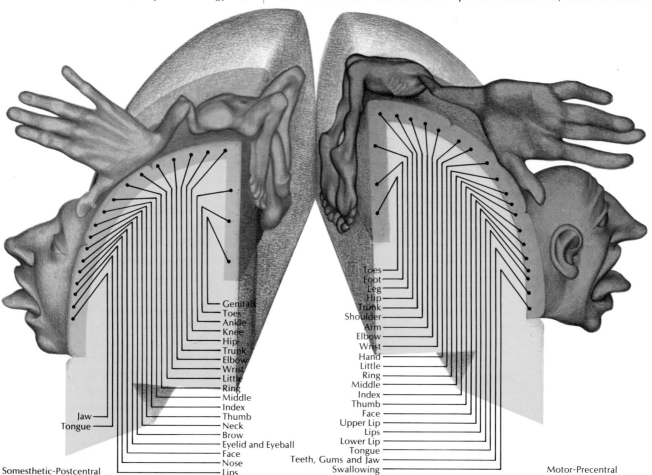

Somesthetic-Postcentral

Genitals
Toes
Ankle
Knee
Hip
Trunk
Elbow
Wrist
Little
Ring
Middle
Index
Thumb
Neck
Brow
Eyelid and Eyeball
Face
Nose
Lips

Jaw
Tongue

Toes
Foot
Leg
Hip
Trunk
Shoulder
Arm
Elbow
Wrist
Hand
Little
Ring
Middle
Index
Thumb
Face
Upper Lip
Lips
Lower Lip
Tongue
Teeth, Gums and Jaw
Swallowing

Motor-Precentral

cortex and the somesthetic cortex show *contralateral control.* In other words, areas in the right hemisphere are concerned with activities and reception on the left side of the body, whereas areas in the left hemisphere control and represent activities in the right side of the body.

Another parallel between motor and sensory representation, illustrated by the drawings of body regions in Figure 12.19, is that the amount of cortex taken up with the motor activities and sensory impressions from different parts of the body is associated with the degree of precise motor control or the sensitivity of the part of the body concerned, not by its size or the mass of its muscle groups. The fingers have much larger representation than the trunk on the motor cortex, for example, and the lips are disproportionately represented on the somesthetic cortex.

The two hemispheres are apparently mirror images of each other, but there must be chemical or structural differences between them in human beings. These differences are necessary to account for handedness and for the localization in one hemisphere of certain functions relating to speech, perception, and intelligence-test performance. Right-handed people have their primary speech center in the left hemisphere, and damage to the left temporal lobe impairs intelligence-test performance severely but leaves perceptual-test performance relatively intact. Damage to the right temporal lobe disrupts perceptual performance more than intellectual performance.

Although there are many routes of communication between the two hemispheres, the primary band of connecting fibers is the *corpus callosum.* Until recently, the functions of the corpus callosum were virtually unknown, but there are some clear indications that this massive group of fibers helps synchronize the activity of the two hemispheres. For example, cats trained to make a visual discrimination on the basis of stimuli that are confined to one hemisphere immediately transfer their discriminatory ability to the opposite hemisphere unless the callosum has been severed. If the callosum is cut prior to learning, the second hemisphere may need total retraining on the discrimination task.

The *thalamus* is the "great relay station" of the human brain. It is composed of a set of nuclei that lie beneath the corpus callosum. Each of the nuclei is a closely packed cluster of nerve cells, which presumably have common functions and connections. In the thalamus the nuclei connect various portions of the brain. Some relay sensory information from the peripheral parts of the sensory system to particular regions of the cortex. Others are involved with connections between different cortical areas or between cortical and subcortical parts of the brain.

The *hypothalamus,* located near the base of the brain, is also composed of nuclei. This structure, although relatively small in the human brain, has enormous importance for the regulation of verte-

Figure 12.20   Control of the involuntary functions of the body by centers in the central nervous system is accomplished by the autonomic nervous system, with its sympathetic and parasympathetic divisions. Reciprocal innervation of most organs by both divisions is the rule. Regulation of body maintenance activity is the primary function of this system.

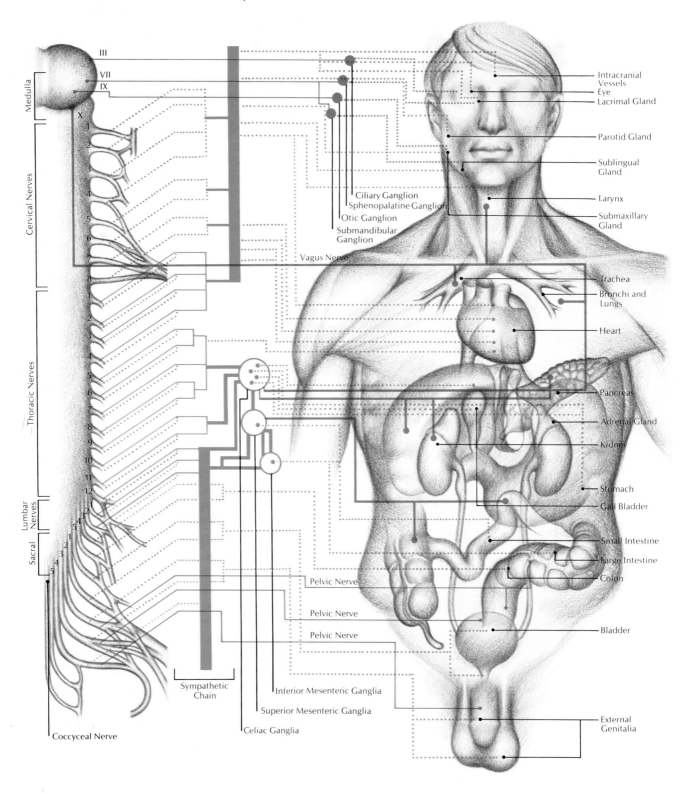

Medulla

III
VII
IX
X

Cervical Nerves

1
2
3
4
5
6
7
8

Thoracic Nerves

1
2
3
4
5
6
7
8
9
10
11
12

Lumbar Nerves

1
2
3
4
5

Sacral

1
2
3
4
5

Coccyceal Nerve

Sympathetic Chain

Celiac Ganglia

Superior Mesenteric Ganglia

Inferior Mesenteric Ganglia

Pelvic Nerve

Pelvic Nerve

Pelvic Nerve

Vagus Nerve

Ciliary Ganglion
Sphenopalatine Ganglion
Otic Ganglion
Submandibular Ganglion

Intracranial Vessels
Eye
Lacrimal Gland
Parotid Gland
Sublingual Gland
Larynx
Submaxillary Gland
Trachea
Bronchi and Lungs
Heart
Pancreas
Adrenal Gland
Kidney
Stomach
Gall Bladder
Small Intestine
Large Intestine
Colon
Bladder
External Genitalia

270

brate behavior. It helps control the internal environment of the body (blood pressure, heart rate, temperature, and so on); it regulates the activity of the pituitary gland; and it controls many basic drives, such as hunger, thirst, and sex.

The hypothalamus controls the so-called involuntary functions of the body through the autonomic nervous system. The autonomic nervous system consists of two divisions: the *sympathetic* and the *parasympathetic* (Figure 12.20). The control centers for these two divisions lie in the brain and in the spinal cord, and every vital organ is connected to both parts. The way that vital organs function is determined by the balance between the two parts, for both send impulses to every organ. The sympathetic system works to speed up body processes; the parasympathetic system works to slow them down. The rate at which impulses arrive from either system determines which system predominates.

In general, the sympathetic and parasympathetic systems can be viewed as working antagonistically. The sympathetic system tends to promote energy expenditure. For example, it mobilizes the body

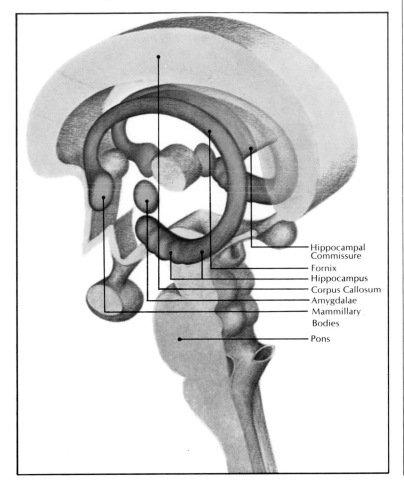

Hippocampal Commissure
Fornix
Hippocampus
Corpus Callosum
Amygdalae
Mammillary Bodies
Pons

Figure 12.21   The structure of the limbic system, which is located deep in the forebrain. In conjunction with the hypothalamus, this system can exert profound control over emotional behavior. Some limbic structures related to the hippocampus have been associated with complex memory function.

Figure 12.22   The reticular activating system serves to inhibit or facilitate the passage of impulses from the sense organs to the cortex. Thus, it acts as a sensory filter and exercises general control over an animal's behavioral state. If an animal suffers damage to the midbrain portion of this system, it will sleep much more than normal and will show a cortical brain wave pattern characteristic of stages of deep sleep.

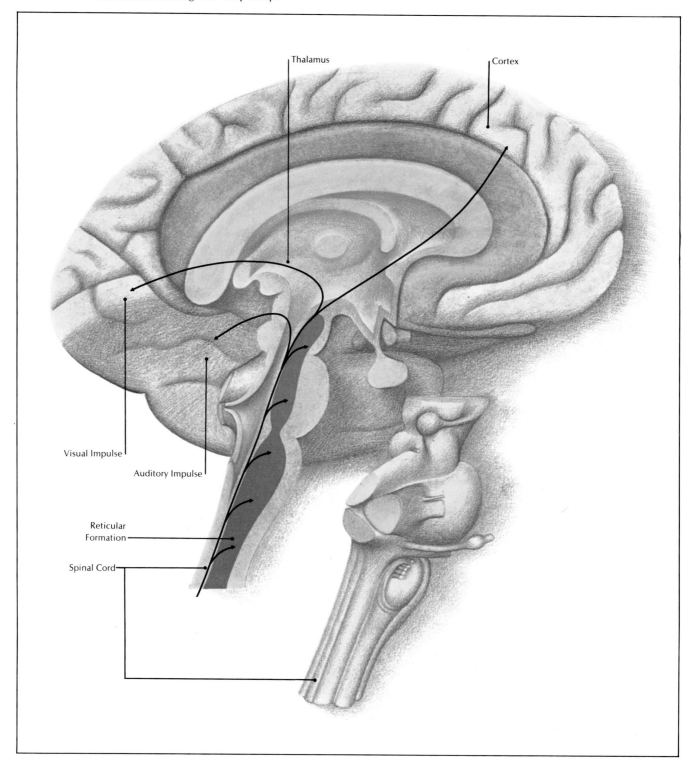

Thalamus

Cortex

Visual Impulse

Auditory Impulse

Reticular
Formation

Spinal Cord

to meet emergency conditions by increasing blood-sugar levels and heart rate, dilating the blood vessels through which oxygen and nutrients reach the skeletal musculature, and inhibiting digestive processes. The parasympathetic system is primarily an energy-conserving system. For example, it slows heart rate and enhances digestive activity.

Two other parts of the forebrain that are important in emotional arousal are the limbic system and the reticular activating system. The *limbic system*, shown in Figure 12.21, is a group of interconnected structures that includes the amygdala, cingulate cortex, hippocampus, and septal area. Damage to these areas has profound effects on emotional reactivity, even to the point of turning usually excitable creatures such as monkeys and mountain lions into docile pets or converting tame laboratory rats into ferocious combatants (at least for a limited time). These highly complicated structures also have other functions; in man, for example, the hippocampus has been implicated in memory storage processes.

Below the forebrain, the central core of the midbrain, and the upper hindbrain is a latticework of nerve cells known as the *reticular activating system*. This primitive system, whose activities are diagrammed in Figure 12.22, exists in the brains of all vertebrate species. It helps direct an organism's attention to incoming sensory stimuli and regulate its level of alertness.

After Charles Darwin published *On the Origin of Species* in 1859, man's conception of himself underwent a profound change. No longer could he regard himself as a special creature, apart from all animals. Instead there was the unsettling contention that he was a part of the animal kingdom—a species that had evolved by natural selection in the same way that all animal species had evolved. To be sure, man was unique in some ways, but at the same time he was similar in many, many ways to other species—especially apes. The stage was set for a controversy that would not end until well into the next century. Eventually most of the issues relating to the theory of natural selection would be resolved, partly on the basis of the findings of the emergent science of *ethology*, a biological approach to the study of the behavior that is characteristic of a particular species of animal. Through ethological studies, one of Darwin's assumptions—that the behavior patterns of an animal are as characteristic of its species as its bodily characteristics—would be confirmed.

INNATE BEHAVIOR AND LEARNED BEHAVIOR

Ethology grew out of Darwin's evolutionary theory, and much ethological research has been devoted to the study of hereditary influences on behavior. In this context, two major patterns of behavior are recognized: innate and learned.

Innate behaviors, or instinctive behaviors, are responses that are exhibited automatically by any individual animal, even when it has been raised without contact with other members of its species. The sucking responses of a newborn human, for example, are invariably made by any normal infant. A young wolf will howl and pounce in a way characteristic of wolves even if it has never seen another wolf; a baby rhesus monkey raised in a human family will produce the calls and facial expressions characteristic of rhesus monkeys even if it has never seen another of its kind. Many complex patterns of mating and nesting behavior in birds are innate, for they are performed unerringly by birds reared in isolation, with no opportunity to observe the behavior of other birds.

In a sense, innate behaviors are the "genetic learning" of a species—individuals with the appropriate behavior patterns tend to survive and reproduce. Through evolution there is a gradual accumulation of genetically determined behaviors appropriate to the conditions under which the species lives.

Learned behaviors, in contrast, are patterns that are developed or modified as a result of experiences. Consider the rapid development of the behavior patterns of the hunting wasp (Figure 13.1). This animal digs a nest, flies once or twice around the nest site, and then heads off in search of food. The wasp can find its way back flawlessly, something that many people would not be able to do. It obviously learns landmarks extremely rapidly. If the landmarks are moved a little, however, the wasp becomes irrevocably

13

Forms of Behavior

Figure 13.1 The hunting wasp learns where it has dug its nest by circling it a few times and observing the arrangement of prominent landmarks. If the landmarks are moved slightly, the wasp will not be able to find the nest, even if it is in plain sight.

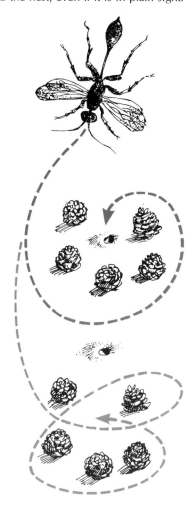

disoriented even though the nest is in clear sight. Here is an instance of a behavior pattern that clearly depends on learning but that, once learned, is extremely inflexible. As a general rule, however, the greater the level of organization of an animal, the more flexible its behavior can be.

A central issue for the definition of innate behavior is the relative importance of hereditary and environmental factors to the development of a behavior. Heredity must contribute a great deal to the behavior of the digger wasp, which can learn nothing from its parents—they die before the young wasp emerges from its pupa in the spring. In contrast, "higher" animals, especially human beings, usually undergo a long period of tutelage by their kind. Despite the apparent implications of extreme examples, however, no behavior can be classed as *completely* hereditary (totally uninfluenced by the environment) or as *completely* environmental (totally uninfluenced by hereditary endowment). Every behavior is influenced in some ways by both factors.

Imprinting is one of the most striking examples of the interaction of complex hereditary factors with learning. The German

Figure 13.2 Imprinting baby chicks to a large moving object—a blue ball. The ball is slowly rotated around the runway, eliciting a following response in the young bird. Imprinting normally takes place within a critical time period soon after hatching, when the young birds are highly receptive to environmental stimuli.

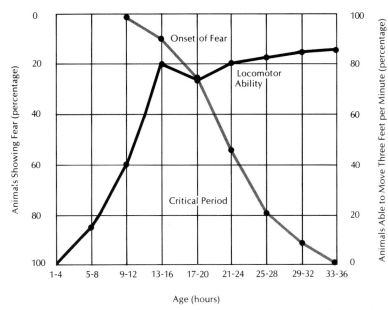

Age (hours)

Figure 13.3 The sensitive (or critical) period for imprinting can only begin when the animal has attained a high enough level of locomotor ability so that it can follow the stimulus. As shown in this graph, chicks gain quickly in locomotor ability until they reach 13 to 16 hours of age. This factor sets one limit on the critical period; animals cannot follow a moving object until they can walk well. The other delimiting factor is the onset of fear of novel stimuli, which occurs when animals have become familiar with a certain environment. Although this period of shyness soon ends, it too limits the critical period for attachment; animals are afraid of a novel moving object during this time. The period of shyness in chicks begins between the ninth and twelfth hour of life and ends at about the thirty-fifth.

ethologist Konrad Lorenz brought to the attention of ethologists the interesting fact that a newly hatched duckling or gosling, at a certain early age, will follow almost any moving object—a human being, a wooden box on wheels, a bird of a different species (Figure 13.2). A slightly older duckling or gosling (sometimes only a few hours older) will not follow these objects. Apparently there are certain times in the life of these animals when they are particularly susceptible to attachments. It is not difficult to see the importance of these attachments for young birds. Birds of a feather *do* flock together, and early attachments to particular moving objects—in the wild, usually other birds of the infant's own species—seem to be one of the first stages of interaction.

Even though the objects that are followed do not have to look like members of the animal's own species, imprinting seems to be a primary mechanism for the early development of social bonds. In fact, Lorenz argued in his early writings that these first social bonds, once formed, are difficult if not impossible to break. For example, he mentions a barnyard goose that not only followed its keeper but, upon reaching sexual maturity, actually attempted to mate with him.

Although it was once believed that imprinting could only occur during a highly restricted time period—known as the *sensitive period*—this view has changed. Experiments have shown that some attachments can be modified or reversed, and some learning relevant to the attachment can occur before or after the sensitive period.

Studies of imprinting have shown that once animals are familiar with a given set of environmental surroundings, they often avoid stimuli that are unfamiliar. Indeed, one factor that defines the end of the sensitive period, during which imprinting can occur, is the development of a fear of novel objects. It is as if an animal builds up an internal representation of certain features of its environment and then avoids stimuli that do not match that representation. Many children and other young mammals, such as the wolf, show this behavior.

SYSTEMS OF RESPONSE

An organism is a living system that can produce many different outputs (responses) in reaction to many different inputs (stimuli). At one level, these responses work to maintain the organism's *internal environment* within a range that permits survival. Water balance must be regulated, for example, and nutrient levels, oxygen levels, acidity, temperature, and so on. To maintain this balance, or homeostasis, the organism must exchange materials and energy with the external environment. These exchanges occur through the respiratory, circulatory, digestive, and excretory systems, as described in Chapter 11.

Most physiological systems work continuously throughout the life of the organism, and generally their activity is rhythmic, as in the beating of the heart. In some cases, the rhythm is maintained by nonneural systems, and the nervous system simply modulates the frequency or amplitude of the rhythm. In other cases, the rhythm is maintained by specialized neurons or neuron systems, which are in turn modulated by other neural structures.

At another level, responses relate to the *external environment*. These responses include orientation and coordination of the body and its parts, movement from one place to another, avoidance of or protection from threats and dangers, rhythmic activities, communication with other organisms, and specific patterned sequences of behaviors.

Classifying response activities into two broad categories does not necessarily mean the two are not interrelated, however. Consider, for example, Vincent Dethier's description of the feeding behavior of the blowfly. As the time since its last meal becomes longer, the blowfly increases its locomotor activity, flying in a random course until it encounters the smell of food. At that point it flies toward the food, lands near it, and begins walking about in a random manner. If it encounters a drop containing sugar (its natural food), it extends its proboscis (a sort of hollow tongue) and begins to suck. Over a period of about 10 minutes, the animal takes one long drink and perhaps a few short ones. It then remains immobile for a few hours before it resumes its random flight.

Dethier began his search for an explanation of this behavioral cycle by examining in more detail what happens when a fly encounters a sugar solution (Figure 13.4). The blowfly, like many other insects, has taste receptors in its legs. When a group of these receptors is stimulated by sugar, the fly turns toward the area of stimulation and extends its proboscis downward. When the sugar

Figure 13.4 (left)   Diagram of the blowfly, an insect used in feeding behavior studies.

Figure 13.5 (right)   Diagram of the three feedback loops that control feeding behavior in the blowfly.

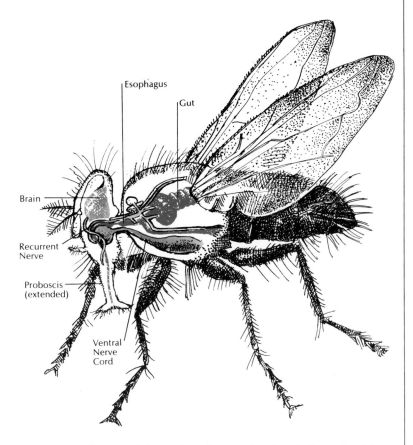

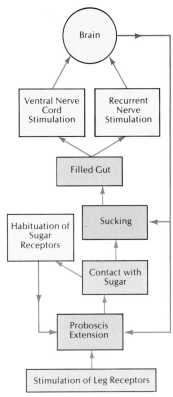

receptors at the tip of the proboscis are stimulated, the animal begins sucking the sugar up into its esophagus and gut. After a minute or two, the receptors lose their sensitivity to sugar and the animal stops feeding.

The receptors become sensitive again a few minutes after feeding has stopped, but the animal does not eat for several more hours. What keeps it from doing so? Dethier found two nerves that affect eating. Severing one of them caused the animal to eat much more than usual; severing the other made the fly eat so much that it actually burst. Later investigation showed that the first nerve, called the "recurrent nerve," receives impulses from receptors that measure the distention of the foregut. This nerve tells the brain when the foregut is full. The second nerve, called the "ventral nerve cord," receives impulses from stretch receptors in the fly's body wall and tells the brain when the body wall is becoming distended with food. As Figure 13.5 shows, the blowfly's feeding system includes three "feedback loops" for controlling the amount of food eaten. Each loop uses a different source of information, but all of them influence the same behavior.

As this description indicates, the fly's eating behavior depends on several *internal* as well as *external* factors: not only on the pres-

Figure 13.6 This map of the Western hemisphere traces the migration patterns of green sea turtles, gray whales, Alaskan fur seals, and barn swallows and nighthawks. Recent research on how migrating birds navigate has shown that no single factor can be said to account for their ability to unerringly make such long journeys. Night-flying birds do use the stars for navigation, but they also learn to take account of the rotational motion of the night sky; fledglings who were reared in a planetarium with artificial star patterns that had been rotated to a new axis chose to migrate north rather than south in the autumn.

ence of a source of food but on the periodic sensitivity of the external sugar receptors, the distention of the foregut (as measured by the recurrent nerve), and the pressure of the gut against the body wall (as measured by the ventral nerve cord).

Orientation and Locomotor Responses

Most animals work to maintain a specific orientation of the whole body to some part of the external environment. Usually this orientation requires the use of energy and coordinated movements. Neural controls work to detect any departure from the animal's normal position with respect to some reference point in the environment, and these controls initiate muscular movements that restore the proper orientation. The most pervasive reference points are the gravitational field (up and down) and the direction of light. Surface contacts, chemical gradients, and wind or water currents also serve to orient some organisms.

Relatively simple animals orient themselves in two different ways. They may sense where more favorable conditions lie and move in that direction, or they may simply move faster under unfavorable conditions and slow down when they encounter favorable ones. Most animals move bodily from one place to another by flying, walking, swimming, or crawling. Such forms of locomotor behavior are usually accomplished through specific sequences of coordinated, oriented activities. Much of the locomotion of any species occurs through only a few motions. In the horse, for example, there are three primary motions: walking, trotting, and galloping. Each of these *fixed action patterns* is a distinct series of movements that varies little from one horse to another, and all normal horses display the patterns.

The most stunning example of orientation and movement patterns is the recurring phenomenon we call *migration*. Many animal species migrate over vast distances. Alaskan fur seals, for example, breed only on the Pribilof Islands in the Bering Sea. During the winter, the females and young seals swim some 3,000 miles to the coast of southern California, while the males travel to the southern Aleutian Islands (a much shorter journey). In the spring, the sexes join each other again in the Pribilofs. Eels born in the Sargasso Sea find their way to the coasts of North America and Europe; salmon return to their home streams to spawn after journeying hundreds of miles into the ocean; and both the nighthawk and the barn swallow of North America fly from as far south as Argentina to breed in the cold reaches of the Yukon and Alaska, a trip of about 7,000 miles (Figure 13.6).

Before an organism can move purposefully from one place to another, it must have knowledge of the direction of its destination and a means of orienting itself toward the proper direction as it moves. The exact cues used by various species during their remarkable trips have been under investigation for many years, but

Figure 13.7   Migration is a behavior that is determined by both genetically controlled physiological rhythms and external stimuli such as climatic changes. The Monarch butterfly *Danaiis plexippus* (shown at top) is thought to be the only species of butterfly that performs a seasonal two-way migration. The same individuals that fly south in the autumn return in the spring, after being inactive from November through March. During their southward migration and while in the south, these butterflies congregate by the thousands each night in certain trees and may remain there for weeks at a time. This massing discourages intruders. Shown in the middle and lower photographs are birds and whales, animals that migrate on a yearly schedule.

Figure 13.8 There is evidence that starlings use the position of the sun to orient themselves as they prepare to migrate. Although the altitude of the sun above the horizon varies through the year, the azimuth of the sun is always the same at a given hour of the day, no matter what the season. If the birds do measure direction by comparison to the sun's azimuth, then they must possess some internal mechanism to help them orient and solve complex navigational problems. This diagram depicts an experiment designed to test this hypothesis. The experiment starts at left, with the starlings always trained to feed at the cup in the compass direction south. After training, the birds choose the feeding cups indicated by dots, showing some error in choice of directions. They are placed in an artificial light-dark cycle (LD:12, 12) that is 6 hours behind the real time—in such a cycle, the onset of light came 6 hours after true local dawn. During the first few days, their hypothetical clocks shift gradually 6 hours behind (black arrows). Consequently, the real time of day when they think it is noon (red arrow) drifts later and later until it reaches 6 P.M. When the birds are briefly exposed to the sun and allowed to look for food, they choose cups west rather than south. During the following days, they are left in constant light; their clocks and thus choices of direction gradually drift. Retrained in the same LD:12, 12 schedule, their clocks and choices of direction sharpen up again. They are then shifted to a normal light-dark cycle and, when retested, look for food in the south. To understand why the "sun-compass" theory explains this behavior, imagine the birds' using the rule "When my clock says noon, the sun should be due south of me and that is where I get food."

because of the difficulties of studying migration in the field, no more than the bare outlines of the behavior have become clear. Birds, the most thoroughly studied migratory animals, seem to use the sun and the stars as guides as well as various landmarks (Figure 13.8). It has also been found that homing and migration do not necessarily follow a fixed course. In some experiments, birds were transported over large distances and were then released; many species still found their way home. What are the mechanisms underlying these behaviors? No one knows exactly, but several ideas have been advanced and will be described in more detail later in the chapter.

Avoidance and Protective Responses

Among the most important behavioral responses are the ones that serve to protect an animal when it is threatened with danger. In most cases, the *avoidance response* tends to take precedence over other possible behaviors. Avoidance behaviors are often quite specific and differ greatly among species. Forest animals flee from an approaching fire. If light from a controlled source is suddenly dimmed, the annelid worm *Sabella* withdraws rapidly into its tube: the sudden dimming of light may be caused by the shadow of an approaching predator. The call of a hunting bat may cause a pursued moth to fold its wings and drop toward the earth. Unusual vibration may cause a caterpillar to stop moving, so that it looks much like a leafless twig. A normally stationary sea scallop that is touched by a predatory starfish may release its hold on a piece of coral and swim furiously away (Figure 13.10). Avoidance activities are quite varied, and many involve orientation and locomotion. Each activity is elicited by a stimulus that somehow has been associated with danger during the evolution of a species. The response has survival value for the individual and, therefore, for the species.

Whereas avoidance responses help an organism elude danger, *protective responses* generally minimize damage after danger strikes. Direct injury requires an immediate response from an organism. The first step in any protective response must be an

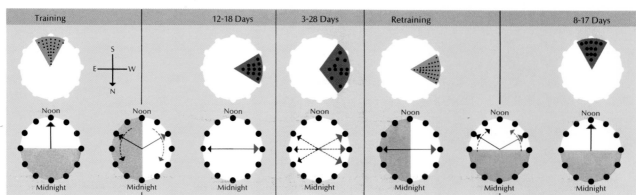

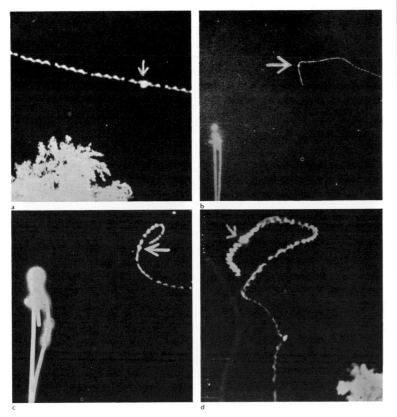

c   d

Figure 13.9   Streak photographs of the avoidance response tracks made by free-flying moths when exposed to a source of ultrasonic pulses mimicking a bat. (A) Track of a moth that did not react to the ultrasonic stimulus (start at arrow). (B) A distant moth turns and flies away from the sound source at left. Turning away from the sound source occurs only when faint sounds reach the moth. At this range, a real bat's sonar could not detect the prey, so this maneuver is a logical one for the slower-flying moth. (C) Turning away as in B. (D) Erratic twisting during power drive in response to intense sounds. The moth eventually stopped flapping and made a free fall to the ground.

awareness that injury has occurred. Sensory receptors usually provide information about the nature of damage. On the basis of this sensory input, the nervous system generates appropriate coordinated behaviors. Protective responses in mammals, such as blinking and tearing, grooming of wounds, and limping, are familiar to everyone. Examples are found in all groups of animals. A strong pinch to a crab's leg may cause the appendage to break off at a specialized joint. A pinched or damaged earthworm may break completely in two and discard the damaged segments. A lizard, if it is grabbed or damaged, may discard its tail. Protective behaviors—even those involving self-inflicted damage—increase an animal's chances for survival.

Again, however, the actual classification of responses may give an illusion of unvarying responses in a given situation. Animals are not that simple. Not only must individual, fixed patterns be examined, but the *relationships among them* must be taken into account. *Displacement activity*, for example, involves several fixed response patterns. Suppose that a tern is sitting on a nest of eggs. When it sees a man approaching, the bird begins to vacillate between escaping and staying on the nest with the eggs. At some point the bird is likely to stop vacillating and do something that appears to be irrelevant to the situation: it preens its feathers (Fig-

Figure 13.10   This scallop, *Chlamys sperculoris*, is shown swimming to escape attack by its predator, the sea star (*Asterias rubens*), an example of an avoidance response.

Figure 13.11   In many species, such as the egret, preening or grooming is a form of instinctual behavior. Preening is an example of a fixed response pattern, which is a relatively stereotyped species-specific behavior. Also, preening commonly occurs as a displacement activity: a bird that is in conflict between two opposing fixed action patterns will suddenly begin to carry out an apparently irrelevant pattern, such as preening.

ure 13.11). Suppose another tern is vacillating between attacking another bird and incubating its eggs. It, too, may show a seemingly irrelevant behavior by making standard nest-building movements. Displacement activities are common to many species. Human beings, under moderate levels of conflict or tension, draw from a repertoire of behaviors that include drumming their fingers, scratching their head, and sighing profoundly. And both people and other primates, in frustrating situations, may displace their aggression onto targets that are not the source of the frustration.

SYSTEMS OF SENSORY STIMULI

We know from our daily lives that the sensory world stimulates and guides much of our behavior. The same is true for the nonhuman animals, but the world they experience through their senses may be quite different from ours. In many animals, only a few stimuli produce observable effects on behavior. The European naturalist J. von Uexküll painted a vivid picture of the behavior of the mated female tick, which will climb into the branches of a tree and wait for weeks, if necessary, for a mammal to pass directly beneath her. As she waits, the tick seems unresponsive to the barrage of sights, sounds, and odors about her. The stimulus that signals an approaching meal and leads the tick to drop on her host is the smell of butyric acid, generated by mammalian skin glands. In Uexküll's words, the tick is like a gourmet who picks the raisins out of a cake.

Sign Stimuli

Ethologists were quick to realize the importance of such observations. Different animal species live in different sensory worlds. Indeed, many species can be classified according to the stimuli they attend to, in much the same way that they can be compared in terms of fixed responses. Highly specific cues in the external environment often seem to serve as signals for complex sequences of behavior. Ethologists call these cues *sign stimuli*.

For example, Niko Tinbergen reports that the male stickleback fish will attack a crude model of another stickleback if the model has a red belly. The general appearance of the model is unimportant as long as it has a red belly. The stickleback ignores models that seem very fishlike to human beings and quickly attacks other models that seem unfishlike but have a red underside (Figure 13.12). For a male robin protecting its territory, red breast feathers are a sign stimulus for an attack.

Although *color* is a key element in many sign stimuli, it is not always important. Tinbergen also reports a remarkable example in which *size* is the key element. Many sign stimuli elicit a response only if they are about the right size. But a bird called the oystercatcher, offered a model egg several times larger than one of its own eggs, prefers the larger egg—even when the model egg is so large that the bird cannot assume a normal brooding position, as shown in Figure 13.13.

Habituation

Sometimes an animal responds in a certain way to a sign stimulus but then gradually stops responding as it sees more and more of the stimulus. This familiarization process is known as *habituation*.

W. Schleidt showed that turkeys try to escape from any unfamiliar object of an appropriate size that is moving overhead, whether it is shaped like a natural enemy or not. When he exposed turkeys to circles or rectangles moving overhead, for example, they showed the same initial alarm reactions that they did to an overhead silhouette of a hawk. As the turkeys became familiar with the shapes, their alarm reactions decreased. In each case, the habituation of the turkeys' fear reaction was specific to the shape that was repeatedly presented. Turkeys that repeatedly saw a hawk silhouette but not a circle eventually stopped reacting to the hawk but did show fear in response to the circle. Conversely, animals that were repeatedly exposed to the circle but not to the hawk silhouette showed no alarm reactions to the circle but many reactions to the hawk.

Although interpreting the meaning of sign stimuli is a difficult task, the general point remains. The sensory apparatus of a particular species is characteristic of that species, and the meaningfulness of objects and events in the world of an animal are determined partly by the capacities and limitations of his sensory apparatus. Man cannot tell whether light is polarized, but bees and many other invertebrates can; they use polarization as an important cue for locomotion and orientation. For the female tick, the significant stimulus is butyric acid, in concentrations that the human olfactory

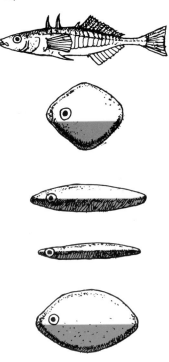

Figure 13.12 The male stickleback fish may ignore the uncolored top model but readily attack any of the lower ones with a painted red "belly." Note, however, that all patches are on the underside of the model. Apparently, both color and position are important in eliciting the attack response.

Figure 13.13 An oystercatcher attempting to brood a superoptimal model egg several times larger than its own egg. The bird's preference remains steadfast even though it cannot assume a normal brooding position.

Figure 13.14   Photograph of the dinoflagellate *Gonyaulax*. This plankton organism, well known for its production of "red tides," also exhibits circadian rhythms in luminescence, respiration, and photosynthesis.

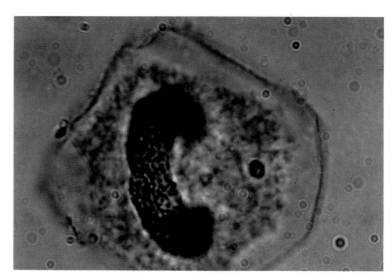

system cannot ordinarily detect. The upper limitation of auditory sensitivity in man is about 20,000 cycles per second, but bats, porpoises, and some rodents can hear up to 100,000 cycles per second. Bats and porpoises navigate by means of echolocation, involving frequencies far too high for human hearing. Man can only speculate about the nature of the auditory worlds of these animals.

RHYTHMIC PATTERNS OF BEHAVIOR

Rhythmic events recur everywhere in the environment. The sun rises and sets, the moon moves through its monthly cycle, and the tides and seasons repeat their inexorable rhythms. Their continuity and precision are proverbial: night follows day; and when winter comes, spring cannot be far behind. Temperature, light, humidity, weather, and other environmental characteristics vary along with these cycles.

Animals must adjust their activities to these cycles in order to avoid adverse conditions, to be able to use their senses to best advantage, and to improve their chances of obtaining enough food and of producing offspring. They show rhythms in response to both external and internal stimuli. The solar day-night cycle or lunar (tidal) cycle commonly is expressed in activities such as metabolism, locomotion, and food getting. Annual or seasonal behavioral change occurs in relation to reproduction or to periods of adverse weather such as winter or drought. In many animals, however, cyclic activity has become somewhat independent of environmental cues, and the cyclic changes of state may continue regularly in the *absence* of these cues. This ability to maintain an independent cycle of behavior has been taken by many biologists as evidence that in addition to direct stimulation by the environment, organisms possess an *internal clock* or *clocks*. The evidence

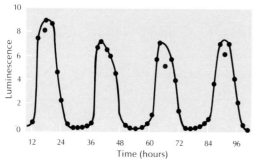

Figure 13.15   Rhythm of luminescence in the dinoflagellate *Gonyaulax polyhedra* reflecting entrainment to a 12-hour light-dark period.

suggests that the timing mechanisms of higher animals are located within the nervous system.

Timing of Biological Clocks

When organisms are removed from their habitat and placed in conditions of constant light intensity, temperature, pressure, humidity, and chemical composition, they may continue to exhibit nearly the same rhythmic fluctuations of behavior they showed in their natural habitat.

There are two possible explanations for this persistence of biological rhythms. According to one view, there are always subtle *geophysical influences* acting on even isolated organisms. These influences vary cyclically and synchronize the organism's behavior. Geomagnetic fields, cosmic rays, or electrostatic fields have been suggested as possible synchronizing stimuli. According to the other view, organisms possess internal *biological clocks* that determine the timing of changes in behavior and behavioral states. External stimuli serve only to reset this clock periodically so that it does not get out of phase with the outside world.

No one knows whether one or the other of these general hypotheses is correct. Some researchers suspect that organisms possess internal timers *and* respond to subtle environmental clues. Biological timing mechanisms have also been suggested as the basis of animal homing and direction finding as well as seasonal changes of behavior and physiology. These behaviors seem to require the animal to determine the time of day, or at least the length of the day or night. In the case of homing, many animals navigate by the position of the sun or stars. It has been suggested that to synchronize behavior to the seasons, an animal needs to measure the length of day or night (days lengthen in the summer) or needs to measure the relation between environmental day and night cycles to an internally generated rhythm. Surely timing mechanisms that control these behaviors interact with some mechanisms that control rhythmic activity, but the proof has yet to be discovered by researchers.

Circadian Rhythms

Activities that are synchronized to the 24-hour day-night cycle are called *diurnal cycles*. Activities that recur with a periodicity of about 24 hours under constant conditions are called *circadian rhythms*. If there are no external rhythmic stimuli, the behavioral cycle is said to be *free-running*. Such cycles vary slightly from the 24-hour cycle, and they also differ slightly from one individual to the next (Figure 13.16).

Circadian rhythms are present in all the mammals that have been studied, including man. Men who have been isolated show 24-hour variations in sleeping and waking cycles, in the concentrations of chemicals in blood and tissues, and in the rate of cell

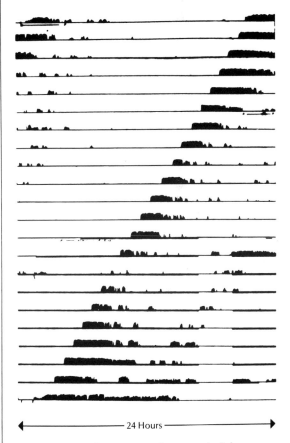

← ——————— 24 Hours ——————— →

Figure 13.16 Twenty-two-day record of the running activity of a small mammal kept in a laboratory darkroom without the daily light-dark cycle. The chart records of each day are placed below one another, making it easy to see that the activity begins earlier each day.

Figure 13.17   The alteration of an innate circadian rhythm controlling the emergence of adult *Drosophila* is illustrated by experimental control of environmental conditions. This diagram shows the distribution of emergence activity in a population of *Drosophila* maintained at 21° Centigrade in an environment where 12 hours of light alternated with 12 hours of darkness. On about day 17 of the 25-day life cycle of the flies, emergence began at about the onset of the light period. Within a few hours emergence stopped, but another peak of activity occurred exactly 24 hours after the first one. The amplitude of the 24-hour rhythm gradually decreased from day to day as the population was "used up," until all flies had emerged by the day 8. Under natural conditions, this species of *Drosophila* is often observed to emerge from the pupa cases a few hours after dawn—a time when humidity tends to be high, a favorable condition for the delicate, emerging fly whose cuticle is still hardening and whose wings are still unfurling as they are pumped up.

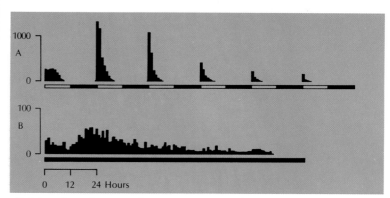

division, heartbeat, and excretion. In such experiments, the ability to estimate time as well as the error rate in problem solving also exhibited a circadian periodicity.

Under controlled conditions, most human subjects settle into a circadian periodicity that differs from the normal 24-hour day only by an hour or two. For most subjects, the period has been greater than 24 hours. After several days of the experiment the subject is out of phase with the external world, although he is unable to ascertain it. In a few individuals, activity periods approximating a 48-hour cycle occurred, without the subject's suspecting that they were living differently from normal.

It has been possible, in other experiments, to synchronize humans to light-dark cycles of periods slightly different from their natural 24-hour cycle. But as anyone who has traveled by jet across several time zones knows, such a shift is sometimes an uncomfortable experience and the effects of a change in schedule may persist for several days.

Such experiments provide insight into the range of environmental conditions that humans tolerate. Efficiency as well as accident

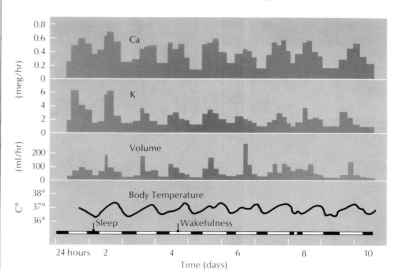

Figure 13.18   Circadian rhythm of urine excretion (calcium, potassium, and water), body temperature, and wakefulness in a human subject kept under constant environmental conditions in complete isolation.

288

and error rates of workers are known to vary with time of day, but little is known about the effects of various shift schedules on these normal circadian rhythms. Airline pilots who cross time zones as well as astronauts on space missions are deprived of normal environmental cycles. Whether this deprivation has any adverse effect is not yet known.

In the process of searching for the biological mechanisms of the mammalian clock, many organs and glands of the body have been found to exhibit circadian rhythms. Particular attention has been focused on neuroendocrine organs (such as the adrenal glands) and the pituitary system. Adrenal glands of hamsters, even if maintained in a test tube, show some metabolic cycles that can be synchronized by light-dark cycles. In living hamsters, a hormone secreted with a circadian rhythm from the adrenal gland controls many body rhythms, such as fluctuations in glycogen in the liver and epidermal mitotic activity. These cycles stop if the adrenal glands are removed, but the hamster's activity rhythm continues. Clearly, the adrenal glands are not the repository of the master biological clock.

Work continues in the search for the basis of the biological clock, even though at times it seems as if, like memory, it is everywhere at once and nowhere in particular. Perhaps the answer will be found in the maze of self-controlled feedback activities of hormones and enzymes of the body as a whole.

BEHAVIOR IN THE ENVIRONMENT

Communication and social behavior in the environment must involve two parties—one to send the signal and one to receive it. And, throughout the animal kingdom, there has been a progression of complexity both in mechanisms designed to send signals to other individuals and in mechanisms designed to recognize and inter-

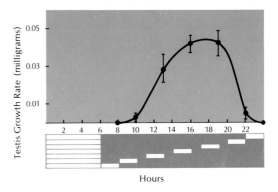

Figure 13.19 Testicular enlargement in male white-crowned sparrows is a circadian rhythm entrained to a long-day photoperiod. An experimental group was kept on a schedule of 6 hours of light and 18 hours of dark, with 2-hour blocks of light interrupting the dark period at various selected times. Introduction of the 2-hour light blocks in effect simulated a long-day photoperiod. Note that the maximal growth rate occurred after the 2-hour blocks of light administered became equivalent to a long day.

Figure 13.20 A schematic representation of the functional relationships between neural and hormonal systems in photoperiodically entrained gonadal cycles in birds.

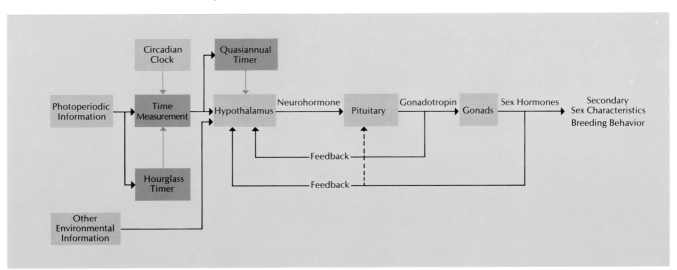

Figure 13.21   Restrictions of chemical communication as illustrated in gypsy moths. The distance and area from which females can recruit males with pheromones is a function of diffusion and air turbulence. At distances of less than 2 meters, males orient themselves by discriminating the diffusion gradient. At distances of greater than 2 meters, they orient themselves by flying into the wind.

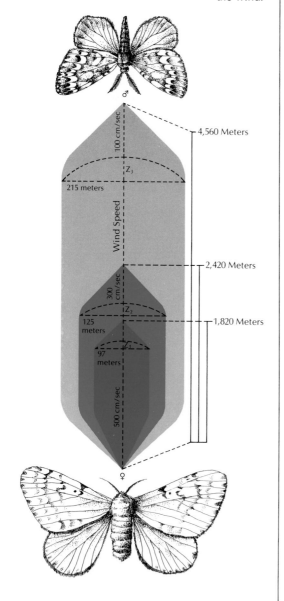

Z_1 = 48.5 Meters for 500 cm/sec Wind
Z_2 = 62.5 Meters for 300 cm/sec Wind
Z_3 = 108 Meters for 100 cm/sec Wind

pret signals sent by others. The signals and the recognition systems are often startling in the intricacy and the efficiency with which they intermesh. The bee, for example, has evolved complex visual recognition capacities that enable it to find a flower. The flower has evolved distinctive shapes and color patterns that the bee can easily recognize. The bee gets food from the flower, and the flower gets the advantages of cross-fertilization from the bee's visits.

The components of behavior that trigger responses of other organisms in the environment can be chemical, visual, auditory, or tactual; or they can be a combination of these and other stimuli. These components are called *social releasers*. They are often marvelously adapted to the properties of the physical world and to the modes of communication available. Chemicals with long lifetimes serve for long-range, low-speed communication. Chemicals with shorter lives deliver signals that must act only over short distances and short times.

A species may possess a repertoire of different chemical signals, or *pheromones*, with different diffusion characteristics, produced by various glands on the body, transmitted in different ways and serving such varied functions as marking trails, inducing mating, encouraging aggregations for nesting, disseminating alarm, permitting recognition of individuals and groups, and attracting others to a food source. The sensitivity of a male moth to the pheromone emitted by a female is a good example of the species specificity of long-distance communication that can be accomplished through use of pheromones. The male can locate the female at distances of less than a meter or two by noting the diffusion gradient in concentration of the pheromone (Figure 13.21). At greater distances, the male can locate the female only by flying into the wind that brings him the scent (Figure 13.22). Females emit the pheromone only when there is enough wind so that it is likely to be detected by males.

The pheromone used for mate location by moths is active over long distances. Ants, on the other hand, have trail-marking substances that fade more rapidly (Figure 13.23). If the source of food at the end of a trail becomes exhausted, returning foragers who normally must follow scent trails between the nest and food sources do not add their own trail substance to the existing trail. Because the trail gradually fades if it is not renewed by successful foragers, the net effect is that foragers looking for food no longer can find that particular trail, and they spend their effort going to better food sources. Ants also release alarm pheromones, which elicit aggressive behavior either by specialized soldier members of their society or by workers. Ants at a short distance from a disturbance are induced to release their own alarm substance, which leads to the spread of the alarm message. The alarm substance must not persist very long, otherwise a minor disturbance at one spot could lead to a chain reaction that would leave the anthill in

Figure 13.22 Chemical communication in the silkworm moth *Bambyx mori.* The male can locate a female at distances of less than a meter or two by noting the diffusion gradient in concentrations of the pheromone. At greater distances, the male can locate the female only by flying into the wind that brings him the scent. In this illustration, the concentric circles are 1 meter apart. Tracks of various males are for 1-hour periods.

an uproar for many hours after the cause of the original disturbance had passed.

Other social releasers such as visual signals give immediate and potentially private communication, and auditory signals give private or general audience, depending on how they are used. In animals that require relatively little information, signals tend to be simple. But versatile systems for communicating many different possible messages have evolved as well. Human language is one such system; the language of the honeybee is another.

Honeybees are social insects. They live in colonies that include one egg-producing "queen," a few males called "drones," and many "workers," which are females that have not developed genitals. The insects eat flower nectar and pollen, fertilizing plants by spreading pollen from one to another as they forage for food. Only a few bees, called "scouts," leave the hive to search for food. When they find it, they collect some on their bodies and in their stomachs and return to the hive. Shortly thereafter, many other workers fly to the place where the scouts found food and begin gathering it. How do the workers find the food? The scouts do not lead them to it; they somehow tell the workers where it is when they return to the hive.

In an early experiment, the Austrian zoologist Karl von Frisch demonstrated the precision with which worker bees locate a food

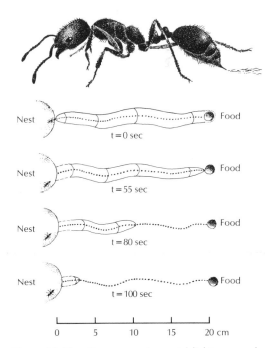

Figure 13.23 Fire ant worker establishing an odor trail with its extended stinger (above) and the history of representative odor space of the trail pheromone (below).

Figure 13.24 Diagram depicting the dance language of bees. A scout bee that finds a rich source of food returns to the hive and performs a specific dance within the hive on the vertical surface of the comb. The scout performs the round dance only when the nectar source is close at hand; this dance signals the workers to search in the general area of the hive.

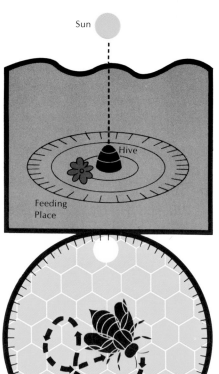

source that the scouts have found. He fed some scout bees with food that he had placed on a particular group of flowers at a botanical garden. The plants were a species not usually visited by the scouts, but within minutes after they had left the garden, many workers from the hive arrived. The workers lighted only on the species of flower where the scouts had been fed, ignoring the 700 other varieties of flowers in the garden.

How could anyone go about determining the nature of a communication process that works so precisely? Ideally, the bees' communication medium—their "language"—would have to convey information about the type of flower the scouts had found, its quality and abundance, and its distance and direction from the hive of the bees.

An obvious possibility is that the first factor, the type of flower, is communicated by odors the returning bees carry. The scout returns with flower odors on its body, and its stomach is filled with fragrant nectar, some of which it gives to the other bees. Through careful experimentation, von Frisch determined that bees use both these odors in finding the food source. He also found that the bees use a specific scent that the scouts themselves leave at the food source by means of a scent gland.

The odor of a flower alone would help a bee locate food near the hive, but it would not be enough to tell workers about the abundance of food or to locate distant sources—and bees often forage over a mile from their hives.

Von Frisch found that scouts returning to the hive from nearby food sources (50 yards or less away) perform a circular movement called a "round dance," illustrated in Figure 13.24. Bees in the hive follow the dancing bee, touching it with their antennas. When von Frisch varied the richness of an artificial food source he found that the bees danced more vigorously when the source was richer. The bees in the hive learned about the richness of the source from the vigor of the dance. As the scouts perform their dance on the vertical face of the comb, other workers cluster about the dancing bees, then leave the hive and fly to the food source.

But the round dance apparently tells nothing about direction, as von Frisch showed in another experiment. He fed scouts from a lavender-scented dish near the hive, around which he had placed a number of similarly scented but empty dishes. After the scouts went back into the hive, he removed the full dish and put an empty scented one in its place. When the bees emerged from the hive to find the food, they did not settle on that dish but distributed themselves equally among all the dishes. Apparently the scouts had told the other bees that food was nearby but had not told them exactly where to look for it.

Later on, von Frisch repeated the experiment but put the dishes farther away from the hive. This time something entirely different happened. At about 200 yards, almost all the bees went to dishes

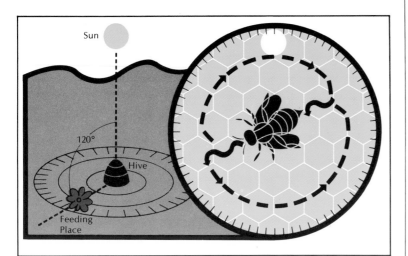

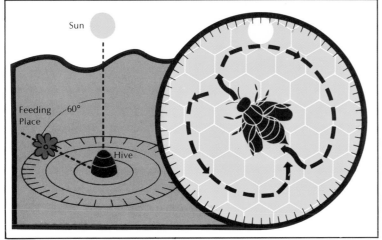

Figure 13.25 In the upper portion of this diagram, the bee's figure-8 dance, with the tail-wagging runs in the middle, is traced at approximately a 120-degree angle to the sun; the scout wags its tail rapidly in this case, signaling that the source is relatively close. In the lower portion of the illustration, the figure 8 is traced at a 60-degree angle to the sun, and the scout wags its tail slowly, signaling a more distant location.

very near \the one where the scouts had eaten. Why? Von Frisch noticed that the scouts did a different dance when the food source was far from the hive than when it was nearby. Instead of the round dance that conveyed no information about direction, they did a "wagging dance." In the wagging dance a bee runs straight forward for a short distance, turns 360 degrees to the left, runs again, turns 360 degrees to the right, and repeats that sequence several times.

The distance from hive to food source, von Frisch managed to show, is indicated precisely by the speed of a scout's 360-degree turn. The direction of the straight run between turns indicates the angle of the path of flight. The bees "calculate" the angle by using the line connecting the hive and the sun (Figure 13.25). The calculation does not require direct sunlight—bees are sensitive to polarized light, which results from the scattering of the sun's rays in the atmosphere and is transmitted by blue sky, indicating the position of the sun even when it is out of sight behind a cloud or a hill.

Figure 13.26 The honeybee, of the arthropod class Insecta, exhibits a complicated pattern of social behavior.

Polarized light is not transmitted when the sky is completely overcast, but bees can see the sun through a thicker cover of clouds than human beings can.

Bees' language, then, is highly complex. It involves at least three senses—smell, sight, and touch—and conveys information not only about the type and quality of the food the scouts have found but about how far the food source is from the hive and, if it is not nearby, in what direction.

In any species where groups of individuals live in proximity to one another and interact, a fairly complex and regular *social organization* can be observed. There may be a number of different roles within the group, each with its own pattern of appropriate social behavior. Males, females, old, young, leaders, foragers—the kinds and numbers of roles vary greatly from species to species. In many cases, the assignment to various roles is based on biological differences—male versus female, old versus young, or, in bees, worker versus drone—but in others the roles result solely from behavioral interactions among biologically similar individuals.

A famous example of social organization among animals is the *pecking order*, first described among hens by T. Schjelderup-Ebbe and later observed in a wide range of other animals. If hens that have no experience with one another are brought together in a barnyard or a large cage, they show aggressive behavior toward one another. At first, there seems to be no pattern to the behavior; any two hens are apt to fight upon an encounter. After a period of time, however, a social organization appears within the flock. Encounters no longer result in random aggressive behavior. Each hen exhibits aggression or submission predictably, depending on

Figure 13.27 This photograph shows how individual rank in a dominance hierarchy among hens can often be ascertained by observing individual precedence over food or water.

which hen she encounters, and a simple pecking order, or dominance hierarchy, is established, as seen in Table 13.1.

Table 13.1
Dominance Hierarchy in a Flock of Hens

Hen 1 pecks	2	3	4	5	6	7	8	9	10	11	12	13
Hen 2 pecks		3	4	5	6	7	8	9	10	11	12	13
Hen 3 pecks			4	5	6	7	8	9	10	11	12	13
Hen 4 pecks				5	6	7	8	9	10	11	12	13
Hen 5 pecks					6	7	8	9	10	11	12	13
Hen 6 pecks						7	8	9	10	11	12	13
Hen 7 pecks							8	9	10	11	12	13
Hen 8 pecks								9		11	12	13
Hen 9 pecks									10	11	12	13
Hen 10 pecks							8			11	12	13
Hen 11 pecks											12	13
Hen 12 pecks												13
Hen 13 pecks none												

Source: Adapted from W. C. Allee, *The Social Life of Animals* (New York: Norton, 1938).

In simple hierarchies, older animals dominate younger ones, larger ones dominate smaller ones, and males dominate females. But there are complications. In breeding seasons, female birds and primates may become dominant. A small group of individuals may cooperate to depose a high-ranking male. In some birds and fishes, a low-ranking male may dominate a higher-ranking one if his own mate is nearby. It is difficult to avoid the impression that human social order is not all new.

Dominance hierarchies diminish the amount and intensity of fighting and thereby allow more time for pursuits of greater value to the flock or the species. A. Guhl in 1956 compared two groups of hens. One group had a stable order, and the other was disrupted by shifting its members. The birds in the disorganized flock "fought more, ate less food, gained less weight, and suffered more wounds." Presumably, in nature, the young of a flock with an established order are more likely to survive and reproduce than if disorder prevails.

Territoriality is also a common form of social organization. Living space and resources such as food and nesting sites are distributed among individuals by this mechanism. In many species, the strongest and most dominant individuals are able to claim territories, whereas weaker individuals fail to do so. They are often unsuccessful in their efforts to obtain a mate and thus fail to produce any offspring.

In migrating populations of song sparrows, the male usually returns to the same area in successive years. In year-around resident populations, the male remains near his territory during the winter. As spring advances, he establishes song posts around the periphery

Figure 13.28   Territoriality is a common form of social organization among prairie dogs—this characteristic and their varied patterns of social behavior has placed them among the most successful of the social rodents. The top diagram schematically portrays a 5-acre prairie dog "town" (each square represents 50 square feet). Family territories, or coteries, are shown in brown, with smaller coteries in the process of being established by emigrating adults. Filled-in dots indicate large occupied burrows; open dots indicate smaller burrows; small dots indicate small holes. The lower series of drawings depicts a few of the key social interactions that take place between adults. At the top, two individuals of the same coterie meet, exchange an identification "kiss," and go on to groom each other. The bottom series depicts two strangers approach-

ing, exchanging an identification kiss, the tail-raising cere-
mony, gland sniffing, and departure. The center diorama depicts various aspects of the prairie dogs' life style. Major
predators are the hawk, badger, coyote, and weasel; occasional burrow neighbors are the burrowing owl and prairie
rattlesnake. The raised mound entrance functions to prevent
flooding of the burrow and as a lookout point for danger.

of the territory. At each perch, he sings for several minutes and then moves on, marking the limits of his territory by singing. Intruders of other species usually are ignored, but other male song sparrows are threatened in a ritual manner. If this threat posture fails to dispatch the intruder, more forceful measures are used to drive him away. Female song sparrows are allowed into the territory, and eventually a pair is formed. The pair join in defense of the territory, where they build a nest, incubate the eggs, and seek food for the young.

The size of territories depends on innate behaviors as well as on external factors. If the size of a population decreases, external pressures are reduced and the size of territories increases—but not without limit. Innate behaviors that influence the behavior of song sparrows keep the territory size from increasing much beyond an acre. Conversely, when the population becomes dense during a series of favorable years, the innate behaviors prevent the territories from being compressed to less than about half an acre. Thus, innate behavior patterns help regulate population size by imposing an upper limit on the number of successfully breeding adults in any one region.

What are the functions of territorial behaviors? P. Klopfer has proposed several. They may serve to increase the efficiency with which material resources are used by restricting food gathering to a limited area well known to the animal. They may limit the intensity of competition for food, because the minimal territory size limits the total population density. It is possible that territorial behav-

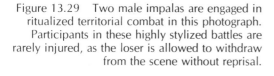
Figure 13.29   Two male impalas are engaged in ritualized territorial combat in this photograph. Participants in these highly stylized battles are rarely injured, as the loser is allowed to withdraw from the scene without reprisal.

iors strengthen pair formation between the sexes. They may also reduce the effect of predation, because an individual is better able to escape a predator in his well-known territory. Finally, such behaviors may reduce the level of aggression between members of the species, because in most cases the interaction between two males is reduced to a ritual display of threatening behaviors, as shown in Figure 13.29.

Social organization in birds has been studied for many years. Recently, there have been a number of studies of social behaviors in primates, insects, and fishes. But there are many species on which relatively little information is available. Although general patterns such as dominance hierarchies and territoriality have been observed in many different animals, an increasing number of studies, and greater attention to quantitative detail in those studies, have tended to reveal more differences among species than similarities. Even among the primates, there is a wide variety of social organizations and patterns of social behavior. Therefore, most ethologists tend to be skeptical of any attempt at this time to explain human behavior in terms of the few generalizations that have been made about behavior in "lower" animals.

Thomas Malthus in 1798 observed that "nature has scattered the seeds of life abroad with the most profuse and liberal hand. She has been comparatively sparing in the room and the nourishment necessary to rear them." Ecologists today point out that the growth of the human population in particular has upset the balance of distribution and resources of the biosphere. Ecologists are the scientists who investigate the interactions among organisms, populations, and the environment. Their units of study are the *population*, a group of individual organisms belonging to the same species and living in the same area; the *community*, which encompasses all populations living in an area; and the *ecosystem*, the community together with the inorganic environment. Thus, the largest ecosystem is the earth. The smallest ecosystem has its limits set by the interest and convenience of the observer; it could be a single tree or a rock in a tidal pool. As in Malthus' time, the concern lies with understanding of populations and how they change in terms of how energy and materials are used by life on earth, as well as in terms of explaining evolutionary processes.

A population is characterized by its range (the area it occupies) and the number of individuals it contains. Also important is its *density*, the number of individuals per unit area, which may vary from place to place within the geographical location. The rate at which new individuals are added to a population through birth and immigration—and the rate at which individuals are lost through death and emigration—determine its overall *growth rate*. Still another characteristic is the distribution of individual traits such as coloring, size, sex, and age. The ways in which this distribution, or population structure, change with time correlate with the way species change and give rise to new species. It is seldom practical to count or measure every individual in a natural population or community, nor is it possible to determine all the interactions between individuals, populations, communities, and their environments. Therefore, the modern ecologist must deal with information gathered by means of sampling and model-building and must work extensively with theories of statistical analysis using modern computers.

POPULATION GROWTH

Consider a population of bacteria living together in a laboratory culture dish. Under ideal conditions, one bacterium can divide to produce two daughter cells, and each of its daughter cells will divide after 20 minutes. The size of the whole population doubles every 20 minutes. Table 14.1 shows the calculated growth of such a population, assuming that no members are lost through death and that the doubling continues to occur every 20 minutes.

In real life, such exponential growth rates cannot continue for long. Even for organisms that reproduce more slowly than bacteria, the exponential growth pattern would soon lead to standing

14

Life and the Environment

room only all over the world. Charles Darwin, for instance, calculated the growth of a theoretical population of elephants, assuming that each elephant pair began breeding at age 30, survived until

Table 14.1
Growth of Hypothetical Bacterial Population*

Time (minutes)	Population Size (number of individuals)	Natural Logarithm of Population Size
0	1	0.00
20	2	0.69
40	4	1.39
60	8	2.08
80	16	2.77
100	32	3.47
120	64	4.16
140	128	4.85
160	256	5.55
180	512	6.24
200	1,024	6.93
220	2,048	7.63
240	4,096	8.32
260	8,192	9.01
280	16,384	9.70
300	32,768	10.40
320	65,536	11.09
340	131,072	11.78
360	262,144	12.47

*It is assumed that each individual divides at 20-minute intervals to give rise to two daughter cells and that no individuals are lost from the population through death.

age 100, and produced six offspring during its life. Within a mere 750 years that single pair of elephants would have given rise to a population approaching 19 million elephants!

In actual populations, the growth rate begins to decrease after relatively few generations. In a population of yeast cells grown in laboratory culture, the exponential growth pattern continues for only a few hours. The growth rate of the population does not continue to be proportional to the size of the population (Table 14.2). Instead, the growth rate approaches zero, and the size of the population becomes nearly constant. The growth curve takes on an "S" shape, which is called a *sigmoid curve* (Figure 14.1).

But this leveling off does not mean that the birth rate has become zero. New members continue to be added to the population even after it has reached its equilibrium, or maximum, size. Equilibrium is reached because individuals are dying at the same rate that new individuals are being born. To make things easier, this discussion will consider only populations where members cannot enter or leave their ecosystem—that is, new individuals are added only through birth and individuals are lost only through death.

The equilibrium size of the population changes with environmental conditions. A change in temperature, available space, available food, chemical composition of the environment, or the presence of other populations may alter the size of the equilibrium

Table 14.2
Population Growth of Yeast Cells in Culture

Time (hours)	Population Size (number of individuals)	Growth Rate (individuals per hour)
0	10	0
2	29	9.5
4	71	21
6	175	52
8	351	88
10	513	81
12	594	40.5
14	641	23.5
16	656	7.5
18	662	3

Source: Adapted from Raymond Pearl, *The Biology of Population Growth* (New York: Knopf, 1925).

population (Figure 14.2). It is often said that the environment has a particular *carrying capacity*—it has a limit to the size of the population it can support.

Both the birth rate and the death rate of the population may be affected by environmental conditions as well as by the size and structure of the population itself. If a stable population size is to be achieved, conditions within the population and in the environment must be such that these two rates can become equal.

The growth curves for most real populations approximate the sigmoid curve, but various deviations are common. For instance, there is often a time delay between a change of conditions and the resulting changes in birth and death rates. Therefore, the popula-

Figure 14.1 The upper graph depicts the growth curve of yeast cells grown in a laboratory culture. Note the S-shape.

Figure 14.2 The lower graph shows the growth curves of yeast cells grown under varying environmental conditions.

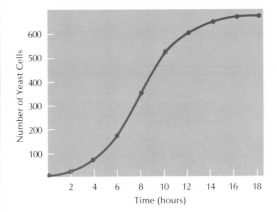

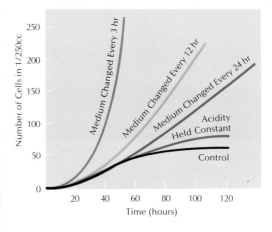

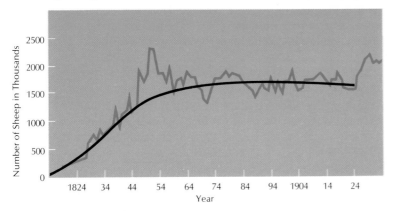

Figure 14.3 Growth curve of sheep following their introduction to an area. Note the initial sigmoid pattern followed by approximate equilibrium.

Figure 14.4   This photograph of a cemetary in the borough of Queens, New York emphasizes the conditions of overcrowded urban centers. In some ways, a cemetary reflects the statistics of a population's life table and therefore may be considered as a rough indicator of the death rate of a population.

Table 14.3
Life Expectancies for Selected Countries

Men	Years		Women
Netherlands	71.4	75.4	Sweden
Sweden	71.3	75.0	Iceland
Norway	71.1	74.8	Netherlands
Israel	70.9	74.8	Switzerland
Iceland	70.7	74.7	Norway
Denmark	70.4	74.2	Canada
Switzerland	69.5	74.1	France
Canada	68.4	73.9	England
New Zealand	68.2	73.8	Denmark
England	68.0	73.4	United States
Spain	67.3	73.0	U.S.S.R.
Puerto Rico	67.3	73.0	Israel
Czechoslovakia	67.2	72.8	New Zealand
France	67.2	72.8	Czechoslovakia
Japan	67.2	72.4	Australia
Australia	67.1	72.3	West Germany
West Germany	66.9	72.1	Japan
United States	66.6	72.1	Puerto Rico

tion may increase beyond the carrying capacity *before* the factors limiting growth can act to decrease the growth rate to zero. In such a case, the size of the population may fluctuate around the equilibrium size (Figure 14.3). If the time lag is such that the oscillations increase in size rather than damping out, the population may disappear on one of the downward turns when the population reaches zero size. In some extreme cases, the population size increases exponentially far past the carrying capacity and then "crashes" toward a very low level. Such a "J-shaped" curve is often observed in the unicellular plant populations that "bloom" in bodies of water in the spring and rapidly exhaust the supplies of nutrients that have accumulated over the winter.

Although the sigmoid curve can be expressed mathematically, the mathematical model has little value in predicting the equilibrium size of a population or in understanding deviations from the smooth sigmoid curve. A more detailed understanding of growth requires a closer look at the opposing factors of birth and death.

Birth Rates

The *maximum birth rate* of a population represents the maximum rate at which new individuals can be added under the best possible conditions. This value is an exponential growth rate in which the rate of births is proportional to the size of the population. In most real populations, the biologist is more concerned with the *actual birth rate* that occurs in some specific environment.

Sometimes birth is most easily measured in terms of a *crude birth rate*, such as "*n*" births per year per 1,000 individuals in the population. But this method excludes the fact that only certain members of the population produce young. For example, a population in which 90 percent of the individuals are male or one in which 90 percent of the individuals are not yet mature will have birth rates markedly different from more normally structured populations. To account for the effects of changes in population structure on the birth rate, biologists prefer when possible to work with the *age-specific birth rate*, which is expressed as the number of offspring produced per unit of time by females of a particular age class. Even more useful for most calculations of population growth is the *gross reproductive rate*, which represents the average rate at which females are born to females of each age group.

Death Rates

Death rates change with age and other individual characteristics even more dramatically than do birth rates. The *crude death rate* represents the number of individuals dying during a period of time for each 1,000 individuals in the population. *Specific death rates* calculated for various age groups are of greater use. Death in a population is commonly represented by a *life table* (a form first developed by insurance companies). It includes a calculation of the *life expectancy* of an individual of a particular age; in other words, the average time that an individual of that age may be expected to remain alive (Table 14.3).

Table 14.4 is a life table for cottontail rabbits. From these data, a graph may be plotted to show the percentage of individuals that

Table 14.4
Life Table for 10,000 Cottontail Rabbits

Age Interval in Months	Number Living to Beginning of Interval	Number Dying During This Age Interval	Death Rate*	Life Expectancy†
0–4	10,000	7,440	0.744	6.5
4–5	2,560	282	0.11	6.6
5–6	2,278	228	0.10	6.5
6–7	2,050	246	0.12	6.5
7–8	1,804	307	0.17	6.4
8–9	1,497	150	0.10	6.4
9–10	1,347	175	0.13	6.3
10–11	1,172	164	0.14	6.3
11–12	1,008	212	0.21	6.3
12–13	796	143	0.18	6.3
13–14	653	98	0.15	6.2
14–15	555	55	0.10	6.0
15–16	500	65	0.13	5.8
16–17	435	31	0.07	5.6
17–18	404	24	0.06	5.3
18–19	380	49	0.13	5.0
19–20	331	36	0.11	4.9
20–21	295	47	0.16	4.6
21–22	248	20	0.08	4.4
22–23	228	39	0.17	4.2
23–24	189	32	0.17	4.0
24–25	157	13	0.08	3.7
25–26	144	7	0.05	3.4
26–27	137	30	0.22	3.1
27–28	107	12	0.11	2.9
28–29	95	13	0.14	2.6
29–30	82	32	0.39	2.4
30–31	50	7	0.14	2.3
31–32	43	9	0.21	2.1
32–33	34	11	0.33	1.9
33–34	23	16	0.70	1.9
34–35	7	3	0.35	2.3
35–36	4	0	0.00	2.0
36–37	4	0	0.00	1.5
37–38	4	0	0.00	1.0
38–39	4	4	1.00	0.5

*Fraction of those alive at the beginning of this interval that die during the interval.
†Average number of months of life remaining at beginning of age interval.

Source: R. D. Lord, Jr., "Mortality rates of cottontail rabbits," *Journal of Wildlife Management* 25 (1961):33–40.

Figure 14.5   The survivorship curves of the animals in this graph are expressed in a variety of different shapes. A majority of large mammals tend to have long life spans.

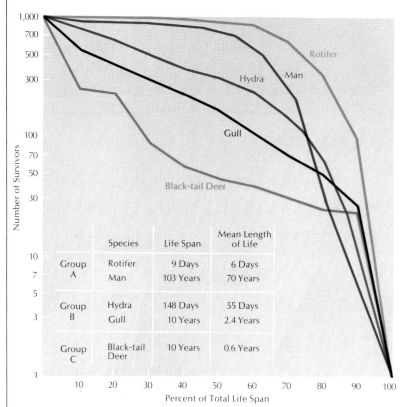

	Species	Life Span	Mean Length of Life
Group A	Rotifer	9 Days	6 Days
	Man	103 Years	70 Years
Group B	Hydra	148 Days	55 Days
	Gull	10 Years	2.4 Years
Group C	Black-tail Deer	10 Years	0.6 Years

survive to any particular age. Such survivorship curves tend to approximate one of three generalized shapes (Figure 14.5). Among cottontail rabbits, nearly three-quarters of the individuals die within the first 3 months of life; for the survivors, the average life span is about 10-1/2 months, but some individuals survive for more than 3 years. Thus, the survivorship curve for cottontail rabbits drops rapidly just after birth and then decreases slowly. Similar curves are typical for many small mammals, fishes, and invertebrates. At the other extreme, a majority of large, mammalian individuals tend to live through a relatively long lifespan, and there is a sudden drop in survivorship at the end of that typical lifespan.

POPULATION CONTROL

In any population, the birth rate cannot long exceed the death rate. The exponential growth resulting from excess births soon causes the population to exceed the carrying capacity of the environment, no matter how large that capacity may be. Various factors then bring about an increase in deaths, a decrease in births, or both, ultimately bringing the population to an equilibrium size or causing it to decrease drastically in size. The mechanisms acting to control population size have been studied extensively, but biologists are not yet in agreement on the nature or importance of many of these mechanisms. It is generally agreed that there are two major categories of control mechanisms. *Density-independent* factors are environmental factors that act directly on members of the population. Their effects are largely unaffected by the density of the population. *Density-dependent* factors result from interactions within the population; their effects vary with the density of the population. Most natural populations show cyclic fluctuations of a greater or lesser extent, and in most cases these fluctuations are controlled by both density-independent and density-dependent factors of varying degrees.

Density-Independent Factors

Extreme cases of population control by density-independent events may be seen in many populations that live in harsh conditions. For example, small insects living in desert areas may be able to survive through dry seasons only as eggs or in a dormant state. During rainy seasons, conditions may be such that the population burgeons exponentially at the greatest rate possible. When the rainy season ends, environmental conditions change dramatically and nearly all of the population dies. The growth curve for this population consists of a series of J-shaped curves occurring at annual or seasonal intervals. The height of each population peak is determined by the number of eggs or dormant individuals left from the last population burst (which, in turn, is a function of the height of that peak) and by the duration of the favorable season.

Some ecologists have argued that such density-independent factors as weather are the major influence on the shapes of population curves that show seasonal peaks. Others have found evidence that, even in extremely seasonal populations, the growth rate is influenced strongly by factors that depend on the density of the population. For example, the number of black-tail deer that die during

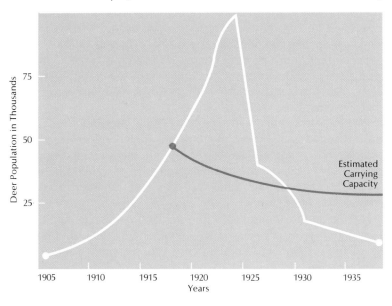

Figure 14.6 As reflected by this graph, the removal of predators from a population generally results in an increase in numbers to the point where the carrying capacity is exceeded and the population eventually becomes drastically reduced.

307

Figure 14.7 The male peacock (*Pavo cristatus*) both before (above) and during (below) courtship display. Certain animals go through elaborate courtship and territorial displays in order to compete for resources in a given area.

a harsh winter may depend not only on the severity of the weather but also on the number of deer competing for available food.

Density-Dependent Factors

Among most populations, the large fluctuations caused by extreme environmental changes are relatively rare. The population size usually does not reach the maximum that could be supported under existing environmental conditions. Instead, as population density increases past some equilibrium value, the growth rate tends to decrease. If the population density drops below this equilibrium value, there is an increase in growth rate. This homeostatic, or "balancing," mechanism tends to keep the population at a size somewhat below the maximum theoretical carrying capacity of the environment. As a result, the population is better able to survive extremes of weather or other potential catastrophes without great increases in mortality.

Various forms of social behavior appear to be among the most important homeostatic mechanisms of population control. By means of dominance hierarchies and territorial behavior, as described in Chapter 13, the available supply of food and space is shared among the stronger members of the population. These individuals can reproduce normally. If the population is too large for the available resources, extra members cannot secure a territory, or mates, or an ample supply of food. Accordingly, the birth rate of the population is adjusted to match available resources and size of the existing population.

There is a possibility that physiological changes may be caused by overcrowding. In many species, births decrease and deaths increase as the population density increases. In some cases, these effects are brought on by increasing competition for food, increasing likelihood for transmission of disease, or lack of space. It has long been known, however, that physiological changes can accompany increasing population densities. During peaks of the population cycle in snowshoe hares, for example, many individuals die from "shock disease," a condition of severe physiological stress characterized by extremely low levels of blood sugar and liver glycogen.

Among many mammalian species, high population densities are associated with various malfunctions of the endocrine system, particularly the adrenal glands. Apparently the increased population density leads to more frequent aggressive encounters between animals, particularly those involved in establishing territories and dominance hierarchies.

In laboratory experiments with mice and moles, definite physiological changes were found to accompany increases in population density. An increase in the size of a population confined to a constant space led to an increase in the weight of adrenal glands and a decrease in the weight of thymus and reproductive glands.

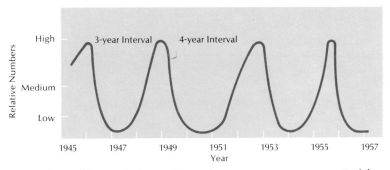

Figure 14.8 Generalized curve of the three- to four-year cycle of the brown lemming population.

As in the wild populations, these changes were accompanied by decreasing birth rates and increasing death rates.

The interactions among different species are important as a mechanism of homeostatic population control. For example, the cyclic fluctuations in lemming populations of the Arctic tundra regions have been explained as the result of interactions between the lemming population and the populations of plants on which it feeds (Figure 14.8). An increase in lemming population size leads to overgrazing and a deterioration of the soil as it is exposed to direct weathering. Overgrazing leads to a food shortage for the lemmings and a lack of cover under which to hide from predators. As the lemming population decreases as a result of the higher death rates, the destruction of the plant populations is lessened. The plant cover is restored in the following growth season, and the stage is set for another population burst among the lemmings.

HETEROGENEITY OF ENVIRONMENT: HABITATS

No environment is completely homogeneous over a large area. Even in a desert, a plain, or a prairie where conditions appear to

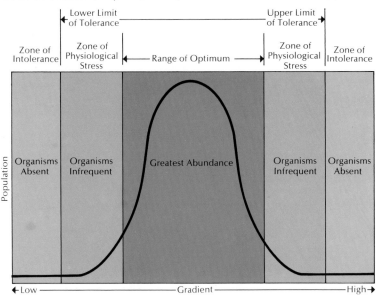

Figure 14.9 The law of tolerance. Each species has a particular range of tolerance within which its life processes can function. The upper and lower limits of tolerance for all environmental conditions of a given species may vary geographically and seasonally.

Figure 14.10 Populations comprising "simple" ecosystems or those with reduced species diversity may sometimes go awry. As an example, the lemming populations of the Arctic tundra and alpine zones may show a marked increase in response to abundant foods. Lemming predators are unable to check this rapid population explosion and thus disease, eventual lack of food, and a unique behavioral phenomenon known as shock disease act to reduce the lemming population. As the population increases, social behavior of an aggressive nature increases. This constantly increasing level of hostile social encounters affects the lemmings' neurophysiology by causing atrophy and final destruction of the adrenocortical complex. These physiological changes produce behavioral changes. The animal first goes into a state of shock during which it may wander aimlessly (giving some credence to the famous "march to sea"), but eventually this wandering leads to coma and death.

be monotonously identical, careful study shows differences in soil, temperature, light energy, and moisture from place to place, even within distances of fractions of an inch. These differences are seen on a larger scale in the distribution of mountains, deserts, oceans, and other macroenvironments over the surface of the earth. For each species, it is possible to describe the environmental conditions that make existence possible. For a particular species, there are a maximum and a minimum temperature that can be tolerated. Within the range of tolerance is a smaller range of optimal conditions within which growth, metabolism, and reproduction occur at the best possible rates. Similar ranges of tolerance and favorable conditions can be established for other conditions such as moisture and food supply (Figure 14.9). Tolerance limits may define the broad outlines of the range of a population. But different conditions often limit the population at different parts of its range. For example, the distribution of a plant may be limited by cold temperatures in the northern part of its range and by a lack of moisture in the southern part of its range. Within the broad limits of this range, individuals or groups tend to cluster in the smaller patches that provide the most favorable conditions. Often, the details of distribution can be interpreted as the effect of competition—as when a water supply is altered unfavorably by other species.

The general conditions under which populations of a particular species are found are described as the *habitat* of the species. In some cases, the habitat is best characterized in terms of temperature, moisture, soil or mineral types, and other abiotic factors. In other cases, the habitat is best characterized in terms of other organisms; for example, the habitat of an insect species may be on the leaves or flowers of a particular plant species. But any description should be based on environmental factors that limit the range of the species and of its constituent populations.

If two species are introduced into a uniform, mutually suitable environment and if all resources—food, light, space, water—are available in unlimited supply, both populations theoretically will continue to grow exponentially without limit. In any real situation, however, the two species are certain to share at least one resource that is in limited supply (space, if nothing else). The two species will have slightly different growth curves. Assume the size of each population approaches the carrying capacity of the environment, whose conditions now include the presence of the other species, which probably allows a lower carrying capacity for each species than either would have alone in the same environment. The result will be that the growth rates of the two populations will decline. Because the two curves are not identical, the growth rate of one population will decline to zero while the other population is still growing, at least slightly. The continued growth of the second population causes it to use still more of the limited common resource, further reducing the growth rate of the first population.

Figure 14.11 Graph illustrating competitive exclusion between two species of floating aquatic plants, *Lemna gibba* and *Lemna polyrrhiza*.

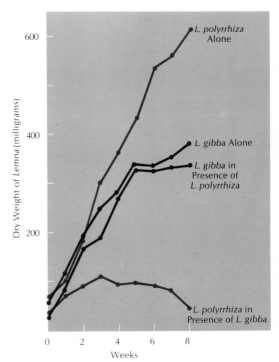

311

Figure 14.12   These growth curves of the protozoans *Paramecium aurelia* and *Paramecium caudatum* demonstrate competitive exclusion in the latter species.

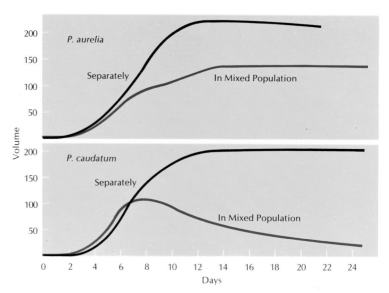

Figure 14.13   The normal feeding position of the shrub-browsing gerenuk. This graceful African gazelle is noted for feeding on shrubs at intermediate height and thus avoids intense competition with neighboring browsers and grazers who feed on grasses.

Now the first population will begin to decline in size while the second population continues to grow (Figure 14.11). This situation is commonly described by saying that the second species has a *competitive advantage* over the first. The more the second population grows, the lower the carrying capacity of the environment for the first population (because both are competing for a single limited resource), and therefore the second population will eventually crowd out the first population entirely.

The principle of *competitive exclusion* summarizes the foregoing argument by stating that under uniform conditions with at least one resource in limited supply, *not more than one species can continue to exist indefinitely.* This principle was first stated by A. Grinnell in 1904 and has since been confirmed repeatedly by laboratory experiments. The coexistence of numerous species in nature is possible only because of the heterogeneity of the environment from place to place or from time to time.

If the conditions vary slightly from place to place, one species may hold the competitive advantage in some places while another holds the advantage in other places. Environmental changes over time may also permit the coexistence of two species. For example, suppose that one species holds the competitive advantage under summer conditions, whereas another holds the advantage under winter conditions. The regular alternation of seasons may prevent either species from completely disappearing before it has an opportunity to rebuild its population at the expense of the other.

In natural environments, the principle of competitive exclusion can be expressed in more general terms: No two species can coexist for long if they live in the same habitat and simultaneously exploit without limit a single resource. Careful examination of coexisting species reveals a slight difference in habitat preference or some behavioral or anatomical difference that will eventually limit the exploitation of common resources of food, space, and so on.

To describe the ways in which organisms exploit the resources of their environment, ecologists have developed the concept of the *niche*. If the habitat of a population is generally described as the place where the population lives, the niche may be similarly described as the way that the population makes its living. A plant-eating animal and an animal-eating animal occupy different niches even though they may occupy very similar habitats. The day feeders and the night feeders also occupy different niches in similar habitats. Thus, no two species can occupy identical niches in identical habitats for long.

COMMUNITIES AND ECOSYSTEMS

In the heterogeneous natural environment, a single population rarely lives alone in a particular area. In most natural macrohabitats, plant populations use solar energy to convert inorganic materials into organic compounds such as carbohydrates; animal populations feed upon the plants; other animal populations feed upon the plant eaters; and fungi and bacteria populations convert the remains of plants and animals back into inorganic materials. A group of populations occupying the same area and interchanging materials and energy is called a *biotic community*.

Obviously, a community occupying a desert floor contains species different from those in a community in a lake or ocean. But some generalizations about the structure of communities can be made. Almost every community contains populations of *producers*, or organisms that use energy from the nonliving environment and produce complex organic molecules from inorganic substances. Photosynthetic plants and algae are the major producers of most communities, but photosynthetic and chemosynthetic bacteria can

Figure 14.14 This club-shaped fruiting body, commonly called a mushroom, is an example of a decomposer in a biotic community. Decomposers break down dead organic matter to obtain their energy.

Figure 14.15 A pyramid of biomass shows the relationships between organisms of certain trophic levels of a biotic community.

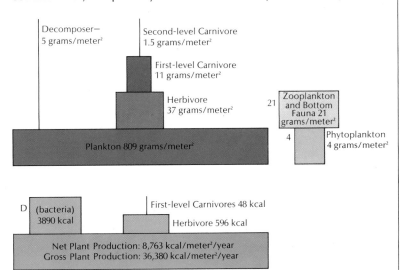

Decomposer—
5 grams/meter²

Second-level Carnivore
1.5 grams/meter²

First-level Carnivore
11 grams/meter²

Herbivore
37 grams/meter²

Plankton 809 grams/meter²

21 Zooplankton and Bottom Fauna 21 grams/meter²

4 Phytoplankton 4 grams/meter²

D (bacteria) 3890 kcal

First-level Carnivores 48 kcal

Herbivore 596 kcal

Net Plant Production: 8,763 kcal/meter²/year
Gross Plant Production: 36,380 kcal/meter²/year

Figure 14.16   This schematic illustration depicts the trophic levels of a community and shows how one trophic level is dependent on another. The sea must produce 1,000 pounds of phytoplankton (d), which is fed upon by tiny herbivores (c), 100 pounds of which is necessary to feed carnivorous fish (b), 10 pounds of which are needed for a man (a) to gain 1 pound. So, for a man to gain a pound the sea must produce a half ton of living matter. The 10 to 1 ratio will vary depending on the nature of the ecosystem.

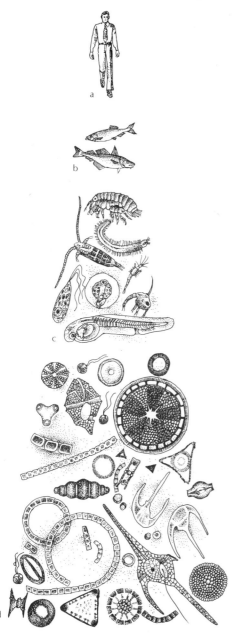

also be producers. For a few communities—such as those in caves or deep in the ocean—the major source of energy and organic molecules may consist of organic material that comes from other communities, and producers may be rare or absent.

Decomposers (bacteria and fungi) are also present in nearly every community. If nitrogen, phosphorus, and other crucial elements are to be constantly available to the producers for synthesis of new organic substances, these elements must be reclaimed from the remains of dead organisms. Carbon, oxygen, and hydrogen are available in water and in the atmosphere in relatively large quantities, but even these materials must be recycled on a global scale if life is to continue. Only in an environment with a constant influx of inorganic nutrients and a constant removal of organic debris would it be possible for a community to exist indefinitely without decomposers.

The *consumers* of the community may be present in greater or smaller numbers, depending on the amount of organic material available from the producers. Although it is theoretically possible to have a community without consumers, rarely does a real community exist in which some organisms have not taken advantage of the food supply from the producers to establish themselves as consumer populations. *Primary consumers,* or herbivores, obtain their supply of organic nutrients from tissues of the producers. *Secondary consumers,* or carnivores, feed on the tissues of primary consumers. In communities with dense producer populations there may be *tertiary consumers,* or second-level carnivores, who feed primarily on the secondary consumers.

A unidirectional flow of energy occurs through the interactions of the community. Energy from the environment is stored in organic molecules by the producers. At each of the other *trophic* levels (primary consumers, secondary consumers, decomposers) of the community, a large portion of that energy is dissipated. The populations of each trophic level consume a portion of the organic material produced in the bodies of the lower level (Figure 14.16). The amount of energy stored in the organic substances of each level is normally less than 10 percent of the total energy stored in the bodies of the next lower level. The remainder of the energy that is obtained through feeding is expended in the life processes of the organisms; most is eventually reradiated from the earth.

As a result of this inevitable energy loss in the community, the structure of the community may be graphed as a pyramid. Because organisms vary greatly in size, it is normally more useful to examine the pyramid of biomass rather than the pyramid of numbers (Figure 14.15). A pyramid of energy is even more informative but normally difficult or impossible to calculate from available information on a natural community.

In most real communities, the actual structure of the trophic levels is quite complicated. Each population feeds on certain combinations of the other populations. For example, a particular population of secondary consumers may feed primarily on a single population of primary consumers but may also occasionally feed upon other primary consumers, other secondary consumers, and even on the producers themselves. The actual trophic relationships form a complex *food web*, whose structure may vary with seasonal and other changes (Figure 14.18). A simplified picture may be obtained by looking at the energy relationships along a single *food chain*, as shown in Figure 14.19.

Few communities have been quantitatively studied in detail. Those that have are relatively small and atypical. In a Georgia salt marsh, for example, about 6 percent of the solar energy falling on the surface is used by photosynthetic organisms in the production of organic materials. Of the energy converted by the producers, 77 percent is used in respiration within the producers themselves, a figure much larger than that typical of forests or other communities with large plant bodies. Of the energy stored in plant tissues, about 47 percent is dissipated by bacteria (decomposers) and about 8 percent by primary and secondary consumers. The remaining 45 percent of the stored energy is lost to the ecosystem as the living and dead organisms are washed out of the salt marsh.

In this community, producers and decomposers play the major role in energy transfer, and a sizeable proportion of the energy fixed by the producers is exported from the community (Table 14.5). In a spring community, on the other hand, the primary source of energy is organic debris washed into the community, and little energy is exported (Table 14.6).

Table 14.5
Energy Utilization in a Salt Marsh Community*

Energy Use Breakdown	Kilocalories per Square Meter per Year	Percentage of Total Energy Supply
Total incoming light energy	600,000	100.000
Reflected, heating environment, etc. (not utilized by community)	563,620	93.937
Gross production	36,380	6.063
Producer respiration	28,175	4.696
Building producer tissues (net production)	8,205	1.367
Bacterial respiration	3,890	0.648
Primary consumer respiration	596	0.099
Secondary consumer respiration	48	0.008
Exported from community	3,671	0.612

*Assuming total yearly income light energy of 600,000 kilocalories per square meter.

Source: Adapted from J. M. Teal, "Energy flow in the salt marsh ecosystem of Georgia," *Ecology* 43 (1962):614–624.

Figure 14.17 Comparative ecosystems—a typical pond ecosystem and a hypothetical urban ecosystem. Insight into the operation of ecosystems can be gained by considering the sources and transfers of energy and food material. The ultimate source of energy for all ecosystems (defined as a community of interacting organisms and the environment in which they live) is the sun. Plants use solar energy to drive the energy-requiring reaction called photosynthesis, which is the process by which simple carbohydrates are manufactured from carbon dioxide and water by chlorophyll-containing cells. Even electricity and gas—important as energy sources in urban ecosystems—can ultimately be traced to the sun. Electricity is often produced by generating plants that use fossil fuels (including coal and oil) to drive generators. Coal and oil are products of the decomposition of ancient plants, which originally obtained their energy from the sun by the process of photosynthesis. Only energy produced by nuclear reactions, such as in nuclear-powered electricity generating plants, cannot be traced to the sun. In the pond ecosystem, plants include pond weeds such as water milfoil (u) and water celery (d) as well as microscopic, unicellular plants including diatoms (t) and other algae (h) that are collectively termed phyto-

plankton (a). The urban ecosystem has far fewer species of plants within its confines. In fact, to supply food for their inhabitants, cities must bring in plant-produced material (including wheat and corn) from outside. One of the most significant cultural achievements of man has been the development of agricultural techniques that maximize the amount of food produced per unit of land area utilized. In a food web, plants are referred to as primary producers. The next level or link in the food web contains the herbivores, the plant-eating organisms. In the pond ecosystem, herbivores include copepods (i) and other members of the zooplankton grouping (which feed on phytoplankton) and ducks (which feed primarily on pond weeds). In the urban ecosystem, the herbivores are cattle, which do not live in the city but certainly supply a good deal of its food. The next position in the food web consists of carnivores— animals that eat other animals. In the pond ecosystem, the carnivores are the larger fish such as catfish (j), bass (l), and sunfish (m); birds such as the heron (p) and osprey (r); water snakes (s); frogs (g); and dragonflies (q). In the urban ecosystem, some species of birds and cats fill the role of carnivores. Animals that eat both plants and other animals are omnivores, which in the pond include fish such as

carp (k) and minnows (n); water beetles (o); and some species of turtles. In the urban environment, man is by far the most significant omnivore in terms of amount of consumption. Dogs, which are almost always supported by man, can also be considered omnivores. Another grouping of organisms that are extremely important in the cycling of organic matter and energy within ecosystems are the scavengers, which feed primarily on dead plant and animal matter. In ponds, this grouping includes crayfish (b) and snails. In cities, cockroaches, silverfish, and rats are scavengers. Again, it is important to realize that these animals depend almost entirely on the waste products of man for their source of food. All of these classifications are somewhat arbitrary. Often animals perform more than one function in any given ecosystem—carp (in ponds) and dogs (in cities) can be both omnivores and scavengers. A final grouping of organisms that perform an extremely important function in any ecosystem are the decomposers. Decomposition is the process by which organic material is broken down into smaller compounds, including carbon dioxide and water. Thus, keeping in mind that photosynthesis requires carbon dioxide and water to manufacture simple carbohydrates, it can be seen that the transfer of

matter and energy within ecosystems is cyclical. Bacteria (f) are, by far, the most important decomposers in any ecosystem. In ponds, they are located either in the water or in bottom mud. In urban ecosystems, the process of decomposition occurs to a large extent in sewage disposal plants where bacterial cultures are maintained by man. By keeping decomposition confined to small, isolated areas, man also strives to minimize the spread of disease bacteria that may be carried with waste products. Every animal must have a place to live that provides shelter from climatic forces and refuge from predators. In ponds, this important function is often performed by larger plants such as water lilies (e), cattails (c), and pond weeds (d) that provide shelter for birds, fish, and other animals. In cities, man has provided homes not only for himself but also for a large proportion of the other animals. Such structures as buildings and sewers provide homes for cats, dogs, birds, silverfish, and cockroaches. In the modification of the environment to suit his needs, man has drastically changed the kinds and numbers of ecological niches, that is, the positions that specific organisms occupy in an ecosystem.

Figure 14.18 The food web of a salt marsh in midwinter. Producer organisms (terrestrial and salt marsh plants) are eaten by herbaceous invertebrates that live on land (grasshopper and snail). Marine plants are consumed by herbivorous marine invertebrates. Fish eat plants from both ecosystems and are in turn eaten by first-level carnivores (great blue heron and common egret). Examples of omnivores that make up a food web are ducks, sparrows, rats, mice and other small rodents, and sandpipers. Another first-level carnivore is the shrew; representative second-level carnivores are the hawk and owl.

Table 14.6
Energy Relationships in a Cold Spring Community

Energy	Kilocalories per Square Meter per Year	Percentage of Available Chemical Energy
Source		
Organic debris entering community	2,350	76.150
Gross photosynthetic production	710	23.006
Immigration of caddisfly larvae	18	0.583
Decrease in producer biomass	8	0.261
Use		
Heat dissipation by community	2,185	70.803
Deposition of organic matter	868	28.127
Emigration of adult insects	33	1.070

Source: Adapted from J. M. Teal, "Community metabolism in a temperate cold spring," *Ecological Monographs 27* (1957):283–302.

An ecosystem chosen to represent a relatively uniform vegetation or animal assemblage is often called a *biome*. Typical biomes include the temperate deciduous forest, the coniferous forest, the desert, the tropical rain forest, and the prairie.

The boundaries of communities, ecosystems, and biomes are drawn arbitrarily. Any portion of the environment could be chosen for study as an ecosystem, but such a system is convenient for study only if the flows of energy and materials across its boundaries are minimal, or easily measured and estimated. Because each species has its own limits of tolerance for various environmental factors, the ranges of any two populations are seldom identical. For that reason it is normally impossible to draw firm lines around a community that unequivocally separate it from neighboring communities. Biomes, similarly, grade into one another so that boundaries must be fixed rather arbitrarily.

CHANGES IN COMMUNITIES

Within any community the sizes of various populations are continually changing. Nearly all populations show sizeable fluctuations in size, usually in regular cyclic patterns. As environmental conditions change or new populations immigrate into the community, the balance of competitive advantage may shift and some old populations may dwindle or disappear. Nevertheless, most communities are surprisingly stable. The balance among populations moves now one way, now another, but over a span of a century the total changes are surprisingly small.

The most rapid changes in communities are seen when a forest fire, flood, or other drastic change in the environment suddenly alters the conditions. The barren land produced by a natural catastrophe, or by human action, is first occupied by a relatively simple community of hardy *pioneer organisms*. This community gradually

Figure 14.19   Flow diagram of energy relationships along a single food chain showing a consecutive pattern of fixation and transfer by the components and considerable respiratory losses at each transfer. As the color key indicates, grass represents a producer; bacteria, a decomposer; the mouse, a primary consumer (herbivore); the skunk, a secondary consumer (first-level carnivore); and the wolf, a tertiary consumer (second-level carnivore). P = gross primary production; P_n = net primary production; and P_2, P_3, P_4, and P_5 = secondary production at the indicated levels.

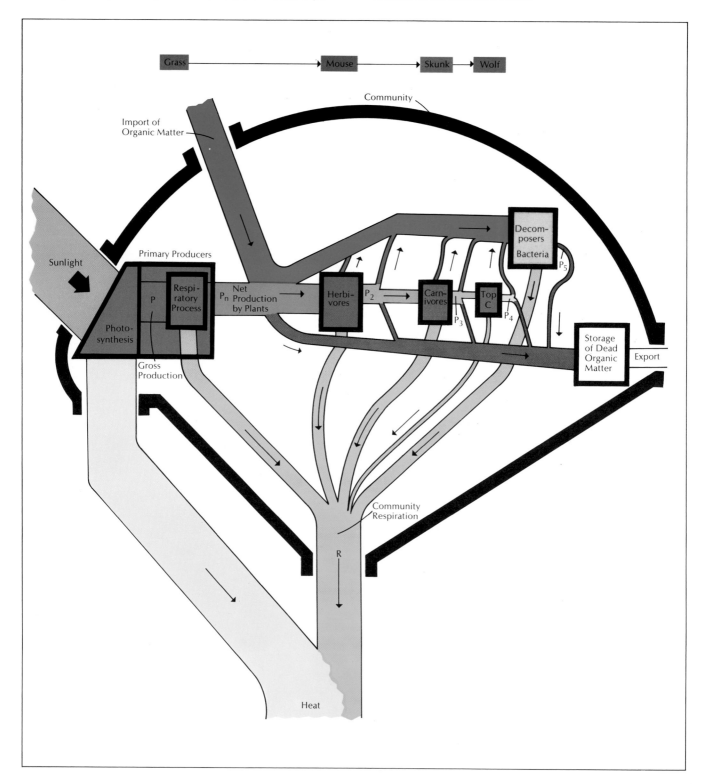

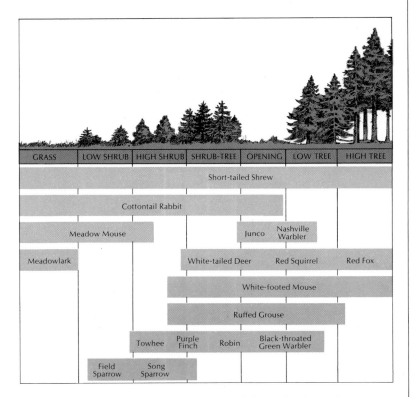

| GRASS | LOW SHRUB | HIGH SHRUB | SHRUB-TREE | OPENING | LOW TREE | HIGH TREE |

Figure 14.20 Community changes as illustrated by wildlife succession in conifer plantations in central New York. Note the disappearance of some species and the appearance of others as conditions of vegetation height and density change.

alters the nature of the soil and establishes shade and moisture within which other organisms can grow. As conditions are gradually changed by the organisms, new populations become established and the pioneer populations lose their competitive advantage and disappear. The end product of this process of *succession* —which may pass through a number of distinctive intermediate communities—is a complex and relatively stable community similar to the one that occupied the region before the catastrophe.

The process of succession can occur only where individuals from neighboring regions are able to immigrate into the region of succession. Presumably such immigration goes on all the time, but only when conditions are just right can individuals become established, compete successfully, and form a growing population.

The legacy of both human greatness and human tragedy was engraved into the genetic program of our species by forces at work millions of years ago. During the period of geologic time known as the Pliocene, a drought that began in Africa was to last for 10 million years. This unrelenting desiccation slowly pushed back the ancient forests that had cloaked the land during the preceding 12 million years of the Miocene Epoch, and in its place there appeared an open, parklike grassland known as a savanna.

The new habitat called for a new behavior, and at least one group of forest-dwellers responded to the challenge. These were the semi-erect, apelike herbivores that may have already begun foraging expeditions beyond the margin of the forest, returning at dusk to the branches far above the ground-dwelling predators of their kind. These creatures, over time, would evolve into all modern-day apes—and man.

Three of the great apes alive today—the chimpanzee, the orang-utan, and the gorilla—are forest dwellers. And, as speculative as the notion may be, the behavior of these species may in some respects afford us insight into the behavior of our early Pliocene ancestors. All modern apes, for example, use rudimentary tools. The chimpanzee and the gorilla crush leaves and use them to soak up drinking water out of otherwise inaccessible hollows. A thin twig is inserted into termite quarters, later to be removed and stripped of the edible soldier termites that have clamped onto the twig with their mandibles. Enraged apes throw rocks and attack with clubs. And perhaps most significantly, the chimpanzee and the gorilla generally band together into social groups. Social patterns are established to provide order, stability, leadership, and predictable defensive behavior when danger threatens.

Logic tells us that our Pliocene ancestors would not have ventured far from the safety of the trees unless they had somehow managed to establish similar patterns of social behavior. And, with the thinning out of the forests, there would have been environmental pressures for them to do so. Immense reproductive advantages would accrue to those groups that could move into, and survive in, the new environment. The fossil record attests to their success. There was a steady increase in brain size of these creatures, particularly in the cerebral cortex, the region associated with conceptual thought. There was the development of an upright stance, which freed the hands for defense; without the formidable teeth of the carnivore, our herbivorous ancestors would need all the adaptive advantages they could muster. And there may have been a concurrent increase in the utilization of rudimentary tools until, some 3 million years ago, the forest herbivores unquestionably had evolved into omnivorous hunter-gatherers of the plains.

Could these creatures be called "man"? If brain size alone were the criterion we would think not; the brain size of these australopithecines, as they came to be called, was scarcely above one-third

15

Man in the Biosphere

Figure 15.1 Schematic diagram illustrating the "spectrum" or "continuum" in the evolutionary development of the hominids. In early times, the hominid family was characterized by diversity as much as by unity. The fossil evidence supports a view that all human evolution should be regarded as a continuum, with a wide variety of forms existing at any time and a gradual shift in genotypic and phenotypic characteristics through time.

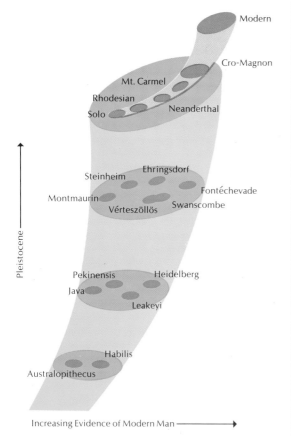

that of the average modern brain. And yet, what is the correlation between brain size, body size (the australopithecines weighed a scant 100 pounds), and the capacity for conceptual thought? The answer to that question has not been resolved. If tool use alone were the criterion, then surely they would be no more or no less manlike than the tool-using chimpanzee today; the early australopithecines left behind only the most rudimentary tools.

But there are in the evolutionary record indications of a steady progression not only in brain size but in increasingly deliberate modifications of "natural" tools. It seems, then, that the australopithecines were transitional beings standing between tool-using animals and tool manipulators, displaying patterned behaviors that were too complex to be accidentally stumbled upon and learned anew in each passing generation. These behaviors would be the indicators of *culture*—the ability to learn from the accumulated experience of others. With the evolution of culture, the natural selection of genotypes would gradually be supplanted by the natural selection of ideas and technology.

By the Middle Pleistocene, undeniably *human* beings had penetrated the extremes of their natural habitat and, with culture as their environmental mediator, they moved into incredibly diversified environmental zones. Even in the face of advancing glaciations they persisted and, with their increasingly refined tools, they exploited the vast game herds that abounded during the Pleistocene Epoch. About 13,000 years ago, during the last recession of the ice sheets that had covered much of Europe, the game herds inexplicably began to disappear. It was in response to that environmental pressure that man turned in a direction that would effectively remove him from the control of nature: he began the domestication of plants and animals. In stabilizing his food resources, he found the means of proliferating unchecked across the face of the earth. And in this stabilization, he effectively terminated the process of natural selection for the speciation of his kind.

HUMAN POPULATION GROWTH

By the year 1650—in less than 15,000 years—the population of the human species reached half a billion. By 1850 it had doubled to 1 billion; and within 80 years it had doubled again. In this decade it will reach 4 billion. Given the present rate of increase, by the year 2000 there will be four human beings on our planet for every human being that existed in the year 1900, or twenty-eight for every person who lived in the first century.

The forces underlying the explosive growth of humanity are not difficult to define. The size of a population, like the size of an individual, depends on the ratio between input and output. For a population, birth is the input and death is the output. The population grows if more individuals are added through births than are removed through deaths. In 1970, for example, about 18 births oc-

Figure 15.2 (left)   Growth rate of the human
population for the past one-half million years. If
Old Stone Age were in scale, its base line would
extend about 18 feet to the left.

Figure 15.3 (right)   A reasonable interpretation of
man's progression up the evolutionary scale.

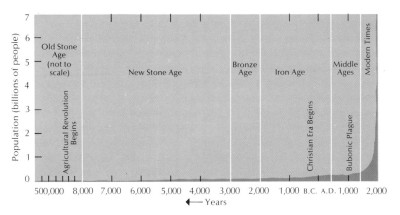

curred for each 1,000 persons in the average total population for
that year; the 1970 death rate was about 10, which means the
population increased by 4/5 percent. If it continues to grow at this
rate it will double every 70 years. Table 15.1 gives birth rates, death
rates, growth rates, and doubling times for various nations.

Not only is the size of the human population increasing; the
rate of growth is also accelerating. What is happening today is a
precipitous decline in the world-wide death rate—especially
among infants and children—largely because of our ability to control microorganisms that cause infectious diseases.

Before medical science intervened, man's natural birth rate was
just another example of what Thomas Malthus described as nature's "profuse and liberal hand." Man was no exception to the
biological law of surplus. All organisms are geared to produce a
surplus of offspring to compensate for everpresent mortality factors—disease, famine, predation, and genetic failure. In parts of
Africa, Latin America, and Asia, Pliocene and Pleistocene levels of
mortality still persist and half of all children born will die before

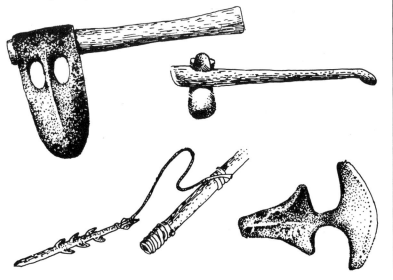

Figure 15.4   Tools from the Upper Paleolithic
included axes, harpoons, and celts. These tools
indicate that the cave dwellers of this period relied
primarily upon hunting and fishing as a way of life,
but there is evidence they may have harvested wild
grains as well.

325

Table 15.1
Birth, Death, and Growth Rates of Selected Modern Nations (1969)

Nation	Birth Rate*	Death Rate†	Growth Rate‡ (*percentage*)	Doubling Time (*years*)
Costa Rica	45	7	3.8	18
Philippines	50	15	3.5	20
Honduras	49	17	3.4	21
Mexico	43	9	3.4	21
Pakistan	52	19	3.3	21
Zambia	51	20	3.1	23
Iran	50	20	3.1	23
Mongolia	40	10	3.0	23
United Arab Republic	43	15	2.9	24
Brazil	38	10	2.8	25
Albania	34	8.6	2.7	26
Laos	47	23	2.6	27
Nigeria	50	25	2.5	28
Turkey	46	18	2.5	28
India	43	18	2.5	28
Indonesia	43	21	2.4	29
Chile	33	10	2.3	31
Cambodia	41	20	2.2	32
Guinea	55	35	2.0	35
Nepal	41	21	2.0	35
Canada	18	7.3	2.0	35
WORLD AVERAGE	34	15	1.9	37
Australia	19.4	8.7	1.8	39
Argentina	23	9	1.5	47
Mainland China	34	11	1.4	50
Uruguay	21	9	1.2	58
Greece	18.5	8.3	1.2	58
Puerto Rico	26	6	1.1	63
Portugal	21.1	10	1.1	63
Japan	19	6.8	1.1	63
United States	17.4	9.6	1.0	70
France	16.9	10.9	1.0	70
Denmark	18.4	10.3	0.9	78
Switzerland	17.7	9.0	0.9	78
Spain	21.1	8.7	0.8	88
Norway	18	9.2	0.8	88
Poland	16.3	7.7	0.8	88
Italy	18.1	9.7	0.7	100
Finland	16.5	9.4	0.6	117
Ireland	21.1	10.7	0.5	140
Czechoslovakia	15.1	10.1	0.5	140
West Germany	17.3	11.2	0.4	175
Hungary	14.6	10.7	0.3	233
Belgium	15.2	12.2	0.1	700
East Germany	14.8	13.2	0.1	700

*Rates in number of individuals per 1,000 total population.
†Low death rates in rapidly growing countries are largely a result of the small proportion of older individuals in those populations.
‡Calculated growth rates include allowance for immigration and emigration, which is very significant in some countries.

Source: Population Reference Bureau, *1969 World Population Data Sheet* (Washington, D.C.: Population Reference Bureau, Inc., 1969).

they are 6 years old. As recently as 1952, 34 percent of all babies died within a year of birth in northeastern Brazil.

Birth rate, too, can change because it is also under environmental control. Lactation, for instance, delays ovulation and helps space babies at 2- to 3-year intervals if the mother nurses. Behavioral and mechanical controls can prevent births, albeit with varying degrees of efficiency (Table 15.2).

Table 15.2
Failure Rates of Birth Control Methods

Method	Pregnancy Rate*		
	High	Average	Low
No contraceptive	80	80	80
Aerosol foam	80	—	29
Foam tablets	43	—	12
Suppositories	42	—	4
Douche	41	—	21
Jelly or cream	38	—	4
Coitus interruptus	38	—	10
Rhythm	38	—	0
Diaphragm and jelly	35	—	4
Condom	28	—	7
Lactation	26	—	24
Steroid hormones ("the pill")			
Progestin alone	2.1	—	0.2
Sequential	1.4	—	0.5
Combined	0.7	—	0.1
Intrauterine devices ("the loop")			
Silicone loop	—	2.2	—
Shell loop	—	0.5	—
Abortion	0	0	0

*Values are equivalent to the percentage of women who will become pregnant during a one-year period while using the specified method. The "low" rates approach the minimum achievable by motivated and medically supervised women.

Source: Data from B. R. Berelson, et al. (eds.), *Family Planning and Population Programs* (Chicago: University of Chicago Press, 1966); S. J. Segal and C. Tietze, "Contraceptive technology: current and prospective methods," *Reports on Population/Family Planning* (New York: The Population Council, October 1969); and C. Tietze, "New intrauterine devices," *Studies in Family Planning No. 47* (New York: The Population Council, 1969).

Abortion is still the most widely used birth-control procedure. In Japan, China, and some East European countries it is a legal and inexpensive method for preventing birth. Where abortion is outlawed it nevertheless continues but is practiced under sorry circumstances. A recent study in Chile showed that 30 percent of pregnancies are terminated by abortion, most of them clandestine and dangerous. Several studies in European cities indicate that about one-half the pregnancies are terminated by illegal abortion, many self-inflicted or performed by untrained persons. It has been estimated that 4 percent of Italian abortions result in maternal death. As steroid hormone pills, the intrauterine device, and other methods become more freely available, abortion will be used less. Until then, however, the practice will continue to be widespread.

THE PROBLEMATIC RISE OF PSEUDO-SPECIES

If birth rate is controllable, why does its decrease lag so far behind that of the death rate? Even as man has proliferated his control over the environment, he has lost control over his own interventions and may be destroying himself in the balance. In the creation of his increasingly artificial world, his conceptual powers have also created pseudo-species that have divided him into struggling social populations.

Today, parents raise about as many children as they want, and they want the number that will maximize their social and economic status. Traditional, space-limited societies may have negative or even nonexistent growth rates. On some Melanesian and Polynesian atolls or on tiny farms in East European countries, family size has been controlled stringently for some time. Relatively low growth rates are characteristic of nearly all industrialized countries. But in some industrialized and nonindustrialized countries, the growth rate is unchecked. Why is there a disparity? The answers are found in the cultural history of these diverse societies.

While industrialization was occurring in certain countries, infant and child mortality was high and explosive population growth was prevented. Even as late as 1900, European death rates were 18 to 20 per 1,000, and in the United States infant mortality was 16 percent. In addition, rising population densities, mechanization of agriculture, and a shift from rural-farming families to urban-worker families made smaller families more desirable. In preindustrial agricultural societies, children—even young ones—contribute to the productivity of the farm. In cities, children are strictly consumers until they become employable, and as long as the immediate family is a stable institution, city parents choose to have only a few children. By the time medicine and sanitation made it possible for nearly all children to survive to adulthood in the industrialized countries, small families were seen to be beneficial. Birth rates fell accordingly.

But the tables have been turned on approximately 90 percent of the world's population. Most people are subsistence farmers. They generally consider large families to be desirable not only because more progeny mean more working hands but because surviving sons are the only form of old-age insurance these people have. But recently these traditional societies have been affected by medical control and rapid change in social patterns.

At the close of World War II, public health programs began employing biocides such as DDT to control malaria-carrying in-

Table 15.3
Infant Mortality Rates for Selected Countries

Country	Infant Deaths*
Sweden	12.9
Iceland	13.3
Netherlands	13.4
Finland	14.2
Norway	14.6
Japan	15.0
Denmark	16.9
Switzerland	17.5
New Zealand	18.0
Australia	18.3
England	18.3
Gibraltar	18.7
Canal Zone	20.1
Luxembourg	20.4
Taiwan	20.6
France	20.6
Scotland	21.0
East Germany	21.2
United States	22.1

*Number of deaths of infants under 1 year of age per 1,000 live births (excluding fetal deaths).

Source: Statistical Office of the United Nations, *Statistical Yearbook* [1968] (New York: United Nations, 1969).

sects. Insect-control, sanitation, and immunization programs brought about a sharp decline in death rates. Before DDT had been introduced into Ceylon, for example, the death rate in that country was 22 per 1,000 per year. Within 8 years it had declined to 10 per 1,000.

Increasing numbers of children are placing severe strains not only on available food resources but on schools and hospitals—at a time when impoverished governments are hoping to divert their limited resources to the support of industrial development and improvement of the standard of living. To make things worse, the affluent industrialized nations—with only about 15 percent of the world's population—are using over 50 percent of the world's annual production of nonrenewable fossil fuels and ores. The implications of this rate of consumption are serious. The required amounts of many metals, including copper and zinc, exceed known and projected reserves. If by some miracle all the world's people were raised to the level of affluency enjoyed by Western societies, the earth would be soon depleted of all ores.

Many countries were not well-endowed to begin with. In fact, many of the so-called undeveloped countries of Asia, North Africa, the Middle East, and Central America exist on the ruins of the most ancient civilizations. More often than not, their high-grade ores have already been depleted, their soils eroded, and their forests cut away. Not only are these people off to a late start in the modern race to affluence, they are also handicapped by inferior economic conditions as a result of the exploitation of their environment by their predecessors.

The accelerating pace of social change aggravates the problem. People from poverty-ridden rural areas look to the cities for alleviation of their conditions. Expectations rise, and untutored masses

Figure 15.5 Changes in the rate of mortality among selected Asian nations. The average rates of 1945–1949 are compared with those of 1960–1961. The decrease in death rate is due largely to "death control" through world health organizations.

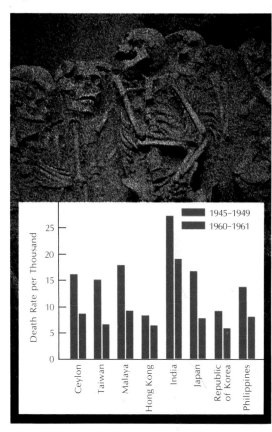

Figure 15.6 Man's short-sighted exploitation of natural resources has left billions of blighted acres unfit for all but marginal subsistence.

329

of people migrate to urban areas, where they converge and add to the existing conditions of poverty, overcrowding, and cultural traditions alien to their own.

Birth rates can decline, but only when people desire smaller families. And small families become desirable only when children are too expensive to feed, clothe, and educate — and when parents need not live in terror that child mortality will leave them alone and without support in their old age. In short, the precondition for lowered birth rates is security — which, paradoxically, is as elusive today as it was for our Pliocene progenitors.

GROWTH AND THE LIMIT ON RESOURCES

The tragedy of the human population explosion is the ultimate finiteness of the planet earth and the absolute authority of the laws of thermodynamics. Whether our planet's carrying capacity is a human population of 3 billion or of 300 billion, there is undeniably a limit. It is, nevertheless, an academic exercise to discuss the ultimate carrying capacity of our planet when a majority of the people alive at this moment are hungry and ignorant. As a species, we presently furnish the materials and facilities to allow about 400 million individuals to reach their full genetic potential. For the other 90 percent of our species, one or more of the following es-

Figure 15.7   A bread sculpture showing the geographical distribution of hunger in the United States. Malnutrition (darkened areas) and deficiency diseases debilitate as many as 20 million people in the United States. Retarded mental development and stunted growth are just a few of the consequences that might result from a diet deficient in requirements such as protein.

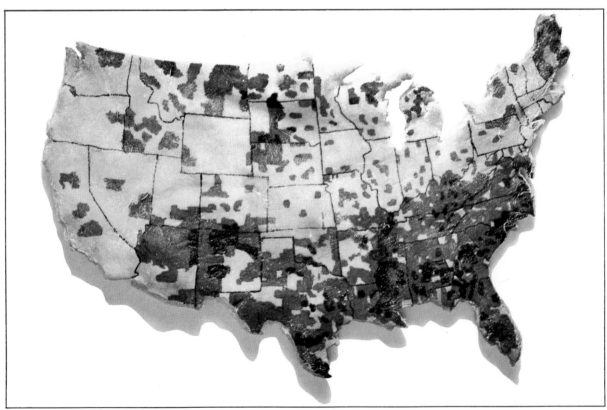

sentials are lacking: good nutrition, quality education, and sufficient medical care.

What are the prospects for feeding another 3 billion people in 30 years? Broad suggestions for increasing food production include breaking new lands, enhancing productivity of lands already exploited, preventing storage losses, and developing agricultural technologies of various sorts.

Three-fourths of the earth's surface is covered with water, most of which is found in the oceans. The oceans will continue to produce important protein supplementation, but seafood is not a significant source of the basic food requirement—calories. By the year 2000 the oceans will probably furnish 30 percent of our minimal *protein* requirements but only 3 percent of our energy needs.

Figure 15.8 shows that the land area itself is not all potential farmland. Between 60 percent and 70 percent is nontillable desert, tundra, or mountains, or is either covered with snow or lacking topsoil. That means 30 or 40 percent is available for farming and grazing. Half of this land is already being farmed. The balance either is used for grazing or lies in the tropics; virtually all of it is considered "marginal" because of low fertility or unpredictable rainfall. In fact, very little of the total land on our planet has the right combination of rainfall, nutrients, temperature, and topography to permit normal agriculture. It is no accident, for example, that Nevada is sparsely populated. Water is so expensive to transport that it is not economical even for rich nations to farm such dry lands. Economically undeveloped nations simply cannot afford large-scale irrigation, even if water were available to transport.

Good lands are being lost to urbanization, erosion, and drying much faster than new land is being broken. Figure 15.9 shows changes in land utilization between 1882–1952. The pace of these changes has been accelerating since then and, at the current rate of population growth and farm productivity, the 950 acres left will provide less than 10 years' supply of food.

In theory, the productivity of present agricultural lands can be doubled or even tripled in many regions. All that is needed is increased input of water and nutrients, or fertilizers. It sounds simple, but it costs money. Why does India with a population of 550 million use the same amount of nutrients as Sweden, with a population of 8 million? Why does the entire continent of South America use no more nutrients than Holland? The answer to this question again lies in economic realities. Agricultural workers would have to be trained and sent to remote villages. The mining of minerals, the building of factories, and the construction of dams, roads, railroads, and aqueducts to move and process the necessary materials would require enormous outlays of capital that, almost invariably, are nonexistent.

Socioeconomic problems beset similar proposed schemes. The development of supergrains, synthetic foods, edible bacteria

Figure 15.8 (above)   Potential farmland on a global scale based on the nature of the soil.

Figure 15.9 (below)   Changes in world land utilization (1882–1952). The net addition of tilled acreage amounts to 210 million hectares during a period when the world population grew by almost 2 billion people. Thus, the gain is reduced to only 0.1 hectare (1/4 acre) per individual.

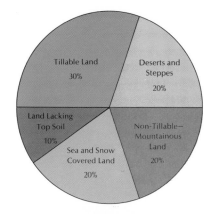

	1882	Percent	1952	Percent	CHANGE 1888–1952	Percent
Forest	5.2	45.4	3.3	29.6	−1.9	−36.8
Desert and wasteland	1.1	9.4	2.6	23.3	+1.5	+140.6
Built-on land	0.87	7.7	1.6	14.6	+0.73	+85.8
Pastures	1.5	13.4	2.2	19.5	+0.7	+41.9
Tilled land	0.86	7.6	1.1	9.2	+0.24	+24.5
Area not especially utilized	9.53	83.5	10.8	96.2	+1.27	+12.9
	1.81	16.5	0.27	3.8	−1.54	−79.9
Total	11.34	100	11.07	100	−0.27	−2.4

Changes in Land Utilization 1882–1952
(billion hectares)

Figure 15.10 This graph shows the decreasing food projection versus the increasing total population to the year 2000.

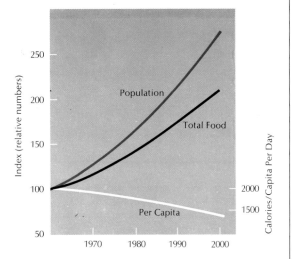

grown on petroleum, nuclear reactor-based agricultural-industrial complexes, and algae farming turn out to be economically impractical or unprofitable. Nor is a solution in sight for our food storage problem: Between 20 percent and 50 percent of the food produced is lost through spoilage while 2 billion people in the world go hungry.

In our effort to extend our control over the environment, it appears we inadvertently have managed to diminish the quality of our food as well. In 1969, J. Kelsay reported that even in the United States diets have deteriorated since 1955. Protein (or its component amino acids) is an essential dietary constituent. To increase productivity, plant breeders have created larger, starchier grains, and the protein percentage in those grains has declined steadily until today most of the high-yield grains provide only 5 to 10 percent protein. Western countries import protein in the form of fish meal and soya bean cakes, largely to supplement the diets of livestock. Corn of the modern hybrid varieties has so little protein that it must be mixed with protein additives before it is adequate for hog consumption. Grains once supplied a balanced, protein-sufficient diet. In many places today, grain must be supplemented with expensive fish, beans, milk, eggs, and meat.

What effect does malnutrition have on our species? There is increasing evidence that the short stature, slow minds, and physical deformities of many of the world's people result in large part from malnutrition. In East Pakistan alone, for example, 50,000 people a year become permanently blind from vitamin A deficiency. Even though only 20 percent of body growth occurs in the first 3 years after birth, it is during this period that the brain grows 80 percent. Therefore, protein deficiency during this period stunts brain growth and now is widely believed to result in permanent mental retardation.

In 1963 a South African study compared a group of undernourished children raised under extremely unfavorable conditions to a group that was well-fed and well-cared for. The groups seemed similar genetically. By the time they were 8 years old, the undernourished group was on the average 3-1/2 inches shorter and 5 pounds lighter than the well-nourished group. Moreover, the average head circumference, which is directly proportional to brain size, was 1 inch less. Other studies have shown that enrichment of diet *after* dwarfing has occurred does not reverse the effect on brain size even though it partially compensates for stunting of body size. In 1966 J. Cravioto wrote about the relation between nutritional stunting and mental retardation in South America. Significant differences in intersensory organization (a major variable in intelligence testing) were found in Guatemalan Indian children and were correlated with their heights. Short stature was largely the effect of malnutrition. According to N. Scrimshaw's 1968 calculation, there are now 300 million members of our species who are the silent victims of dietary retardation. About 20 million of these people occur in the United States alone.

The future holds no miracles. The ocean's production of fish, now about 60 million metric tons per year (40 percent of the world's protein), can be increased to above 100 million metric tons by the year 2000. But because of the projected population increase, this amounts to a per capita *loss* in protein. One recent development that could be very beneficial is the breeding of grain varieties that have higher than normal amounts of the essential amino acids lysine and tryptophane. This breeding will substantially increase the protein quality of grains, assuming they can be developed for each crop and climate.

The food shortage is now chronic for over half the world's people; about 10 to 20 million die annually of starvation and its effects now. In 30 years, the existing levels of food resources will have to be doubled merely to keep the average amount of food per person the same as it is today. To provide a minimal healthy diet for the entire population, it will be necessary to triple the present food supply. From where will this food come?

OUR ARTIFICIAL ECOSYSTEM AND THE WORLD'S FOOD CHAINS

Complex ecosystems such as forests and savannas are homeostatic systems. A change in the numbers of a few species or a gradual shift in climate can be compensated for. These ecosystems normally can survive for eons. In searching for ways to feed our numbers, we have been destroying the world's complex ecosystems and in their place we are putting a single-species ecosystem—a crop. The crop needs constant protection and care; it has none of the homeostasis of a natural ecosystem. It can be devoured by an invasion of a single insect species. Winds can knock down the feebly-rooted, quickly-grown plants. Irrigation may be needed to supplement precipitation. Early rains may wash away the seedlings, late rains may cause the plants to rot. If the crop is successfully harvested and stored, up to one-third or even one-half will be consumed by insects and rodents. And finally, nutrients removed from the soil by harvesting must be replenished.

The farmer is never more than a stride ahead of calamity; neither is the world he feeds. The loss of a single summer's crop in the Northern Hemisphere would exhaust humanity's slim margin of survival. Only North America has enough food stored to withstand a disaster. The possibility of disaster is not hypothetical, nor is it necessarily something we can plan for and alleviate. In 1815, for example, the volcano Mount Tambura on the Indonesian island of Sumbawa threw 150 cubic kilometers of ash into the atmosphere. There was no summer sun to nurture crops in much of the Northern Hemisphere the following year.

Attempts to feed today's population and the 72 million added annually have led to more strains on the shaky foundations of the

Figure 15.11 A crop that took an entire season to cultivate can be destroyed in a matter of minutes by a swarm of migratory cicadas (commonly known as locusts).

simplified ecosystem. Modern agriculture now is totally dependent on artificial additives—pesticides and inorganic fertilizers.

Pesticides are powerful agents of natural selection. When they are applied, the most susceptible insects die, leaving the least susceptible to reproduce. Because selection promotes the evolution of resistance in pesticide targets, it becomes necessary to increase the dose and frequency of spraying, thus intensifying the selection on the target organisms and increasing environmental contamination. Resistance to the three major groups of insecticides—chlorinated hydrocarbons, organic phosphates, and carbamates—now is so widespread that crops of many kinds are on the verge of collapse. Based on this knowledge, it is clear that the withdrawal of pesticides is the best solution in some areas.

Natural pest control *can* reassert itself, as G. Conway has reported for the cocoa crop in the state of Sabah, Malaysia. Insecticide sprayings were not preventing the defoliation and death of many trees. Because the worst insect outbreaks began *after* the introduction of a heavy spraying program, the application of pesticides was abandoned.

It is the indirect effect of pesticides that is more in the public consciousness. One group of pesticides, the chlorinated hydrocarbons (including DDT, DDD, dieldren, and lindane), is highly resistant to oxidation and enzymatic attack and is relatively insoluble

Figure 15.12   Schematic representation of the path of DDT residues through the food chain. Note the increasing concentration of the residue as one proceeds up the food chain to the carnivore levels.

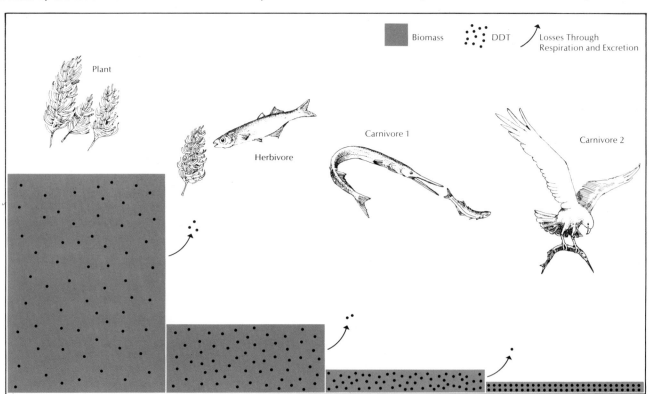

LOCATION		ORGANISM	TISSUE	CONCENTRATION (parts per million)
U.S. (average)		Man	Fat	11
Alaska (Eskimo)				2.8
England				2.2
West Germany				2.3
France				5.2
Canada				5.3
Hungary				12.4
Israel				19.2
India				12.8–31.0
U.S.	California	Plankton		5.3
	California	Bass	Edible Flesh	4–138
	California	Grebes	Visceral Fat	Up to 1,600
	Montana	Robin	Whole Body	6.8–13.9
	Wisconsin	Crustacea		0.41
	Wisconsin	Chub	Whole Body	4.52
	Wisconsin	Gull	Brain	20.8
	Missouri	Bald Eagle	Eggs	1.1–5.6
	Connecticut	Osprey	Eggs	6.5
	Florida	Dolphin	Blubber	About 220
Canada		Woodcock	Whole Body	1.7
Antarctica		Penguin	Fat	0.015–0.18
Antarctica		Seal	Fat	0.042–0.12
Scotland		Eagle	Eggs	1.18
New Zealand		Trout	Whole Body	0.6–0.8

Figure 15.13   DDT residues, including the derivatives DDD and DDE, have been reported in most ecosystem food webs. Note the especially high concentrations in the eggs of the carnivorous birds.

in water (see Figure 15.13). As a consequence, these substances accumulate in the tissues of organisms exposed to them. With some important exceptions, including insect strains that have evolved resistance, organisms cannot dispose of chlorinated hydrocarbons as they do natural wastes and toxins. Predators and filter feeders at the top of food chains naturally accumulate the most. Many predatory and oceanic bird populations are now on the verge of extinction. It is still too early to predict what effects will accrue from the high pesticide levels in whales, porpoises, sea birds, fishes, crabs, shellfishes, and men. Further, no one knows what the effect will be of eliminating top predators from the ocean.

Pesticides are not the only problem. Inorganic nitrate and phosphate fertilizers are responsible for a growing number of ecological imbalances. First, inorganic fertilizers "loosen" the nitrogen cycle in the soil by shortcutting certain steps. As with pesticides, the more they are used, the more they are needed. Second, like pesticides, these fertilizers do not remain where they are applied but are dispersed and concentrated in unexpected places. Lakes are the most seriously affected because they are nutrient traps. Great "blooms" of algae are promoted by the nitrates and phosphates. Oxygen is depleted by bacterial decomposition of the masses of algae, and oxygen-starved fish die and wash ashore. Only the hardiest fish, such as carp, can withstand much oxygen depletion.

Lakes age prematurely under such conditions. In the absence of oxygen, the decomposer organisms cannot keep up with the rain

Figure 15.14   Inorganic nitrates and phosphates often fertilize monstrous growths of algae in fresh-water ponds and lakes. When these algae "blooms" subside, the subsequent bacterial decomposition of their bodies depletes the surface water of oxygen and a "kill" of other organisms occurs.

of algal corpses. Organic sludge accumulates rapidly on the bottom. C. Powers and A. Robertson have indicated the "eutrophication" of Lake Erie has caused it to age the equivalent of 1,500 years in the last 50 years. Biologists used to believe that once the lakes were no longer fertilized by entering nutrients, the nutrients would be forever "locked up" in the sludge at the bottom of the lake, and eutrophication would therefore cease. Now, however, there is considerable evidence that anaerobic conditions promote the recycling of nutrients in such a lake. Thus, even if nutrient pollution were to cease, eutrophication would continue.

Reliance on nitrates already is affecting man directly. High nitrate levels are showing up in wells in agricultural areas throughout the United States. Nitrate itself is relatively innocuous; but bacteria in the intestines of infants convert it to nitrite. In the blood, nitrite combines with hemoglobin to produce methemoglobin. Methemoglobin does not have the affinity for oxygen of hemoglobin, and its presence can lead to respiratory distress and suffocation. The disease is becoming a serious public health problem in the agricultural valleys of California as well as in other states.

If we are able to envision the disastrous consequences of widespread pesticide and nutrient applications, why don't we simply prohibit their use? The reason is that agricultural collapse and widespread famine immediately would follow a sudden termination of pesticides and nutrients. Insect-borne disease also would increase significantly. According to statistics of the World Health Organization, about 10 million people have been saved from fatal effects of malaria by the use of DDT in antimosquito campaigns. Again we are locked in an unworkable ecosystem of our own making.

Can man and other higher animals evolve resistance to pesticides and other chemical pollutants? Perhaps, but the penalties are

Figure 15.15   The accumulation of large masses of organic matter has accelerated the eutrophication process to the extent that Lake Erie has "aged" the equivalent of 1,500 years in the last 50.

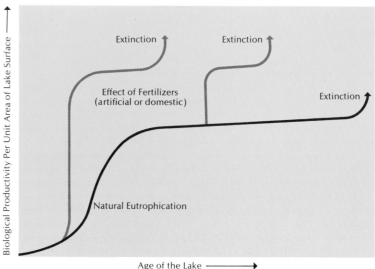

forbidding. Insects have evolved resistance by virtue of natural selection. Usually, only a fraction of a percentage of the population survives a heavy application, and many generations of massacres precede the appearance of significant numbers of resistant individuals. The cost of quickly increasing the frequency of the genes that might confer resistance in man is beyond anything our species has suffered yet, at least if resistance is to be acquired in a few generations.

THE FUTURE FOR MAN

In the evolutionary perspective, man rarely has been physically secure. With the exception of the last few decades in a few countries, half his children have died before reaching maturity. He has always lived under the shadow of famine, drought, and disease. The disease- and danger-ridden eons of human existence provide the context for the present crises. Those who read this book are recent escapees of man's precarious past—part of a small human elite for whom life is not always a battle for mere survival. Man's future will resemble the past more than the present. Once again, survival is the issue.

This time, however, man is competing not against an unrelenting environment but against himself. And this is the paradox of his own perception: He has the capacity to envision the apocalypse of his own self-destruction—his final tragedy—yet at the same time he has the capacity to envision an alternative response before the error is irreversible. As in the dawn of his past, he is faced with a new habitat that calls for a new life style. At least some of his kind will respond to the challenge.

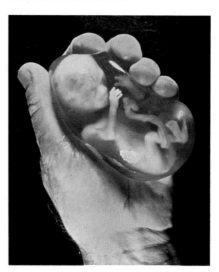

Figure 15.16   In a very real sense, man holds the future in his hand. If he is able to exercise restraint in some areas of endeavor while pushing ahead to meet more and more difficult challenges in other areas, he may be able to do more than just ensure the survival of mankind.

Epilogue

"What is life?" The question has been asked innumerable times but has been answered to the satisfaction of few. Science is based on the experience that nature gives intelligent answers to intelligent questions. To senseless questions, nature gives senseless answers—or no answers at all. If nature has never provided an answer to this question, perhaps something is wrong with the question.

The question is wrong indeed. It has no sense, for life in itself does not exist. No one has seen or measured life. Life is always linked to material systems; what man sees and measures are living systems of matter. Life is not a thing to be studied; rather, "being alive" is a quality of some physical systems.

A look at the living world reveals an incredible variety of shapes, sizes, forms, and colors. There seems to be an infinite variability among living systems. How can man approach such complexity? How can he ask intelligent questions?

One key to an intelligent approach may be the simple fact that things can be put together in two different ways: randomly or meaningfully. Things put together in random fashion form a senseless heap. Nine persons selected at random and placed together probably will form nothing more than a slightly puzzled collection of nine individuals. Nine persons selected and combined in a meaningful fashion may form a championship baseball team. The whole in this case is more than the sum of its parts—it is what is called organization.

If an atomic nucleus is combined with electrons, an atom is formed. This atom is something entirely new, quite different from electrons or nuclei alone. When atoms are combined, molecules are formed. Again, a new thing is generated with strikingly different qualities. Smaller molecules—say, amino acids—may be combined to form a "macromolecule"—perhaps a protein. This macromolecule has a number of amazing qualities. It demonstrates self-organization—the ability to create more complex, higher structures. It may act as an enzyme to speed up a particular chemical reaction, or it may act as an antibody to neutralize the effects of some other specific protein molecule. Proteins can be created in a literally inexhaustible variety of forms, each with its own qualities.

Macromolecules may be combined to form small "organelles," such as mitochondria or muscle fibrils. When they are combined, the result is a cell—the unit of life, the miracle of creation—capable of reproduction and of independent existence.

The more complex the system, the more complex its qualities. Organs may be built from cells; from organs may come an individual organism, such as a human being. Individuals in turn may be combined to form societies or populations, which again have their own rules. At each level of complexity are new qualities not present in the simpler levels. The study of each level yields new information for the biologist.

The history of biology has been marked by a penetration into ever smaller dimensions. In the sixteenth century, Vesalius was dependent on his unaided eyesight for his study of the human body. In the following century, the optical microscope led to the discovery of many new details of structure. Marcello Malpighi observed the capillary vessels that complete the cycle of blood circulation and showed that even such tiny insects as the silkworm have an intricate internal structure. Anton van Leeuwenhoek described blood cells and the compound eyes of insects. Robert Hooke described the cellular structure of plants.

As microscopes were improved, more and more details of structure were described. By the nineteenth century, it was becoming clear that all complex organisms are composed of semi-independent units called cells. The major structural features of cells were established. Bacteria were discovered and studied.

In this century, the electron microscope has taken the scientist down to molecular dimensions, and he has learned to observe with x-rays as well as with visible light. Organic chemistry was established in the nineteenth century, and by the beginning of this century, it was clear that this approach could be applied to the study of living systems. Biologists have had to learn a new anatomy — the anatomy of molecules. Chemists and physicists have penetrated the atom, first finding the elementary particles and then moving still deeper into the realm of wave mechanics. The discovery of the wave properties of the electron has given a deep insight into the nature of biological reactions.

As scientists attempt to understand a living system, they move down from dimension to dimension, from one level of complexity to the next lower level. I followed this course in my own studies. I moved from anatomy to the study of tissues, then to electron microscopy and chemistry, and finally to quantum mechanics. This downward journey through the scale of dimensions has its irony, for in my search for the secret of life, I ended up with atoms and electrons, which have no life at all. Somewhere along the line, life has run out through my fingers. So, in my old age, I am now retracing my steps, trying to fight my way back toward the cell.

I have concluded that life is not linked to any particular unit; it is the expression of the harmonious collaboration of all. As I descended through the levels of complexity, I studied simpler units and found myself speaking more and more in the language of chemistry and physics.

J. F. Danielli has shown that the subcellular organs of various cells are interchangeable. They can be transferred from one cell to another, much as organs can be transplanted from one human individual to another. The parts of the cell have no individuality. The quality of individuality resides in the higher organization — in the cell or the individual.

No one yet knows the higher principle that holds a cell together. Perhaps the answer will be found in irreversible thermodynamics. The good working order of a living cell may correspond to a stable state with a high probability of occurrence. Perhaps some new principle — as yet undiscovered — keeps the cell together. Living systems do not only maintain their good working order but they all tend to improve it, to make the working structure more complex. When the fundamental principle that holds the cell together is found, perhaps we will then also understand what brought together the first living system and understand what drives living systems toward self-perfection.

Scientists know today that rather complex molecules — amino acids, nucleic bases, even macromolecules — can under certain conditions be built without intervention of living systems. They are still seeking the principle that brought them together for the first time and that makes these systems improve themselves by building more complex structures, capable of more complex functions.

Biology is a very young science. It has called itself a separate science for only some eight decades. No one can expect it to find the answers to all questions. The most important questions are yet unanswered — perhaps unasked. What biologists can do — what they are doing at present — is to ask questions that seem answerable with present techniques. They ask questions about structure and function, from the nature of consciousness down to the behavior of electrons, hoping that some day all of this detailed knowledge will come together in a deeper understanding.

Perhaps some day they will find a new way of looking at things. The best scientists, with the aid of giant computers, cannot yet fully explain the behavior of three electrons moving within an atom. Yet those three electrons — even dozens of electrons — know exactly what to do and never miss. In the essence, nature may be far simpler than is believed.

To see the solutions, scientists must preserve a certain naïveté, a childish simplicity of the mind, an ability to recognize a miracle when they see it every day. The solution may be far closer than it seems. It was a hundred years ago that H. P. Bowditch, one of the first American physiologists, showed that after a frog's heart has been stopped for a while, its first beats are rather weak. The heart gradually regains its original strength, with the record of the heartbeat rising like a series of stair steps. Bowditch called this phenomenon "the staircase."

One might expect the heart to be stronger after a rest. Yet what is observed here can be considered a general quality of living systems. Life generates life: rest or inactivity causes life to fade away. Muscles weaken if they are not used; they become stronger if exercised regularly.

This principle is one of self-organization, one of the most striking differences between a living system and a nonliving one. A machine is worn out by usage. A living system is worn out by inactivity. Living systems are able to organize and improve themselves.

S. J. Hajdu and I tried to discover the mechanism that produced the Bowditch staircase. We found that potassium leaks out from the heart fibers into the surrounding fluid when the heart is not beating. When the heart resumes its activity, the potassium is pumped back into the heart fibers. The change in strength of the heartbeats is caused by this change in the distribution of potassium. The movement of the potassium back into the fibers, against the potassium concentration gradient, increases the amount of organization or order in the system. The entropy of the system is decreased. Ernst Schrödinger in 1944 suggested that the ability to decrease entropy is the most characteristic feature of living systems. In a living system, the decrease of entropy leads to further decrease of entropy, to greater order. The increase of entropy leads to further increase of entropy, the maximum state of entropy being death.

Further pursuit of these thoughts would lead into abstract speculation. I would like to point out, however, that these abstruse questions are not at all far from the sickbeds of suffering patients. One of our most important drugs is digitalis, which is used to stimulate failing hearts. Hajdu and I showed that digitalis helps the heart to pump potassium back into its fibers and to retain it. The inability of a heart to maintain its potassium concentration may be one cause of its failure.

Living systems are clearly different from nonliving systems. There must therefore be a difference in the way these two kinds of systems are thought about. Physics is undoubtedly the most basic science. In a way, biology is only an applied science, applying physics as a tool for the analysis of living systems. However, there are distinctions between physical and biological events.

Suppose a process, left to itself, is likely to occur 999 times in one way for each time that it occurs in another. The physicist concerns himself primarily with the first way. Physics is the science of the probable.

Biology is the science of the improbable. On principle, biological reactions must be improbable. If man's cells worked only through the probable reactions, they would soon run down. In order to regulate itself, a cell must use improbable reactions and must make them take place through very specific tricks. The cell may make use of just that one way in a thousand that the physicist ignores.

The cell finds a way to make the reaction go at just the right moment and at the desired rate. The reaction may be improbable, but if it is thermodynamically possible, the cell will find a way to use it.

Physically, all of us—you and I—are improbable. The probability of atoms happening to come together in the complex structure that makes up my body is so tiny that it is practically equal to zero.

Another difference between the physical and biological approaches to the study of a reaction lies in the matter of isolation. The physicist is apt to attempt to isolate the reaction he wishes to study. In biology, single reactions are rarely encountered. Most biological reactions are parts of complex chains. They cannot be fully understood except as members of the chain, or even as parts of an entire living system—the living biological entity.

For example, one of the most important biological reactions is the "electron flow" that underlies photosynthesis and biological oxidation. These processes generate the energy that keeps living systems going. In these reactions, electrons "flow" from molecule to molecule. If an electron moves from molecule A to molecule B, it leaves a positive electric charge on molecule A. This charge tends to pull the electron back toward A. Electrons could not move against such a strong electrostatic attraction. However, if molecule A is a member of a chain and simultaneously receives an electron from a third molecule, its positive charge will be neutralized. There will then be nothing to pull the electron back toward A. Thus, where electrons cannot move from A to B in an isolated system, they can "flow" without difficulty if A and B are members of a chain.

Like many biological researchers, I have often worked for long periods, using all the tricks of chemistry, wave mechanics, and mathematics to understand a certain reaction. In the end, I have found that the cell carries out this reaction in the only way that it could be accomplished. In my long research career, one of the greatest mysteries to me was the way in which a living cell — without the aid of computers or even a brain — could find this single path to the necessary result.

In one way, the discoveries of genetics have made our understanding of evolution even more difficult. A cell does not directly alter the molecules in evolution. Instead, it alters the code of the nucleic acid in its chromosomes. Then all of the descendants of that cell make the appropriate changes in the molecules involved in the reaction. In most cases, a number of genes must be altered to accomplish a meaningful change in a chemical reaction. If all of these genes were not changed simultaneously, only confusion would result.

According to present ideas, this change in the nucleic acid is accomplished through random variation. The nature of protein molecules formed in the cell is determined by the code of the nucleic acid. The protein cannot alter the nucleic acid. If I were trying to pass a biology examination, I would vigorously support this theory. Yet in my mind I have never been able to accept fully the idea

that adaptation and the harmonious building of those complex biological systems, involving simultaneous changes in thousands of genes, are the results of molecular accidents.

The feeding of babies, for example, involves very complex reflexes. These reflexes require extremely complex mechanisms, both in the baby and in the mother, which must be tuned to one another. Similar mechanisms are involved in the sexual functions of male and female animals. These mechanisms must be tuned precisely to one another in order to achieve successful copulation. Thousands of genes must be involved in the coding of these mechanisms. The probability that all of these genes should have changed together through random variation is practically zero, even considering that millions or billions of years may have been available for the changes.

I have always been seeking some higher organizing principle that is leading the living system toward improvement and adaptation. I know this is biological heresy. It may be ignorance as well. Yet I think often of my student days, when we biologists knew practically nothing. There was then no quantum theory, no atomic nucleus, and no double helix. We knew only a little about a

few amino acids and sugars. All the same, we felt obliged to explain life. If someone ventured to call our knowledge inadequate, we scornfully dismissed him as a "vitalist."

Today also we feel compelled to explain everything in terms of our present knowledge. Identical twins are often exactly alike in the smallest details of physical appearance, indicating that the instructions for building this entire structure must have been encoded in the genetic materials that they share. All the same, I have the greatest difficulty in imagining that the extremely complex structure of the central nervous system could be totally described in the genetic codes. Thousands of nerve fibers grow for long distances in order to find the nerve cell with which they can make a meaningful junction. Surely the nucleic acid did not contain a blueprint of this entire network. Rather, it must have contained instructions that gave the nerve fiber the "wisdom" to search for and locate the only nerve cell with which it could make a meaningful connection. Perhaps this guiding principle also is related to the way in which the first living system came together.

I do not think that the extremely complex speech center of the human brain, involving a network formed by thousands of nerve cells and fibers, was created by random mutations that happened to improve the chances of survival of individuals. I must believe that man built a speech center when he had something to say, and he developed the structure of this center to higher complexity as he had more and more to say. I cannot accept the notion that this capacity arose through random alterations, relying on the survival of the fittest. I believe that some principle must have guided the development toward the kind of speech center that was needed.

Walter B. Cannon, the greatest of American physiologists, often spoke of the "wisdom of the body." I doubt whether he could have given a more scientific definition of this "wisdom." He probably had in mind some guiding principle, driving life toward harmonious function, toward self-improvement.

Life is a wondrous phenomenon. I can only hope that some day man will achieve a deeper insight into its nature and its guiding principles and will be able to express them in more exact terms. It is this mysterious quality of life that makes biology the most fascinating of sciences. To express the marvels of nature in the language of science is one of man's noblest endeavors. I see no reason to expect the completion of that task within the near future.

—Albert Szent-Györgyi
Woods Hole, Massachusetts

Albert Szent-Györgyi, who considers himself to be an American, was born into a Hungarian family of prominent scientists 77 years ago. He was unwillingly involved in his early life in the two World Wars and in the political intrigues of the big powers engaged in these wars. Throughout his life, he has been quietly searching for the principles on which nature and all of life are organized. In 1937 Szent-Györgyi's search resulted in the isolation of vitamin C, for which he was awarded the Nobel Prize. His protest against man's "idiocy," which he believes is evident in the irrational pursuit of war and politics that characterizes our,Western culture, has been capsulized in his two short books, The Crazy Ape *and* What's Next. The Crazy Ape *warns youth against the gerontocracy that rules the world and urges mankind to take advantage of technological skills in order to create a psychologically and socially progressive world where humanistic values are paramount. To him, research is "not a systematic occupation but an intuitive artistic vocation."*

Tell Us What You Think...

Students today are taking an active role in determining the curricula and materials that shape their education. Because we want to be sure **Biology: An Appreciation of Life** is meeting student needs and concerns, we would like your opinion of it. We invite you to tell us what you like about it—as well as where you think improvements can be made. Your opinions will be taken into consideration in the preparation of future editions. Thank you for your help.

Your name_____ Institution_____

City and State_____ Course title_____

How does this text compare with texts you are currently using in other courses?

☐ Excellent   ☐ Poor
☐ Good   ☐ Very poor
☐ Adequate

Name other texts you consider good and why._____
_____
_____

Do you plan on selling the text back to the bookstore, or will you keep it for your library?   ☐ Sell it_____ ☐ Keep it_____

Circle the number of each chapter you read because it was covered by your instructor.

1 2 3 4 5 6 7 8 9 10 11 12 13 14 15

What chapters did you read which were not assigned by your instructor? (Give chapter number.)_____
_____

Please tell us your overall impression of the text.	Excellent	Good	Adequate	Poor	Very Poor
1. Did you find the text to be logically organized?	_____	_____	_____	_____	_____
2. Was it written in a clear and understandable style?	_____	_____	_____	_____	_____
3. Did the graphics enhance readability and understanding of the topics?	_____	_____	_____	_____	_____
4. Did the captions contribute to a further understanding of the material?	_____	_____	_____	_____	_____
5. Were difficult concepts well explained?	_____	_____	_____	_____	_____

Can you cite examples which illustrate any of your above comments?_____
_____
_____
_____

Which chapters did you particularly like and why? (Give chapter number.)_____
_____
_____

Which chapters did you dislike and why?_____
_____
_____

After taking this course, are you now interested in taking more courses in this field?   ☐ Yes   ☐ No

Do you feel that this text had any influence on your decision?   ☐ Yes   ☐ No

Comments:_____
_____
_____
_____

Wilfred J. Wilson, a major advisor and key contributor to this book, is a professor of zoology at California State University, San Diego. He was a recipient of the California State College Distinguished Teaching Award and, as a Shell Merit Fellow at Stanford University, was one of twenty scientists selected from the United States and Canada to design and implement science curricula in colleges and universities. Dr. Wilson has been a visiting lecturer for the Fulbright Foundation in Taiwan and Burma. His research interests lie in comparative animal development; his teaching interests involve the teaching of relevant general and developmental biology to undergraduate students.

Theodore Friedmann, a contributor and advisor for this book, received his M.D. degree from the University of Pennsylvania in 1960 and is currently a faculty member of the Medical School at the University of California, San Diego, in the Department of Pediatrics. Dr. Friedmann, whose major research interests lie in the areas of human genetics and virology, belongs to several associated societies and has recently published related articles in *Science* and *Scientific American.* Before coming to UCSD, he held positions at the Salk Institute for Biological Studies and at the National Institutes of Health, Bethesda, Maryland.

Paul D. Saltman, an advisor on this book, is a professor of biology and vice-chancellor of academic affairs at the University of California at San Diego. Prior to coming to UCSD in 1967, he served 14 years on the faculty at the University of Southern California. Dr. Saltman earned both his B.S. in chemistry and his Ph.D. in biochemistry at the California Institute of Technology. As a professor, Saltman has won several teaching awards and has been active in developing meaningful teaching programs for both undergraduate and graduate students. One of his primary research interests in the area of biochemistry is the chemistry of iron metabolism, especially as it relates to human disease conditions, and he has published numerous papers in his field.

Gail R. Patt, presently acting chairman of Boston University's Department of Natural Science, received her Ph.D. from that institution in 1963. Her interests lie in the field of developmental biology. One of the writers for this textbook, Dr. Patt has also recently published two books in comparative vertebrate biology and in genetics.

Advisory Board

Contributing Consultants

Susan Bryant is an assistant professor in the Department of Developmental and Cell Biology at the University of California at Irvine. She received her early training and her Ph.D. in biology from the University of London. At Case Western Reserve University, she spent two years as a postdoctoral fellow in the laboratory of Professor Marcus Singer, a renowned student of regeneration. Dr. Bryant's research interests lie in the problems involved in the regeneration of vertebrate appendages.

Leland N. Edmunds, Jr. is an associate professor of biology at the State University of New York at Stony Brook. After obtaining his Ph.D. from Princeton University in 1964, he spent a summer at the University of Costa Rica. His research interests include biological clocks, cell biology, and control of the cell cycle. Dr. Edmunds belongs to several professional organizations, including the American Society for Microbiology and the American Society of Plant Physiologists.

Peter H. Hartline received a B.A. from Swarthmore College in physics and later pursued his interest in animal behavior through an M.A. at Harvard University. Because he believed the base of behavior to be the nervous system, he did his doctoral work in the field of neuroscience and received the Ph.D. from the University of California at San Diego in 1969. His main research and scientific interests are in the processing of information in sensory systems and its relation to the perception and behavior of animals.

Jonathan Hodge received his doctorate in the history of science from Harvard University in 1970. He has taught at the University of Toronto and has been a visitor for two years at the University of California at Berkeley. An historian of biology and philosophy, he has been especially concerned with the origins of the Darwinian theory. He has papers on Jean Lamarck and Robert Chambers forthcoming and is working on a book analyzing different traditions in the treatment of origins and species from Plato and Aristotle to the present.

Yashuo Hotta received the Ph.D. in biology from Nagoya University in Japan. He was a postdoctoral fellow at the Plant Research Institute of the Canadian Department of Agriculture and a research associate at the University of Illinois. Currently, Dr. Hotta is an associate research biologist at the University of California at San Diego. His interest in the chemistry of nucleic acids is reflected in his numerous publications.

William A. Jensen, chairman of the Botany Department at the University of California at Berkeley, received both his M.S. and Ph.D. from the University of Chicago. He has done considerable research abroad, and his primary botanical interests lie in the areas of histochemistry, cytology, and cell development.

Lee H. Kronenberg is currently a predoctoral fellow in molecular biology and biochemistry at the University of California at San Diego. He received his undergraduate training in biophysical chemistry at the University of California at Berkeley. His present research interests include the regulation of gene expression, ribonucleic acid metabolism in eucaryotic cells, biological aspects of the secondary structure of nucleic acids and interferons, and related problems in virology. He has taught introductory physics and chemistry at Berkeley, and biochemistry, molecular biology, and chemical evolution at San Diego. He participated in the development of the Contemporary Natural Sciences course at Berkeley and remains deeply committed to communication of science to nonscientists and novices.

Robert D. Lisk has contributed various review chapters and numerous research articles on neural regulation of reproduction to professional journals. He received his Ph.D. from Harvard and is now a professor of biology at Princeton University. He is a member of the Society for Study of Reproduction, the International Brain Research Organization, the Endocrine Society, and is on the editorial boards of several journals.

Vincent T. Marchesi received his Ph.D. from Oxford University in 1961 and his M.D. from Yale University in 1963. After an internship and residency in pathology at Washington University, Dr. Marchesi was a research associate at Rockefeller University. He is presently chief of the Section on Chemical Pathology at the National Institute of Arthritis and Metabolic Diseases. His research interests include cell interactions in inflammatory reactions, the chemistry and structure of cell membranes, and the properties of tumor cells.

David M. Phillips is an assistant professor of biology at Washington University in St. Louis, Missouri. He received his Ph.D. in zoology at the University of Chicago and was a postdoctorate fellow at Harvard Medical School. His research interests include the mechanisms of cellular motility and some aspects of cellular nucleic acid metabolism. His studies on motility involve cinematographic analysis of the swimming movement of spermatozoa from various species of insects and mammals, and the correlation of these swimming movements with differences in structure.

David M. Prescott, who has been an exchange scientist to the USSR for a one-month lecture tour, is a professor in the Department of Molecular, Cellular, and Developmental Biology at the University of Colorado. After receiving his Ph.D. from the University of California at Berkeley, Dr. Prescott was an American Cancer Society Fellow at Carlsberg Laboratory in Copenhagen and a Markle Scholar in Medical Science. His research focuses on mechanisms of chromosome replication and function, chromatid exchanges during the cell cycle, exchange of proteins between the nucleus and cytoplasm, and the factors regulating the initial synthesis of DNA during cellular reproduction.

Roberts Rugh has taught and conducted research in embryology for 44 years. He received his Ph.D. from Columbia University, where he was, until his recent retirement, a professor of radiology in the College of Physicians and Surgeons. His research for the last 23 years has focused on the effects of radiation on the embryo and fetus, and the 221 titles and 6 books he has had published are almost exclusively concerned with embryonic development. Dr. Rugh's most recent publication, *From Conception to Birth: The Drama of Life's Beginnings,* which he coauthored with Dr. L. B. Shettles, is illustrated with color pictures he has taken of human fetuses.

Howard A. Schneiderman is dean of the School of Biological Sciences, director of the Center for Pathobiology, and professor of biological sciences at the University of California at Irvine. Dr. Schneiderman received his Ph.D. in physiology from Harvard University in 1952. He was assistant and then associate professor of zoology at Cornell University between 1953 and 1961. In 1961 he joined Case Western Reserve University as professor and chairman of the Department of Biology and director of the Developmental Biology Center. His principle research interests have been the physiology and development of insects. He has published more than 150 papers, particularly in the field of insect endocrinology and developmental biology. His recent research has been on mechanisms of determination, pattern formation, and intercellular communication in imaginal discs and embryos of *Drosophila,* and mode of action of juvenile hormones and molting hormones.

Michael Soulé, an assistant professor of biology at the University of California at San Diego, received his Ph.D. from Stanford University in 1964. He was a research associate in population biology at Stanford University and studied under Paul Ehrlich. Dr. Soulé is primarily interested in reptilian thermoregulation, insular biogeography, and ecological and evolutionary theory with emphasis on the significance of intraspecific variation.

Albert Szent-Györgyi was born in Budapest, Hungary, and received his M.D. degree from the University of Budapest. In 1927 he took a Ph.D. at Cambridge University. He started research as a medical student in histology, then turned to physiology, pharmacology, bacteriology, and chemistry. In 1937 he was awarded the Nobel Prize for the elucidation and discovery of the catalytic functions of C_4- dicarboxylic acids and the isolation of vitamin C. Dr. Szent-Györgyi has taught at universities in Holland, Hungary, and England. He is also a fellow of the National Academy of Sciences. He is currently director of the Institute for Muscle Research at the Marine Biological Laboratory in Woods Hole, Massachusetts.

J. Herbert Taylor received his Ph.D. from the University of Virginia. He has been a consultant for the Oak Ridge National Laboratory and the Brookhaven National Laboratory and has taught biology courses at Columbia University, University of Tennessee, and University of Oklahoma. He is now a professor of biological sciences at the Institute of Molecular Biophysics at Florida State University. Dr. Taylor's research interests include chromosome structure and behavior, DNA replication, autoradiographic studies of macromolecular synthesis in cells, and characterization of the molecular units of chromosomes.

Robert H. Whittaker obtained his Ph.D. in ecology at the University of Illinois. He has been a weather observer and forecaster for the Army Air Forces, a senior scientist in the Radiological Sciences Department at Hanford Laboratories in Richland, Washington, a visiting scientist at Brookhaven National Laboratory, and is currently a professor of biology in the Section of Ecology and Systematics at Cornell University. He is the author of *Communities and Ecosystems* and *Classification of Natural Communities* and of articles on vegetation analysis, forest productivity, and other research problems. His interests range from the structure and function of natural communities to the evolution and broad classification of organisms. He is also vice president of the Ecological Society of America.

Special Consultation

BIOLOGY: AN APPRECIATION OF LIFE could not have been developed without the assistance, counsel, and encouragement of many individuals whose names are not listed above. In particular, we would like to express our gratitude to the following persons, whose help has been greatly appreciated:

Frank Macfarlane Burnet
John C. Eccles
Edmund J. Fantino
Kurt W.D. Fischer
Terrell H. Hamilton
Maryanna Henkart
Matt Sands

Bibliography

1

The Unity and Diversity of Life

Darwin, Charles. 1859. *On the Origin of Species by Means of Natural Selection, or the Preservation of Favoured Races in the Struggle for Life.* London: John Murray.
    Darwin's classical work, but not the easiest of reading.

————. 1959. *The Voyage of the Beagle.* Millicent Selsam (ed.). New York: Harper & Row.
    An abridged edition of the original journal kept by Charles Darwin on his scientific expedition to South America.

Dobzhansky, Theodosius. 1962. *Mankind Evolving: The Evolution of the Human Species.* New Haven, Conn.: Yale University Press.
    The evolutionary origins of man — from both the genetic and cultural viewpoints — are explored in this highly readable book.

Eiseley, Loren. 1956. "Charles Darwin," *Scientific American*, 194 (February):62 – 72. Also Offprint No. 108.
    An excellent and readable history of the life and work of Darwin.

————. 1958. *Darwin's Century: Evolution and the Men Who Discovered It.* New York: Doubleday.
    A clear and well-written account of the rise of the evolutionary theory.

Lack, David. 1953. "Darwin's finches," *Scientific American*, 188 (April):66 – 72. Also Offprint No. 22.
    Self-explanatory title and there is an interesting evolutionary discussion.

Mayr, Ernst. 1963. *Animal Species and Evolution.* Cambridge, Mass.: Harvard University Press.
    An excellent modern statement of the theory of evolution.

2

Life As a Product of the Chemical Earth

Baker, J. J. W., and G. E. Allen. 1970. *Matter, Energy, and Life.* Reading, Mass.: Addison-Wesley.
    A well-written account of the chemical bases of life processes.

Doty, Paul. 1957. "Proteins," *Scientific American*, 197 (September):173 – 184. Also Offprint No. 7.
    An interesting and important account of the principle substance of living cells.

Fieser, Louis. 1955. "Steroids," *Scientific American*, 192 (January):52 – 60. Also Offprint No. 8.
    The story of one of modern chemistry's major efforts to study the effects of steroid hormones in animals.

Fruton, Joseph. 1950. "Proteins," *Scientific American*, 182 (June):32 – 41. Also Offprint No. 10.
    A noteworthy article on the essential constituent of protoplasm.

Wald, George. 1954. "The origin of life," *Scientific American*, 190 (August): 44 – 53. Also Offprint No. 47.
    An interesting, though old, exposition on the origin of the earth and life.

3

Energy Flow and Early Living Systems

Arnon, Daniel. 1960. "The role of light in photosynthesis," *Scientific American*, 203 (November):104 – 118. Also Offprint No. 75.
    A treatment of photosynthesis as a mechanism of energy transduction.

Cloud, Preston, and Aharon Gibor. 1970. "The oxygen cycle," *Scientific American*, 223 (September):110 – 123.
    Discusses the origin of oxygen in the early atmosphere, and the evolution of higher plants and animals.

Lehninger, Albert. 1961. "How cells transform energy," *Scientific American*, 205 (September):62 – 73. Also Offprint No. 91.
    Discusses energy transformations and their interrelations in mitochondria and chloroplasts.

————. 1970. *Bioenergetics: The Molecular Basis of Biological Energy Transformations.* 2nd ed. New York: Benjamin.
    The classic book on energy with a good introduction to metabolic processes.

Siekevitz, Philip. 1957. "Powerhouse of the cell," *Scientific American*, 197 (July): 131 – 140. Also Offprint No. 36.
    A highly readable article on the nature and function of mitochondria.

Stumpf, Paul. 1953. "ATP," *Scientific American*, 188 (April):85 – 92. Also Offprint No. 41.
    Discusses the formation and activities of the currency of energy.

Wald, George. 1959. "Life and light," *Scientific American*, 201 (October):92 – 108. Also Offprint No. 61.
    A highly readable account of the narrow band in the electromagnetic spectrum upon which life depends.

4

The Fundamental Unit of Life

Allen, Robert. 1962. "Amoeboid movement," *Scientific American*, 206 (February):112 – 122. Also Offprint No. 182.
    Contraction of streaming protoplasm at the front of the cell is used to explain motility in amoebae and in similar cells.

Allison, Anthony. 1967. "Lysosomes and disease," *Scientific American*, 217 (November):62 – 72. Also Offprint No. 1085.
    Discusses how lysosomes, which contain digestive enzymes, could be involved in pathological as well as normal processes.

Brachet, Jean. 1961. "The living cell," *Scientific American*, 205 (September): 50 – 61. Also Offprint No. 90.
    The complex and organized activities carried on by the cell as seen from anatomical and chemical standpoints.

Burnet, Sir Macfarlane. 1951. "Viruses," *Scientific American*, 184 (May):43 – 51. Also Offprint No. 2.
    An old but interesting article on the obscure nature of viruses.

Crick, Francis. 1957. "Nucleic acids," *Scientific American*, 197 (September): 188 – 200. Also Offprint No. 54.
    An article on the structure and function of the genetic polymers, written by the co-discoverer of the DNA double helix.

Hokin, Lowell, and Mabel Hokin. 1965. "The chemistry of cell membranes," *Scientific American*, 213 (October):78–86. Also Offprint No. 1022.
    Concerned with how the fatty substances that make up part of the cell membrane play an active role in the transport of cellular substances.

Loewy, A. G., and Philip Siekevitz. 1969. *Cell Structure and Function*. 2nd ed. New York: Holt, Rinehart and Winston.
    A fine presentation of cell structure and formation at an understandable level.

Margulis, Lynn. 1970. *Origin of Eukaryotic Cells*. New Haven, Conn.: Yale University Press.
    An authoritative but difficult monograph on the symbiotic theory of eucaryotic cells.

Morowitz, Harold, and Mark Tourtellotte. 1962. "The smallest living cells," *Scientific American*, 206 (March):117–126. Also Offprint No. 1005.
    Asks the question: "What are the smallest dimensions necessary to be compatible with life?"

Moscona, A. A. 1961. "How cells associate," *Scientific American*, 205 (September):142–162. Also Offprint No. 95.
    Discusses some of the specific physical and chemical factors that are responsible for holding cells together in a multicellular structure.

Neutra, Marian, and C. P. Leblond. 1969. "The Golgi apparatus," *Scientific American*, 220 (February):100–107.
    Discusses the role of the Golgi apparatus in the packaging of secretions and in the synthesis of large carbohydrates.

Racker, Efraim. 1968. "The membrane of the mitochondria," *Scientific American*, 218 (February):32–39.
    Discusses the major processes of energy metabolism in the cell in relation to ribosomal reassembly.

Raff, Rudolf, and Henry Mahler. 1972. "The nonsymbiotic origin of mitochondria," *Science*, 177:577–582.
    The question of the origin of the eucaryotic cell and its organelles is reexamined.

Robertson, J. D. 1962. "The membrane of the living cell," *Scientific American*, 206 (April):64–72.
    The story of the unit membrane by the originator of the unit-membrane concept.

Satir, Peter. 1961. "Cilia," *Scientific American*, 204 (February):108–116. Also Offprint No. 79.
    A close look into the structure and function of cilia in eucaryotic cells.

5

The Cell As Information Center

Beerman, Wolfgang, and Ulrich Clever. 1964. "Chromosome puffs," *Scientific American*, 210 (April):50–65.
    Discusses the regions believed to be genes at work that appear on the giant chromosomes of some insects.

Benzer, Seymour. 1962. "The fine structure of the gene," *Scientific American*, 206 (January):70–84. Also Offprint No. 120.
    The mutations and their order within one gene of a bacterial virus and how they can be detected, thus permitting analysis within a gene.

Britten, Roy, and David Kohne. 1970. "Repeated segments of DNA," *Scientific American*, 222 (April):24–31.
    Discloses that although the origin and function of repetitive DNA remain unknown, a significant fraction of the giant molecule appears in as many as a million identical or nearly identical copies in higher organisms.

Clark, Brian, and Kjeld Marcker. 1968. "How proteins start," *Scientific American*, 218 (January):36–42. Also Offprint No. 1092.
    Discusses how the initiation of the synthesis of each protein molecule begins with a specific amino acid carried by a specific transfer RNA.

Crick, Francis. 1954. "The structure of the hereditary material," *Scientific American*, 191 (October):54–61. Also Offprint No. 5.
    An account of the research that led to an understanding of the reproductive processes of DNA.

Deering, R. A. 1962. "Ultraviolet radiation and nucleic acid," *Scientific American*, 207 (December):135–144. Also Offprint No. 143.
    The damaging effects of ultraviolet radiation are explained by discussing its specific effects on DNA.

Dulbecco, Renato. 1967. "The induction of cancer by viruses," *Scientific American*, 216 (April):28–37. Also Offprint No. 1069.
    Discusses how "model systems" of cells transformed by cancer are studied to learn how a virus can produce the change.

Edgar, R. S., and R. H. Epstein. 1965. "The genetics of a bacterial virus," *Scientific American*, 212 (February):70–78. Also Offprint No. 1004.
    Discusses how the genes of the T-4 virus are mapped to show the precise architecture of the virus.

Hanawalt, Philip, and Robert Haynes. 1967. "The repair of DNA," *Scientific American*, 216 (February):36–43. Also Offprint No. 1061.
    Discusses the cell's remarkable ability to repair the double helix when damaged.

Hurwitz, Jerard, and J. J. Furth. 1962. "Messenger RNA," *Scientific American*, 206 (February):41–49. Also Offprint No. 119.
    Discusses the role of messenger RNA as the intermediary in protein synthesis.

Ptashne, Mark, and Walter Gilbert. 1970. "Genetic repressors," *Scientific American*, 222 (June):36–44.
    An account of genetic repressors, which have been isolated confirming the hypotheses of Jacob and Monod.

Rich, Alexander. 1963. "Polyribosomes," *Scientific American*, 209 (December):44–53. Also Offprint No. 171.
    Relates how the assembly line of ribosomes works in protein synthesis.

Sinsheimer, Robert. 1962. "Single stranded DNA," *Scientific American*, 207 (June):109–116.
    The method of replication of single-stranded DNA is discussed.

Watson, James. 1968. *The Double Helix*. New York: Atheneum.
    An autobiographical account of the exciting discovery of the double helix. Nontechnical and extremely interesting.

————. 1970. *Molecular Biology of the Gene*. 2nd ed. New York: Benjamin.
    An excellent book for those who want to learn more about molecular biology.

6

Cell Division

Brachet, Jean (ed.). 1961. *The Cell. Mitosis and Meiosis*. New York: Academic Press, Vol. III.
    Well-written account of cellular division, detailing the many steps involved and the chemical changes.

Keosian, J. 1964. *The Origin of Life*. New York: Reinhold.
    A readable book concerned with the possible beginnings and subsequent evolution of organic life.

Lehninger, A. L. 1970. *Bioenergetics: The Molecular Basis of Biological Energy Transformations.* 2nd ed. New York: Benjamin.
  A book for college biologists; comprehensively discusses the energy involved in chemical, thermal, and other biological processes.

Mazia, Daniel. 1953. "Cell division," *Scientific American,* 189 (August):53–63. Also Offprint No. 27.
  A discussion of the many stages and types of cell division occurring at definite stages of cellular life.

———. 1961. "How cells divide," *Scientific American,* 205 (September): 100–120. Also Offprint No. 93.
  An account of the cellular processes, including biochemical secretions, that are initiated and progress through the stages of division.

Oparin, Alexandr (ed.). 1963. *Evolutionary Biochemistry.* New York: Macmillan.
  A set of papers dealing with how cells evolved to their present state, including possible origins. Slightly difficult reading.

Ramsay, J. 1965. *The Experimental Basis of Modern Biology.* New York: Cambridge University Press.
  Provides insight into the techniques needed to determine how and why cellular processes occur.

Spratt, N. T. 1964. *Introduction to Cell Differentiation.* New York: Reinhold.
  A well-written book on the processes of cellular action, including chapters on the action of specific cells in different organs.

Swanson, C. P. 1969. *The Cell.* 3rd ed. Englewood Cliffs, N.J.: Prentice-Hall.
  A comprehensive review of cell division, with numerous fine prints and diagrams.

7

Patterns of Inheritance

Beadle, G. W. 1948. "The genes of men and molds," *Scientific American,* 179 (September):30–39. Also Offprint No. 1.
  A classic study of the red bread mold *Neurospora crassa* showing how genetic material controls enzymes.

Mendel, Gregor. 1965. *Experiments in Plant Hybridisation.* Cambridge, Mass.: Harvard University Press.
  A translation of Gregor Mendel's original paper on his genetic studies of pea plants.

Srb, A. M., R. D. Owen, and R. S. Edgar. 1965. *General Genetics.* 2nd ed. San Francisco: Freeman.
  A clear discussion of the principles of Mendelian and classical genetics, with attention to details.

Sturtevant, A. H. 1965. *A History of Genetics.* New York: Harper & Row.
  A good, readable review of the problems and discoveries of the first scientists who studied genetic processes.

8

Mutation and the Genetic Code

Allfrey, Vincent, and Alfred Mirsky. 1961. "How cells make molecules," *Scientific American,* 205 (September): 74–82. Also Offprint No. 92.
  Discusses the progress being made in understanding the synthesis of macromolecules by the cell.

Carlson, E. A. 1966. *The Gene: A Critical History.* Philadelphia: Saunders.
  A well-written and documented review of the vital characteristics of genes and the problems associated with understanding them.

Crick, Francis. 1962. "The genetic code," *Scientific American,* 207 (October):66–74. Also Offprint No. 123.
  An account of how the chromosomes follow a definite sequence in order to transmit the information of the genetic material to the structure of the organism.

Dobzhansky, Theodosius. 1950. "The genetic basis of evolution," *Scientific American,* 182 (January):32–41. Also Offprint No. 6.
  A synthesis of Darwin's ideas on the language of genes.

Ingram, V. M. 1958. "How do genes act?" *Scientific American,* 198 (January): 68–74. Also Offprint No. 104.
  Discusses how recent discoveries in the structural and chemical composition of genes have shed light on the patterns of gene action.

———. 1965. *The Biosynthesis of Macromolecules.* New York: Benjamin.
  Discusses how and why large molecules, such as DNA, are formed and how they function in the body.

McKusick, Victor. 1971. "The mapping of human chromosomes," *Scientific American,* 224 (April):104–113.
  Relates information on the ability to manipulate and locate all of the genes on the chromosomes.

Uhl, C. H. 1965. "Chromosome structure and crossing over," *Genetics,* 51:191–207.
  A college-level article on the nature and effects of crossing over.

9

Reproduction and Development

Arey, L. B. 1965. *Developmental Anatomy.* 7th ed. Philadelphia: Saunders.
  An advanced textbook that traces the anatomical features in numerous species and discusses how various species differ.

Berrill, N. J. 1961. *Growth, Development, and Pattern.* San Francisco: Freeman.
  Good reading on how the functional development of organisms may control their behavioral characteristics in later life.

Bonner, D. M, and S. E. Mills. 1964. *Heredity.* 2nd ed. Englewood Cliffs, N.J.: Prentice-Hall.
  Discusses the basic laws of population growth as derived from studies in genetics.

Crick, Francis. 1954. "The structure of the hereditary material," *Scientific American,* 191 (October):54–61. Also Offprint No. 5.
  A discussion of how genetic control is adjusted by altering the chemical structure of the chromosome.

Edgar, R. S., and R. H. Epstein. 1965. "The genetics of a bacterial virus," *Scientific American,* 212 (February):70–78. Also Offprint No. 1004.
  Discusses questions on the transfer of information by the genetic material in viruses.

Fischberg, M., and A. W. Blackler. 1961. "How cells specialize," *Scientific American,* 205 (November):124–140. Also Offprint No. 94.
  An account of the intricate machinery of cells and how they are regulated.

Holley, R. W. 1966. "The nucleotide sequence of a nucleic acid," *Scientific American,* 214 (February):30–39. Also Offprint No. 1033.
  Describes how the bases in nucleotides regulate basic functions in protein synthesis.

Markert, C. L. 1964. *Developmental Genetics.* Englewood Cliffs, N.J.: Prentice-Hall.
  This well-illustrated textbook is concerned with how chromosomes control the development of all living organisms.

Ramsay, J. A. 1965. *The Experimental Basis of Modern Biology.* New York: Cambridge University Press.
  A discussion of the kinds of experimental apparatus used in modern research and the results obtained.

Stern, Herbert, and D. L. Nanney. 1965. *The Biology of Cells.* New York: Wiley.
A discussion of how cell processes and biochemistry are integrated at every level.

Watson, James. 1970. *Molecular Biology of the Gene.* New York: Benjamin.
A useful review of the basics of biochemistry; includes information on how the molecular structure and biochemical apparatus of the gene and chromosome systems are directly related to the physiological output of the body.

10

Levels of Integration

Barth, L. G. 1964. *Development: Selected Topics.* Reading, Mass.: Addison-Wesley.
Fairly easy reading on how genetically controlled development can be contrasted with environmentally controlled development.

Burnet, Sir Macfarlane. 1961. "The mechanism of immunity," *Scientific American*, 204 (January):58–67. Also Offprint No. 78.
Discusses how an antibody works to neutralize an antigen in an antigen-antibody reaction.

Ebert, J. D., and I. M. Sussex. 1970. *Interacting Systems in Development.* 2nd ed. New York: Holt, Rinehart and Winston.
Discussion of the stages and stimuli involved in the development of systems in organisms.

Flickinger, R. A. (ed.). 1966. *Developmental Biology.* Dubuque, Iowa: Wm. C. Brown.
A collection of sixteen papers covering the development of organisms from simple to complex.

Needham, A. E. 1964. *The Growth Process in Animals.* Princeton, N.J.: Van Nostrand.
An interesting book on how environment and genetics combine to influence an animal's structure.

Rudnick, D. (ed.). 1960. *Developing Cell Systems and Their Control.* New York: Ronald Press.
Set of papers on the enzymatic and chemical regulation of intercellular development.

Telfer, William, and Donald Kennedy. 1965. *The Biology of Organisms.* New York: Wiley.
A discussion of how the structural integration of parts within an organism equip the organism to cope with the many internal and external stimuli it encounters daily. Good illustrations.

11

Integration of Function

Adolph, E. F. 1967. "The heart's pacemaker," *Scientific American*, 216 (March):32–37. Also Offprint No. 1067.
Shows how the rhythmic beat of the heart is easily upset by manipulations of the pacemaker, which is controlled by enzymatic and feedback cycles of the body.

Barrington, E. J. W. 1968. *The Chemical Basis of Physiological Regulation.* Glenview, Ill.: Scott, Foresman.
Discusses how biochemical compounds, particularly enzymes, act in the body to control not only development but the ability to react to environmental, internal, or external change. Difficult reading.

Biale, J. B. 1954. "The ripening of fruit," *Scientific American*, 190 (May):40–44. Also Offprint No. 118.
An older paper showing the chemical and biological activities involved in the development of seed-plant embryos.

Brachet, Jean. 1960. *The Biochemistry of Development.* New York: Pergamon.
Textbook concerned with genetic material and enzymatic chemicals, and the role they play in an organism's development.

Burnet, Sir Macfarlane. 1961. "The mechanism of immunity," *Scientific American*, 204 (January):58–67. Also Offprint No. 78.
Discusses how an antibody works to neutralize an antigen in an antigen-antibody reaction.

Frieden, Earl. 1963. "The chemistry of amphibian metamorphosis," *Scientific American*, 209 (November):110–118. Also Offprint No. 170.
Relates how metamorphosis in lower animals can follow spectacular pathways due to biochemical and enzymatic changes within the developing organism, resulting in offspring that look nothing like the parents.

Grobstein, Clifford. 1964. "Cytodifferentiation and its controls," *Science*, 143:643–650.
An advanced book on how chemical activity determines the formation of parts and bodies within the cell.

Levey, R. H. 1964. "The thymus hormone," *Scientific American*, 211 (July):66–77. Also Offprint No. 188.
Concerned with the thymus hormone, a physiologically important hormone that causes the generation of other enzymatic substances that are fed directly into the body's biochemical cycles.

Salisbury, F. B., and R. V. Parke. 1970. *Vascular Plants: Form and Function.* 2nd ed. Belmont, Calif.: Wadsworth.
The biology of plant forms explored in terms of the development of function and chemical control of behavior.

Stein, H. 1965. *The Biology of Cells.* New York: Wiley.
Discusses how cells control body physiology through the formation of layer masses, which in turn are controlled by the internal make-up and secretions of individual cells.

Sutherland, Earl. 1972. "Studies on the mechanism of hormone action," *Science*, 177:401–408.
Discusses the general role of cyclic AMP in living systems including vertebrates and bacteria.

12

Nervous Systems

Benzinger, T. H. 1961. "The human thermostat," *Scientific American*, 204 (January):134–147. Also Offprint No. 129.
Discusses the human being's regulatory mechanism for maintaining a fairly constant temperature, which seems to be the direct result of gene action and biochemical balance.

Bloom, William, and D. W. Fawcett. 1962. *A Textbook of Histology.* 8th ed. Philadelphia: Saunders.
An advanced textbook on the control of cellular organisms in terms of body structures and chemistry. Good illustrations.

Eccles, J. C. 1957. *The Physiology of Nerve Cells.* Baltimore: Johns Hopkins Press.
A detailed discussion of nerve cells—how they are controlled biochemically and how they in turn control the behavior of many parts of an organism's body.

———. 1965. "The synapse," *Scientific American*, 212 (January):56–66. Also Offprint No. 1001.
A good discussion of the junction between two neurons or between a neuron and a receptor or effector cell.

Fisher, A. E. 1964. "Chemical stimulation of the brain," *Scientific American*, 210 (June):60–69. Also Offprint No. 485.
States that the electrical discharge found in the brain, and perhaps the processes of thought, are stimulated and maintained by the continuous generation of chemical reactions.

Magoun, H. W. 1963. *The Waking Brain.* 2nd ed. Springfield, Ill.: Thomas.
States that although the workings of the brain are poorly understood, chemical controls and cellular contents can be investigated to provide evidence for the many theories of thought operation.

Nachmansohn, D. 1959. *Chemical and Molecular Basis of Nerve Activity.* New York: Academic Press.
Discusses how nerve activity within the body is governed by reactions to both internal and external stimuli on the molecular level.

Wooldridge, D. E. 1963. *The Machinery of the Brain.* New York: McGraw-Hill.
Experiments and theories on how the brain works and the mechanisms that keep it functioning.

13
Forms of Behavior

Cold Spring Harbor Symposium on Quantitative Biology. 1960. *Biological Clocks.* Cold Spring Harbor, New York: Cold Spring Harbor Biological Laboratory.
Paper on the rhythmic cycles in organisms, and the environmental characteristics and internal chemical changes that cause biological clocks to act.

Dethier, V. G. and Eliot Stellar. 1970. *Animal Behavior.* 3rd ed. Englewood Cliffs, N.J.: Prentice-Hall.
Discusses animal behavior in light of both environmental and genetic influences.

Deutsch, J. A. 1960. *The Structural Basis of Behavior.* Chicago: University of Chicago Press.
Describes the mechanisms involved in the internal control of behavior.

Etkin, William (ed.). 1964. *Social Behavior and Organization Among Vertebrates.* Chicago: University of the Chicago Press.
A series of papers on the formation, maintainance, and progression of social controls among higher vertebrate animals, as compared to humans.

Hess, E. H. 1958. "Imprinting in animals," *Scientific American*, 198 (March):81–93.
Discusses the fascinating phenomenon of imprinting, in which intense, irreversible associations are formed in early stages of development.

Hockett, C. F., 1960. "The origin of speech," *Scientific American*, 203 (September):88–96. Also Offprint No. 603.
Discusses human vocal development in relation to that of other animals, including apes; effects of organ specialization and advanced brain structure on vocal development.

Roe, A., and G. G. Simpson (eds.). 1958. *Behavior and Evolution.* New Haven, Conn.: Yale University Press.
Well-written book that describes behavioral changes as a direct result of coping with the environment.

Shaw, Evelyn. 1962. "The schooling of fishes," *Scientific American*, 206 (June): 128–138. Also Offprint No. 124.
Discusses the survival instinct and how it is manifested by many forms of animals in such ways as grouping for mutual protection.

14
Life and the Environment

Andrewartha, H. G. 1961. *Introduction to the Study of Animal Populations.* Chicago: University of Chicago Press.
Discusses how animal groups generally follow definite patterns of behavior; shows how these patterns are primarily a function of environmental stimuli.

Farber, P. 1963. *Face of North America: The Natural History of a Continent.* New York: Harper & Row.
Narrative view of the development of land, plants, and animals in North America.

Kormondy, E. J. (ed.). 1965. *Readings in Ecology.* Englewood Cliffs, N.J.: Prentice-Hall.
A collection of passages from the original versions of sixty important papers in ecology.

Odum, E. P. 1963. *Ecology.* New York: Holt, Rinehart and Winston.
Covers concepts of systems interactions and environmental controls on population behavior and growth.

Slobodkin, L. B. 1961. *Growth and Regulation of Animal Populations.* New York: Holt, Rinehart and Winston.
Contains a good discussion of time scales relevant to different types of biological investigations.

Wecker, S. C. 1964. "Habitat selection," *Scientific American*, 211 (October):109–116. Also Offprint No. 195.
A discussion of the necessity for populations to adapt to a given environment before a more competitive population has the chance to take it over.

Woodwell, G. M. 1963. "The ecological effects of radiation," *Scientific American*, 208 (June):40–49. Also Offprint No. 159.
Discussion of how radiation can easily change the nature of an ecological niche by causing mutations and creating new environmental conditions.

Wynne-Edwards, V. C. 1964. "Population control in animals," *Scientific American*, 211 (August):68–74. Also Offprint No. 192.
A discussion of nature's ways of controlling animal populations, ranging from mass migrations to mass suicide.

15
Man in the Biosphere

Emerson, Ralph. 1952. "Molds and men," *Scientific American*, 186 (January):28–32. Also Offprint No. 115.
An account of man's relationships with lower forms of plant and animal life.

Gates, D. M. 1962. *Energy Exchange in the Biosphere.* New York: Harper & Row.
Ecological systems are often governed by the ability to exchange various forms of energy within the system.

Leeds, A. 1965. *Man, Culture, and Animals.* Washington, D.C.: American Association for the Advancement of Science.
Traces how man and animals have been intimately related since the beginning of known human culture.

Newell, Norman. 1963. "Crises in the history of life," *Scientific American*, 208 (February):76–92.
A highly readable discussion of various life crises that have occurred throughout history.

Oparin, Alexandr. 1964. *Life: Its Nature, Origin, and Development.* Translated by Ann Synge. New York: Academic Press.
An advanced text that considers the study of life's origins and development as necessary to the study of life's essential nature.

Romer, A. S. 1966. *Vertebrate Paleontology* 3rd ed. Chicago: University of Chicago Press.
Traces in detail the anatomical changes that vertebrates have undergone through time.

Shepley, H. 1963. *The View From a Distant Star.* New York: Basic Books.
An astronomer's view of what has happened and what is now happening on the earth, with predictions of things to come.

Storer, T. I., and R. L. Usinger. 1965. *General Zoology.* 4th ed. New York: McGraw-Hill.
A well-illustrated textbook on the basic principles of zoology.

Washburn, S. L. 1960. "Tools and human evolution," *Scientific American*, 203 (September):62–75.
Correlates the formation of human culture and man's control of the environment with the development of tools.

Young, J. Z. 1962. *The Life of Vertebrates.* 2nd ed. New York: Oxford University Press.
Life histories of vertebrate animals that indicate the wide diversity of environmental adaptations that have occurred and continue to occur.

a

abiotic: not living.

abortion: premature explusion of a fetus, especially before it is independently viable, either by miscarriage or by surgery.

absolute refractory period: see refractory period.

acetylcholine: a chemical secreted at the end of neurons; responsible for the transmission of a nerve impulse across a synapse.

acetyl-coenzyme A: a compound formed as a result of glycolysis and frequently protein breakdown.

ACTH: see adrenocorticotrophic hormone.

actin: the protein in muscle that, along with myosin, is responsible for contraction and relaxation of the muscle.

action potential: a rapid change in the membrane potential, which generally signifies excitation in the cells.

active site: the part of an enzyme molecule that recognizes and interacts with the substrate.

active transport: the movement of molecules across a semipermeable membrane against a concentration gradient, with an expenditure of energy.

adaptation: an inheritable modification in structure or function that enables a species to better survive and reproduce in its environment.

adaptive radiation: the development of groups or species differing in various characteristics although originating from a common ancestral species.

adaptive value: the value that a given characteristic has in helping an organism to survive and reproduce as its environment changes.

adenine: a purine base and a constituent of nucleic acids that is involved in cellular metabolism.

adenosine diphosphate: organic nucleotide compound containing two phosphate groups, which can capture or release energy via high energy bonds. (Abbreviated: ADP)

adenosine triphosphate: organic nucleotide compound, containing three phosphate groups, which can capture or release energy via high energy bonds. (Abbreviated: ATP)

adhesion: the property of remaining in close approximation, such as that due to the physical attraction of unlike molecules.

ADP: see adenosine diphosphate.

adrenal cortex: the part of the adrenal glands that secretes hormones controlling certain aspects of metabolism.

adrenal glands: a pair of vertebrate endocrine glands, one associated with each kidney; these glands secrete adrenalin and other hormones that control certain aspects of metabolism, blood pressure, and sympathetic nervous system functions.

adrenal medulla: the part of the adrenal gland that is involved in the secretion of adrenalin.

adrenocorticotrophic hormone: the hormone secreted by the anterior lobe of the pituitary that causes the adrenal cortex to produce hormones. (Abbreviated: ACTH)

adsorption: adhesion of molecules to a solid body, with the formation of a unimolecular surface layer.

aerobic: requiring oxygen for growth or survival.

aerobic respiration: a series of processes in which pyruvate is broken down to carbon dioxide and water, accompanied by the synthesis of ATP and the consumption of oxygen. (See Krebs citric acid cycle; electron transport.)

afferent neuron: a neuron that carries impulses toward the central nervous system.

afterbirth: the placenta and fetal membranes expelled from the uterus after the birth of a child.

age-specific birth rate: the number of offspring produced per unit of time by females of a particular age class.

agglutination: the aggregation or clumping together of organisms or particles.

albinism: an absence of pigmentation, especially the absence of melanins in an animal.

albumin: common type of protein found in most animals and in many plants.

alcohol: any of a group of organic compounds formed from hydrocarbons by the substitution of one or more hydroxyl ($-OH$) groups for hydrogen atoms.

allele: one of two or more alternate forms of a gene found at the same location (locus) in homologous chromosomes.

allergy: excessive sensitivity or pathological reaction to a common substance, such as pollen; the external manifestation of an antigen-antibody reaction.

amino acid: a building block of proteins, containing an amino group (NH_2) and a carboxyl group ($COOH$).

amino acid activating enzymes: a specific class of enzymes that assist in reactions of amino acids, causing them to react at a faster rate.

amniocentesis: insertion of needle into the uterus through the abdomen to remove amniotic fluid for investigation during pregnancy.

Glossary

351

amniotic sac: the innermost layer of transparent fetal membranes forming fluid-filled sac for protection of embryo.

anabolism: constructive metabolism; the build-up, or synthesis, of complex organic molecules from simple ones—opposite of catabolism.

anaerobic: oxygen not needed for growth or survival.

anaerobic respiration: fermentation; the partial enzymatic breakdown (catabolism) of of organic fuels by an organism in the absence of oxygen; results in end products such as alcohol or lactic acid.

anaphase: the stage of cell division (mitosis) when chromosomes move apart toward opposite poles.

anaphylactic shock: an uncommon hypersensitive reaction to certain proteins or antigens that is so pronounced as to cause violent circulatory, nervous, and respiratory symptoms, frequently resulting in unconsciousness and death.

anisogamy: union of gametes that differ in form or size.

annual: seed plant that reproduces only by seeds and dies at the end of each growing season; the seeds germinate to provide a new generation.

annulus: any ringlike structure. (Plural: annuli)

antibiotic: a substance produced by some fungi and bacteria that diffuses into surrounding area and is toxic enough to prevent the growth of certain other species in that area.

antibody: a specific serum globulin produced in the body fluid in response to the introduction of a specific antigen.

anticodon: a set of three nucleotides of transfer RNA that corresponds to a particular amino acid.

antidiuretic hormone: a hormone released by the posterior pituitary, causing an increase in reabsorption of water from the urinary tubules and a decrease in the amount of urine voided. (Abbreviated: ADH)

antigen: a foreign substance (a protein) capable of inducing the formation of antibodies when introduced into an organism.

anus: the terminal opening of the digestive system through which the solid refuse of digestion is excreted. (See rectum.)

aorta: the thick artery that leaves the left ventricle and supplies blood to all of the body except the lungs via a system of smaller branched arteries.

appendix: the small fingerlike projection of the caecum in certain mammals; produces enzymes that break down cellulose.

arginine: one of the essential amino acids that make up plant and animal proteins.

artery: a relatively thick-walled, elastic, vessel carrying blood away from the heart to the tissues. (See vein.)

arteriole: a small artery that tends to be less than 0.3mm in diameter in humans.

asexual reproduction: reproduction without the production of gametes and subsequent fusion of gametes.

aster: a fiberlike structure that radiates toward the plasma membrane from the centrosome in animal cell mitosis.

asthma: frequent episodes of difficult breathing.

atom: the basic unit of matter; consists of protons and neutrons in a nucleus, surrounded by electrons equal in number to that of the protons, thus being electrically neutral.

atomic number: the number of positive electric charges, or protons, carried in the nucleus of an atom.

atomic weight: the average weight of an atom of an element; units of weight are based on 1/12 the mass of the carbon-12 atom.

ATP: see adenosine triphosphate.

atrioventricular node: a small concentration of special tissue in the heart of higher vertebrates, located between the auricles just above the ventricles.

atrium: a chamber of the heart that receives the blood from the veins. (Plural: atria)

autoimmune diseases: diseases that are thought to be due to attacks by the immune system upon normal body tissues.

autonomic nervous system: the part of the central nervous system that controls the involuntary or vegetative functions of the body.

autosome: any chromosome other than a sex chromosome. (See X chromosome; Y chromosome.)

autotroph: an organism having the means to make its own food out of inorganic materials using simple precursors.

avirulent strain: a nonpathogenic strain of bacteria, virus, or fungi.

avoidance response: a form of behavior that serves to protect an animal from danger or elude an unfamiliar stimulus.

axial: lying along a major axis, as in the vertebrate skeleton.

axon: a nerve fiber that conducts nerve impulses away from the cell body and toward a synaptic junction.

b

bacillus: a rod-shaped bacterium. (Plural: bacilli)

bacterial transformation: the induction of a new inheritable characteristic in bacteria by DNA transmitted from other bacteria.

base: a substance that releases hydroxyl ions when dissolved in water or that can accept protons in chemical reactions; bases have a bitter taste, turn red litmus blue, and unite with acids to form salts.

base pairing: the combination of a purine base with a pyrimidine base in DNA; possible base pairs are adenine and thymine, and guanine and cytosine.

bilaterally symmetrical: the characteristic of having two sides that are symmetrical about an axis, with most organs occurring in pairs—one organ on each side of the mid-sagittal plane.

bile: a yellow secretion of the vertebrate liver; contains organic salts, acids, and other compounds stored temporarily in the gall bladder.

biocide: a substance such as a herbicide or insecticide that kills living organisms.

biological clock: some internal mechanism of a living organism that produces a physiological or behavioral rhythm, which is synchronized with daily, lunar, or annual cycles of such conditions as tides and light.

bioluminescent: light production due to the metabolic activities of a living organism.

biomass: the mass of a population of organisms.

biome: a large geographical area with a characteristic flora and fauna; for example, a grassland or a tropical forest.

biosphere: the part of the earth that is inhabited by living organisms.

biotic: pertaining to life.

biotic community: a group of populations occupying the same area and interchanging materials and energy.

blastocyst: a blastula; a mass of specialized cells on one area of a thin-walled, hollow sphere that becomes the embryo or a blastodermic vesicle.

blooming: a rapid increase in population size, which frequently exhausts the supply of nutrients in an area.

Bowman's capsule: a double-walled, cup-shaped capsule surrounding a knot of capillaries (glomerulus) in the kidneys of many vertebrates.

brain: a mass of nervous tissue in animals at the anterior end of the spinal cord; the center of the nervous system.

bronchiole: one of the many smaller branches of a bronchus that, after considerable rebranching, eventually terminates in the many alveoli of a lung. (Plural: bronchia)

bronchus: one of two tubes that connect the trachea (or air passage) with the lungs. Plural: bronchi)

budding: a form of asexual reproduction in which a small part of a parent's body "buds off," and grows and develops into a new individual; for example, yeasts.

C

Caesarean section: delivery of the fetus through an abdominal incision.

calorie: as a unit of heat, the amount of heat required to raise the temperature of one gram of water 1°C; a dietary calorie consists of 1,000 of these units.

Calvin cycle: a cyclic pathway of glucose biosynthesis from carbon dioxide, occurring during the dark reactions of photosynthesis.

cancer: a tumorous growth resulting from the abnormal and uncontrolled growth of cells in plants or animals.

capillary: a microscopic network of vessels through which the blood moves from the arteries into the veins.

capsule: a fibrous, fatty, cartilaginous or membranous structure surrounding another structure, organ, or part; for example, the protective sheath surrounding the cell wall in certain bacteria or the capsule of the kidney.

carbohydrate: an organic compound of carbon, hydrogen, and oxygen that can be used to supply energy; common examples are sugars, starches, celluloses, and glycogen.

cardiac muscle tissue: a type of involuntary muscle tissue specific to the vertebrate heart; consists of cylindrical, branching, and interconnecting fibers that contain delicate cross-striations.

carnivore: an animal that is solely or chiefly dependent upon the flesh of other animals for its food.

carotene: a red or orange crystalline hydrocarbon that is synthesized by plants; occurs in the chloroplasts and other plastids of plants.

carrier molecule: a postulated protein that can selectively bind to molecules and transport them across a membrane in a process termed facilitated transport.

carrying capacity: the upper limit to the size of a population that a particular environment can support.

cartilage: a semiopaque, nonvascular connective tissue composed of proteins (collagen), polysaccharides, and occasional cells.

catabolism: destructive metabolism; the reactions in which complex organic molecules are broken down into simpler ones. (See anabolism.)

catalyst: a substance that stimulates the rate of a chemical reaction by lowering the activation energy required for the reaction; it is not consumed by the reaction and does not form part of the final product.

cell: the basic subunit of any living system; the simplest unit that can exist as an independent living system.

cell-borne antibody: an antibody associated with the lymphocytes that originate in the thymus gland during early infancy.

cell culture: the technique of artificial growth of cells or tissues *in vitro*.

cell division: a division of the cytoplasm and the nucleus of a cell to produce daughter cells, or gametes.

cell plate: a structure in plant cells that forms along the equatorial plate of the spindle during telophase; it is composed of membranes from the Golgi complexes and forms the new plasma membranes that separate the two daughter cells.

cell respiration: conversion of chemical energy into usable forms of energy in plant and animal cells.

cell wall: the extracellular material surrounding the cell membrane; it consists principally of cellulose in higher plants and chitin in most fungi; it produces a rather rigid shell-like structure around the cell.

cellular fusion: the fusion of a pair of gametes (frequently derived from different parental cells) to form a new diploid individual cell.

cellulose: the principal carbohydrate constituent of the cell walls of green plants.

central nervous system: one of the two traditional divisions of the vertebrate nervous system; it includes the neural structures (interneurons) encased in the skull and in the vertebral column. (Abbreviated: CNS)

centriole: a dark-staining, dense organelle in the centrosome area of animal cells and certain lower plant cells; it forms the spindle pole during mitosis and meiosis.

centromere: a clear, constricted area at the bend or angle of many chromosomes; usually the point at which spindle fibers attach during mitosis and meiosis.

cephalization: in the evolution of animals, a tendency toward the concentration of important organs of the body in the head region.

cerebral cortex: the external layer of neural cells or "gray matter" of the forebrain (cerebrum); it functions in sensory and motor integration and is highly developed in humans.

cerebellum: inferior part of the brain lying below the cerebrum and above the pons and medulla, consisting of two lateral lobes and a middle lobe; it controls voluntary movements and coordinates mental actions.

chemical bonding: any of several means of joining and holding atoms and molecules together.

chemical compound: two or more different kinds of atoms bound together to form a substance.

chemical element: a simple substance formed of one type of atom; it cannot be decomposed by chemical means.

chemical energy: the energy stored by one or more types of chemical bonds.

chemical evolution: the origin, ancestry, and differentiation of chemical compounds from simple molecules to the highly complex forms found in living organisms.

chemical molecule: chemical combination of one or more like atoms bound together, which then acts as a unit.

chemosynthesis: the biosynthesis of carbohydrates from carbon dioxide and water by means of the energy derived from chemical oxidations, rather than from absorbed light; carried out by certain bacteria and algae.

chemosynthetic bacteria: bacteria that are autotrophic but not photosynthetic; for example, the purple bacteria.

chitin: the structural material that forms the outer skeletons of arthropods (occurring in invertebrates); also is a principal component of the cell walls of many fungi.

chlorinated hydrocarbon: a hydrocarbon that has had one or more hydrogens replaced by chlorine atoms, tending to make it toxic to many life forms.

chlorophyll: any of several green, light-absorbing pigments essential as electron donors in photosynthesis; found in green plants and certain bacteria.

chlorophyll *a*: a yellowish-green pigment found in all plants.

chlorophyll *b*: a blue-green pigment found in green land plants and algae.

chloroplast: a type of plastid found in certain plant cells and in some protozoans that contains chlorophyll.

chromatid: one of the two coiled strands of a chromosome.

chromatin: the DNA-containing material of chromosomes (a nucleoprotein), which is characterized by its affinity for basic dyes; the carrier of the genes of inheritance.

chromoplast: a plastid that contains a pigment.

chromosome: a microscopic, threadlike body that is composed of chromatin and that appears in the nucleus of a cell at the time of cell division; chromosomes contain the genes and normally are constant in number within the species.

cilium: a short, hairlike structure in some types of cells; cilia are capable of beating in unison to produce locomotion or to move particles along the surface of the cell. (Plural: cilia)

ciliate: a protozoan with cilia for locomotion and feeding; cell surface covered with a thick, secreted pellicle.

circadian: pertaining to rhythmic activities that recur with a periodicity of about 24 hours under constant conditions. (See biological clock.)

citric acid cycle: see Krebs citric acid cycle.

class: the major subdivision (or subtaxon) of a phylum in the classification of plants and animals; usually consists of several orders.

clone: a strain of cells derived from a single cell by asexual reproduction.

coccus: a spherically-shaped bacterium.

codominance: a situation in which the alleles of a pair are not identical but neither allele is completely dominant over the other. (See allele.)

codon: a sequence of three nucleotide bases on a DNA polymer acting as the code for a specific amino acid.

coelom: the body cavity between the body wall and the digestive tract of higher metazoan animals.

coenocyte: a multinucleate mass of protoplasm that has arisen as a result of the fusion of two or more cells.

collagen: the fibrous, protein material found in bones, tendons, and other connective tissues.

collecting tubule: one of the larger tubules in the medullary portion of the kidneys of higher vertebrates, formed by the confluence of numerous small urinary tubules.

collenchyma: an elongated plant cell with thickened walls; it is frequently present as support in maturing plant tissues.

colloid: a system of finely divided matter suspended in a liquid medium.

colon: the large intestine of vertebrates, exclusive of the anus.

competitive advantage: under conditions of limited resources, the continued growth of one population, resulting in a reduced growth rate of another group, such that the reproductive rate of the first group becomes favored over the second.

competitive exclusion: under uniform conditions with at least one resource in limited supply, not more than one species can continue to exist indefinitely.

complement: the nonspecific protein material in blood plasma that, in combination with specific antibodies in the blood, causes the destruction of the corresponding antigen.

compound: see chemical compound.

conductivity: the property of a body or substance that transmits energy by direct molecular transfer.

connective tissue: any animal tissue containing an abundance of dead, secreted, intercellular material.

consumer: an organism in an ecosystem that feeds upon other organisms.

contractile vacuole: a clear, fluid-filled vacuole in the protoplasm of certain one-celled organisms; it gradually increases in size, then contracts to expel the fluid through a pore in the cell membrane.

contralateral control: acting in unison with a similar part of the opposite side of the body; for example, the right hemisphere of the brain is concerned with the activities of the left side of the body and vice versa.

convergent evolution: evolutionary processes resulting in similarities of form or structure among distantly related forms of organisms; for example, the wings of bats, flying reptiles, and birds.

copulation: sexual union in which the male injects sperm into the body of the female.

cortex: the peripheral portion of an organ (such as the brain or the kidney).

covalent bond: a state in which electrons are shared between atoms so that each atom has a fairly stable electron configuration.

cretinism: a congenital, pathological condition caused by a deficiency in the secretion of thyroxin during childhood; results in physical and mental retardation.

crista: a ridge; a projection or projecting structure; for example, the inner membrane of the mitochondrion is folded into sheets or tubules called cristae. (Plural: cristae)

cross-bridges: links between thick and thin filaments making up banded regions in skeletal muscle tissues.

cross-pollination: an exchange of genetic material between two hermaphroditic plants.

crossing over: an exchange of material between adjacent chromatids of an homologous pair of chromosomes; it is caused by a chromosome break and produces genetic combinations different from those of the parents.

crude birth rate: proportion of births to the total population in a given year.

crude death rate: proportion of deaths to the total population in a given year.

cytokinesis: a series of biochemical and structural changes in the cytoplasm occurring during cell division and fertilization.

cytoplasm: all of the cellular material inside of and including the plasma membrane but not including the nucleus or the nuclear membrane.

cytosine: a pyrimidine base and a constituent of nucleic acids that is involved in cellular metabolism.

cytoskeleton: microtubules serving to give a cell structural integrity.

d

dark reactions: the chemical reactions involved in photosynthesis that do not require light to proceed. (See Calvin cycle.)

DDT (dichloro-diphenyl-trichloroethane): a crystalline, relatively stable, organic compound that is very effective as an insecticide.

deamination: partial disintegration of amino acids to oxidatively remove NH_2 groups, especially in the mammalian liver.

decomposer: an organism in an ecosystem that converts dead organic materials into more simple organic materials and inorganic substances.

dehydration synthesis: the formation of a large molecule from two or more smaller ones with the associated production of water.

deme: an interbreeding population within a species or an aggregate of single cells.

demographic change: a change in human populations in terms of physical environment and geographic distribution.

dendrite: the process of a neuron that transmits impulses from a synaptic junction toward the nerve cell body.

deoxyribonucleic acid: the chief carrier of genetic information; a nucleic acid composed of long chains of phosphate and sugar molecules, with several bases arranged as side groups. (Abbreviated: DNA)

diakinesis: the pronounced thickening and contraction of chromatids following the formation of chiasmata during prophase of the first meiotic division.

diaphragm: the sheet of muscle separating the thoracic and abdominal cavities in mammals that performs movements essential to breathing.

e

differential reproduction: reproduction at a variable rate depending on environmental conditions and stimuli.

differentiation: the biochemical and structural changes in cells, tissues, and organs during development; usually involves diversification and changes from general to specialized conditions.

diffusion: net movement of molecules from an area of higher concentration to an area of lower concentration.

diploidy: the state of having two full sets of chromosomes.

diplotene: in the prophase of meiosis, the stage in which homologous chromosomes begin to pair and in which their division into chromatids becomes visible.

directional selection: selection of a variant of a species that represents the best adaptation at that time.

disease syndrome: the combination of characteristics exhibited by a diseased organism that distinguishes it from an uninfected individual.

displacement activity: an activity of an organism that serves to focus attention away from the main activity or problem.

divergence: in evolution, the spreading out from a given point, resulting in the formation of new characteristics by species.

DNA: see deoxyribonucleic acid.

DNA polymerase: the enzyme that catalyzes the formation of DNA; it forms the polymer bond between sugar-phosphate groups on a complementary strand.

dominance: a genetic principle; when parents differ in one characteristic, their offspring usually resemble one of the parents rather than a blend of the two characters.

dominance hierarchy: the organization of social animals in a natural situation in which the more dominant animals control the less dominant ones by threats and aggressive action.

dominant character: a genetic character that is evident in the phenotype of an individual regardless of the presence of a particular allele; a dominant allele will determine the phenotype without regard to its allelic mate.

Down's syndrome: a condition resulting in mongolism; usually caused by the presence of an extra sex chromosome.

duodenum: the first portion of the small intestine; contains glands that secrete an alkaline mucus.

ecosystem: a system formed by the interactions between the living organisms of a community and their abiotic environment within a definable unit of space or volume.

ectoderm: the outermost layer of cells of an organism.

effector: the motor or secretory nerve ending in an organ, gland, or muscle that stimulates the structure to perform a certain task.

efferent neuron: a neuron that conducts discharges away from the central nervous system.

egg: ovum, or female gamete.

electric charge: a basic unit of charge, either negative or positive, found on the electron and proton.

electron: an elementary particle having a negative charge; it has a mass of 1/1840 of the proton and can exist independently or as the orbiting particle outside an atom.

electron cloud: the smeared out pattern of an electron moving about an atom, due to the probability associated with its position.

electron shell: the path of an electron's orbit about the nucleus.

electron transport: the motion of electrons used to reduce molecules by being passed along a series of molecules called oxidation-reduction carriers. (See oxidative phosphorylation.)

element: see chemical element.

embryo: an organism in the early stages of development.

endergonic reaction: a chemical reaction requiring the intake, or absorption, of energy to proceed.

endocrine gland: a ductless gland whose hormone secretions are released into the circulatory system and play important physiological roles elsewhere in the body.

endocytosis: a process by which large molecules are transported into a cell by being enfolded into a small cytoplasmic sac.

endoderm: the innermost tissue layer in an embryo that forms the primitive gut; also the inner lining of body cavities in coelenterates such as corals and jellyfish.

endoplasmic reticulum: extensive double-membrane network of tubules, vesicles, and sacs, often connected with the nuclear membrane of the cell; the surface may be studded with ribosomes.

endosperm: a multicellular tissue formed inside a developing seed and serving in the nutrition of the embryo.

entropy: a measure of the amount of disorder or randomness in a system; that portion of the energy of a substance not available for the performance of useful work.

environment: the sum of all physical, chemical, and biological factors to which an organism is subjected.

enzyme: an organic protein catalyst that speeds up the rate of a metabolic chemical reaction without itself being changed. (See catalyst.)

epidermal tissue: the protective, outermost layer of an organism; the outer layer and protective covering of roots, stems, and leaves in most plants and of the skin in most animals.

epilepsy: a disease characterised by intermittent convulsions and loss of consciousness, usually due to an organic disfunction of the various nervous systems of the body.

epinephrine: a hormone produced by the adrenal medulla that raises blood pressure, increases heart rate, and liberates glycogen in the liver.

epithelial tissue: animal tissue composed of tightly packed contiguous cells covering internal and external surfaces.

equilibrium: a state of balance; the condition existing when a chemical reaction and its reverse reaction proceed at equal rates.

erythrocytes: flat red blood cells found in vertebrate blood; they are formed in the bone marrow, contain hemoglobin, and carry most of the oxygen transported by the blood.

ethology: the scientific study of animal behavior, in relation to the natural environment of the animal.

eucaryotic: having a true nucleus; said of cells that have an organized membrane-enclosed nucleus in which much of the cellular DNA is found; includes most organisms with the exception of bacteria and blue-green algae.

eutrophication: a process in which the productivity and organic matter of a lake increases, eventually resulting in algal blooms with a subsequent condition of decreased oxygen levels, bacterial profusions, and the death of most aquatic vertebrates.

evaporation: a process of converting a solid or a liquid into a vapor or gaseous state.

evolution: the continuous genetic adaptation of organisms or species to the environment by selection, hybridization, inbreeding, and mutation.

excitability: readiness of a response to a stimulus.

excitatory response: a response to a stimulus applied to an organism.

exergonic reaction: a chemical reaction that liberates energy.

exocrine gland: any gland that secretes its product through a duct leading to a target organ.

exponential growth: a characteristic growth pattern (shown in bacteria and other organisms) where the number of individuals increases exponentially with time.

extinction: the process in which an entire species or other taxon becomes extinct.

f

facilitated transport: a means of transport in which some form of carrier molecules are used to convey the transported material across a membrane. (See carrier molecule.)

fallopian tube: see oviduct.

family: the major subdivision of an order or suborder in the classification of plants or animals; it usually consists of several genera.

feedback control: a process by which the presence of one substance results in the production of a second substance that, after reaching a certain level, suppresses the level of the original substance.

fermentation: see anaerobic respiration; substrate-level phosphorylation.

fertilization: the union of male and female gametes to form a diploid zygote.

fetus: the unborn offspring of any live-bearing animal; in man, refers to the period between the end of the second month of development (when the first true bone tissue is produced) and birth.

fibroblast: an irregular branching cell found in connective tissues, forming and maintaining the fibers in such tissues.

filter feeder: an animal that obtains its food by filtering it from water.

first filial generation: first generation offspring of a particular mating. (Symbol: F_1)

first polar body: a small cell formed during oogenesis; it contains a haploid nucleus and little cytoplasm.

fission: asexual reproduction in animals by a division of the body into two or more parts.

fitness: the ability of a genotype to reproduce its alleles in a particular environment or after some environmental change.

fixation: a process of rapid killing and chemical hardening of tissues with minimum distortion so that subsequent study and interpretation is possible.

flagellate: a protozoan that moves using flagella.

flagellum: a thin, fiberlike extension of the cell membrane that is used in propulsion by some forms of protozoa and certain fungal spores. (Plural: flagella)

flavin adenine dinucleotide: a chemical whose molecules can accept energy by becoming reduced; this energy can subsequently be released for ATP formation via the electron transport system. (Abbreviated: FAD)

follicle cell: a cell, found in the ovary, that helps to nourish the developing egg cell; a different kind of follicle cell produces animal hair.

food chain: the natural transfer of food energy from producers (green plants) to successive consumers (herbivores and carnivores); certain species characterize a given food chain.

food web: energy interrelationships in any community with special reference to feeding habits; a typical food web includes green plants, herbivores, carnivores, omnivores, and detritus feeders.

frame-shift mutation: a mutation caused by the insertion of or depletion of bases in a gene.

free-running cycle: a behavioral cycle in which there are no external rhythmic stimuli.

fundamental tissue: tissue that is composed largely of a single kind of specialized cell.

fungivore: an organism that feeds on the tissues of fungi.

fusion (biological): the act of uniting cells, or the cohesion of two cells.

g

G_1 period: a metabolically active period preceding the S-phase after the end of cell division, during which time protein and RNA is synthesized.

G_2 period: a gap in cell division of variable length, following the S-phase and preceding cell division, during which no new DNA synthesis occurs.

gamete: a mature, haploid, functional sex cell capable of fusing with an alternate sex cell to form a zygote.

gametogenesis: the process in which male and female reproductive cells are formed.

ganglion: a discrete mass of nervous tissue whose nerve cell bodies are located outside the central nervous system. (Plural: ganglia)

gene: A discrete unit of inheritance carried on a chromosome, transmitted from one generation to the next.

gene flow: gradual changes in the genetic make-up of a population as a result of emigration and immigration of organisms.

gene frequency: relative occurrence of a given gene in any population.

gene linkage: the concept that genes are linked together in various degrees on chromosomes and therefore do not assort independently.

gene pool: all of the alleles of all the genes in a given population.

genetic code: the sequence of genes in a chromosome and the associated chemical make-up that directs the functional and structural organization of an organism.

genetic drift: the random fluctuations in the gene pool caused by chance mortality of individuals.

genetic equilibrium: a state in which the population genes are fairly stable; new changes are being counteracted by others that tend to keep the population gene types at the same level.

genetic learning: the innate behavior of an organism thought to be genetically controlled.

genetic material: the DNA molecules in every somatic cell that hold all of the information required for differentiation and ultimate function.

genetic variation: a change in chromosomal constituents due to mutational influences.

genetics: the study of heredity and variation in organisms.

genotype: the assortment of genes that comprise the fundamental hereditary constitution of an individual; the category to which an individual is assigned on the basis of its genetic make-up, as opposed to only its external appearance. (See phenotype.)

genus: the major subdivision of a family or subfamily of plants or animals; it usually consists of more than one species. (Plural: genera)

glomerulus: small rounded cellular mass; for example, the cluster of capillaries within the cup of Bowman's capsule in the kidney.

glucose: a six-carbon sugar occurring in many fruits, animal tissues, and fluids; a common source of metabolic energy resulting from the breakdown of starch or glycogen.

glycogen: an animal and fungal carbohydrate composed of many united glucose molecules (a polymer).

glycolysis: a complex series of biochemical steps in the cell protoplasm preceding the Krebs cycle, beginning with glucose and ending with pyruvate, during which time energy is released without requiring oxygen.

Golgi complex: an internal reticular apparatus; a network of fibers of canals, apparently involved in cellular synthesis and secretion. (Also called Golgi bodies or Golgi apparatus.)

gonad: a gamete-producing (reproductive) organ in an animal; the ovary in a female animal and the testes in a male animal.

gonocyte: primary reproductive cell formed by repeated meiotic divisions of the germ cell in a male animal.

granum: a layer in the chloroplast upon which is found chlorophyll molecules. (Plural: grana)

gray matter: nervous tissue consisting chiefly of nerve-cell bodies in vertebrates, mostly concentrated in the outermost layers of the brain.

green plant: a plant that is a photosynthetic autotroph and acts as an energy source for consumers (heterotrophs).

guanine: a nitrogenous purine base and a constituent of nucleic acids.

h

habitat: the specific place or environmental situation in which an organism lives. (See niche.)

habituation: a condition of tolerance to the continued use or presence of a drug, poison, or set of behavioral stimuli.

haploid: having a single set of chromosomes in each nucleus; as in gametes and in the somatic cells of most fungi.

Hardy-Weinberg law: a generalization describing the equilibrium established between gene frequencies in a population after a generation of random mating and interbreeding.

heart: a specialized muscular organ or blood vessel used for pumping blood.

heme: a nonprotein, organic, iron-containing fraction of the hemoglobin molecule and certain other respiratory pigments.

hemoglobin: any of a large group of respiratory pigments occurring in the erythrocytes of vertebrates and dissolved in the blood plasma of many invertebrates.

hemoglobin S: sickle-cell hemoglobin; hemoglobin in which the erythrocytes are shaped like crescents or sickles.

hemolytic anemia: destruction of red blood cells and the resultant escape of hemoglobin.

herbivore: an animal that relies chiefly or solely on vegetation for its food.

heredity: the process by which physiological, morphological, and innate behavioral qualities are transmitted from parent to offspring.

hermaphrodite: a single individual that possesses both male and female reproductive organs.

heterogamy: a condition in which two uniting gametes are unlike in size or structure. (See anisogamy.)

heterotroph: an organism that feeds on food materials that have originated in other plants and animals.

heterozygous: having different members (alleles) on a particular gene; in this condition, one allele of a pair is dominant and the other allele is recessive. (See homozygous.)

histamine: an organic base or amine occurring in animal and plant tissues, causing the dilation and increased permeability of capillaries.

histone: a basic protein that is associated with chromosomes.

homeostasis: a tendency toward stability in normal internal fluids and metabolic conditions.

homeostatic mechanism: organ systems integrated by automatic adjustments to keep within narrow limits disturbances excited by, or directly resulting from, changes in the surroundings of the organisms.

homolog: one of a pair of similar chromosomes that are found in all somatic cells of diploid organisms.

homozygous: having identical members (alleles) on a particular gene; both alleles of the pair are either dominant or recessive. (See heterozygous.)

hormone: an organic secretion of endocrine glands that is released into the bloodstream and transported to some other site where it exercises a variety of regulatory roles.

hybridization: the production of a new organism from parents of two different species or genotypes.

hydrogenation: the chemical process in which a substance is combined with hydrogen.

hydrogen bond: the type of bond formed between two molecules when the nucleus of a hydrogen atom is attracted to a highly electronegative atom of a second molecule.

hydrolysis: the chemical decomposition, or splitting, of a compound into simpler substances by the addition of water.

hypothalamus: the vertical portion of the vertebrate brain just below the cerebral hemispheres; it affects control of basic drives, regulation of pituitary secretion, and control of the body's internal environment.

i

immune response: the reaction of an organism triggered by an antigen.

immunization: the process of making an organism immune to infections of specific types.

imprinting: behavior attached to stimuli very early in life and generally not reversible; it occurs at critical stages of development.

inclusions: in cells, the bodies of inactive materials, such as droplets or crystals, enclosed in the cytoplasm.

incomplete dominance: the occurrence of F_1 individuals intermediate between the homozygous dominant character of one parent and the homozygous recessive character of the other, due to the incomplete masking of the recessive character.

independent assortment: the random behavior of genes on separate chromosomes during meiosis; the result is recombination, or a new mixture of genetic material in the offspring. (See recombination.)

indeterminate growth: growth that is not limited by genetic information.

innate behavior: a mode of behavior dependent upon genetic constitution.

inorganic: pertaining to chemical substances that do not contain carbon and are mostly not derived from living organisms.

intercellular cement: a mucopolysaccharide secreted by epithelical cells in a thin layer between adjacent cells.

internal clock: the timing mechanism in some animals and plants that allows for cyclic behavior or changes of state even in the absence of environmental cues.

interneuron: a short nerve cell with its cell body in the dorsal part of the spinal cord; interneurons form a connective link between the sensory cells and motor cells; they process the sensory input and command the motor output.

interphase: the interval in a cell when the nucleus is not undergoing any mitotic changes and the chromatin material is dispersed.

interstitial cell: a small, undifferentiated connective-tissue cell; for example, the glial cell of the nervous system that binds, insulates, and nourishes the neuron.

ion: an atom or group of atoms that have lost or gained one or more orbital electrons.

ionic bond: a bond between atoms in which an electron from one atom is completely transferred to the other, resulting in attraction between the resulting different electrical charges.

irritability: the ability of a system to become sufficiently altered when confronted by an environmental change that it responds in some dynamic way to the change.

isogamy: a condition in which two uniting gametes are similar in size and structure.

k

kidney: an organ of excretion and water balance in vertebrates.

kilocalorie: the amount of heat needed to raise the temperature of one kilogram of water from 15° to 16°C (1000 calories).

kingdom: any of five categories into which organisms are usually classified: Monera, Protista, Fungi, Plantae, and Animalia; each consists of two or more phyla.

Krebs citric acid cycle: a sequence of catabolic events occurring in mitochondria in which the pyruvate produced as the end product of glycolysis is degraded to carbon dioxide and water, producing energy.

l

lactation: the production of milk by mammary glands.

lactic acid: a simple organic acid formed in many animal cells during preliminary enzymatic anaerobic oxidation of glucose via pyruvic acid.

lacuna: a small space in the matrix of bone or cartilage that contains the living bone or cartilage cells; in many invertebrates, a space in the tissues serving in place of vessels for the circulatory fluid. (Plural: lacunae)

large intestine: the colon, a large posterior part of the digestive tract of mammals.

learned behavior: patterns that are developed or modified as a result of experience.

leptotene: preliminary stage of meiosis during which the chromosomes first become apparent as individual beaded filaments in the nucleus.

leucocyte: a colorless, nucleated white blood cell found in the blood or in other body fluid; functional in combating infections.

leucoplast: a colorless plastid.

life cycle: a series of morphological and physiological changes and activities in an organism from the time of zygote formation until death.

life table: the statistical table for a particular species showing such data as mortality rate, survival rate, and life expectancies for given age groups.

light reactions: the chemical reactions involved in photosynthesis that require light for their processing. (See photosynthetic phosphorylation.)

lipid: one of the large variety of fats and fatlike organic compounds that occur in living organisms.

liver: in vertebrates, a large glandular organ that is primarily responsible for glycogen balance, deamination of proteins, formation of urea, and other nutritional functions.

lung: a thin-walled, elastic respiratory organ capable of being filled with air.

lymph: a clear, yellowish fluid, containing white blood cells, that is derived from the tissues of the body and conveyed to the bloodstream by the lymphatic vessels.

lymph node: an organ of the lymphatic system that strains foreign bodies from the lymph and contributes lymphocytes to it.

lymphatic system: a complex system of thin-walled ducts, lymph nodes, and fluid-filled intercellular spaces.

lymphocyte: a variety of nongranular, white blood cells originating in vertebrate lymphoid tissue and found in blood and lymph fluid.

lysis: the destruction of cells by a specific antibody that causes the rupture of the cell membrane.

lysosome: a minute, cytoplasmic particle consisting of several enzymes that are involved in the digestion and synthesis of certain large cellular molecules.

m

macromolecule: a molecule of high molecular weight; examples are proteins, polysaccharides, and nucleic acids.

macrophage: an enlarged, transitional monocyte that helps to relieve inflammation by ingesting foreign material and cellular debris.

maturation: the process by which a cell or organism grows and differentiates to some genetically predetermined state.

meiospore: a spore formed by meiotic division that develops into a haploid organism.

meiosis: the process of cell division by which chromosome number is halved during the formation of sperm and egg cells.

meristem: the undifferentiated plant embryonic tissue that is the principal site of active cell division.

messenger RNA: a ribonucleic acid, produced in the cell nucleus, that transcribes the genetic code of DNA in chromosomes; when attached to ribosomes, it serves as the template for determining amino acid sequence in protein synthesis. (Abbreviated: mRNA)

metabolism: the sum total of physical and chemical processes by which living organisms obtain and utilize energy.

metamorphosis: a period of abrupt transformation from one distinct stage in the life history to another; for example, the transformation of a larva into an adult insect.

metaphase: the intermediate stage in mitosis during which the duplicated chromosomes are arranged on a flat plane in the center of the mitotic spindle midway between the two asters.

methemoglobin: an oxidized form of hemoglobin in which the iron atom is trivalent.

microtubule: a tubelike structure that is commonly found in the cytoplasm and in some cylindrical organelles of eucaryotic cells; serves as a cytoskeleton to give structural integrity.

microvillus: a minute, fingerlike irregularity in the cell membranes of some tissues; each microvillus increases the surface area and absorptive capability of the cell. (Plural: microvilli)

mis-sense mutation: a mutation resulting from changes in DNA base composition, which lead to errors in amino acid incorporation in protein formation.

mitochondrion: a minute, oval-shaped organelle in the cytoplasm of most cells, consisting of protein and lipid material; it is in the mitochrondria where most ATP is produced via the electron transport system. (Plural: mitochondria)

mitosis: the process of nuclear cell division in which there is an equal division of chromosomal material between the two resulting daughter cells so that both receive identical genetic material.

mitospore: a spore formed by mitotic division that develops into an individual similar to the parent plant.

mitotic apparatus: the combination of cell parts involved in mitosis: the spindle, the asters, and the centrioles in cells that possess them.

mole: the amount of material containing 6.02×10^{23} atoms or molecules, equal in grams to the molecular weight of an element or compound.

molecule: see chemical molecule.

mongolism: see Down's syndrome.

monocyte: a large, mononuclear leucocyte or white blood cell.

monomer: a simple molecule; an individual molecular subunit of a polymer.

morphogenesis: the processes by which tissues of germ layers of organisms are shaped into organs with adult shape and form.

morula: a solid mass of dividing cells resulting from cleavage of the ovum; it develops into the blastocyst.

motor neuron: a neuron carrying stimuli away from the central nervous system and toward an effector tissue or organ.

mucopolysaccharide: a mucuslike polysaccharide that forms a protective covering over the outer surface of a cell.

mucus: the slime or viscous secretion of mucous cells or membranes; it consists of proteins and/or polysaccharides (carbohydrates).

muscle tissue: a type of tissue consisting of highly contractile cells.

mutation: a new, abrupt genetic feature in any individual produced as the result of a change in genes or chromosomes.

myelin sheath: the fatty cell membranes surrounding the axon of some nerve cells.

myofibril: a long, contractile protein fibril of cardiac and skeletal muscle tissue.

myofilament: one of the filaments that make up the protein myofibril in the skeletal muscle cell.

myosin: the most abundant protein in vertebrate muscle, largely responsible for contraction and relaxation.

n

natural selection: a process of evolution resulting in the continuation of only those forms of plant and animal life that are best adapted to reproduction and survival in the environment.

nematocyst: a minute stinging structure with which certain coelenterates, such as jellyfish, inject a paralyzing poison into their prey or enemies.

nephron: a microscopic unit of kidney structure in reptiles, birds, and mammals; usually includes a network of blood capillaries and a renal tubule.

nerve fiber: a general term for any process of a neuron, whether a dendrite or an axon.

nerve impulse: an electrochemical impulse that travels from one part of an animal to another by means of nerve fibers.

nerve net: a diffuse network of simple branching nerve cells, whose dendrites and axons cross and intermingle in a netlike fashion; for example, the nervous system of coelenterates.

nervous system: all of the nerve cells of an animal.

neuroglia: a type of interstitial tissue found in association with nerve cells.

neuron: a nerve cell consisting of a cell body, dendrites (which receive impulses from a synaptic junction), and axons (which transmit nerve impulses to an adjoining synaptic junction).

neutron: an elementary particle of nearly the same mass as a proton but lacking a charge; found in all atomic nuclei except hydrogen.

neutrophil: a type of phagocytic leucocyte formed in bone marrow and found in vertebrate blood.

niche: the distinctive way in which organisms of a species use the resources of their habitat or environment; where resources are limited, no two species can long occupy the same niche in the same habitat.

nicotinamide adenine dinucleotide phosphate: an organic compound used to capture energy obtained as the result of light reactions in photosynthesis. (Abbreviated: NADP or NADPH)

nitrate: a salt of nitric acid that is quite commonly used as a fertilizer and is subsequently incorporated into a plant to form proteins.

nitrogenous base: a nitrogen-containing molecule having the properties of a base (the tendency to acquire a hydrogen atom); for example, the purines and pyrimidines.

nondisjunction: failure of homologous chromosomes to separate during meiosis.

notochord: a rodlike cord of cells forming the chief axial structure of the chordate body; in mammals, it is present only in the embryonic stage and is supplanted by the vertebral column.

nuclear membrane: the double-layered membrane surrounding a cell nucleus.

nucleic acid: one of several complex organic acids such as RNA or DNA that combine with proteins to form the nucleoproteins of chromatin; carries genetic information.

nucleolar organizer: a small piece of chromosomal material found in the nucleolus that contributes to the formation of the nucleolus and of the ribosomal RNA.

nucleolus: a spherical body in the nucleus of nondividing cells that contains large amounts of RNA and protein and is the site of protein synthesis. (Plural: nucleoli)

nucleoprotein: one of a group of chemical constituents of chromatin composed of proteins united with a nucleic acid.

nucleotide: an organic molecule that acts as a chemical unit, or a building block of DNA and RNA; consists of a sugar, a phosphate, and a purine or pyrimidine base.

nucleus: a central body present in many cells; an organelle bounded by a double nuclear membrane and containing the chromatin material and certain other discrete bodies; in an atom, the positively charged mass composed of neutrons and protons occupying only a small fraction of the volume but possessing most of the mass. (Plural: nuclei)

nurse cell: a cell that helps provide nourishment for developing gametes.

o

omnivore: an animal that uses a variety of living and dead plants and animals in its diet.

one gene–one enzyme hypothesis: the postulate that each enzyme is produced under the direction of a single gene.

oocyte: a cell derived from a mature oogonium, which undergoes meiosis and eventually produces an ovum.

oogamy: fertilization of a nonmotile female gamete (ovum) by a motile male gamete (sperm).

oogenesis: the process of egg formation from sex cells in plants or animals.

oogonium: a cell that is the precursor of one or more mature oocytes; it divides twice to form four primary oocytes. (Plural: oogonia)

open steady state: a state of dynamic equilibrium in which a system is open to the environment for matter and energy exchanges.

operon: a group of two or more adjacent structural genes and a controlling gene of this group.

order: the major subdivision of a class or subclass in the classification of plants or animals; it usually consists of several families.

organ: a discrete structure composed of tissues and having one or more specific functions; morphologically distinct from adjacent or adjoining structures.

organelle: a discrete specialized part of a cell having one or more specific functions.

organic: pertaining to carbon-containing substances that may be derived from living organisms.

organic phosphate: an organic compound containing phosphate, used as a pesticide in many cases.

organic soup: a watery environment, filled with floating organic compounds, thought to have been the nature of the early oceans.

organism: any living individual.

ossification: the formation of bone.

ovary: the female gonad in which the ova are formed.

oviduct: a tubule used to carry eggs or egg cells away from the ovary; for example, the fallopian tube of mammals.

ovum: a mature but unfertilized female egg cell, typically containing a haploid number of chromosomes. (Plural: ova)

oxidation: an increase in the positive valence of an element occurring as a result of the loss of one or more electrons in a chemical reaction.

oxidation-reduction carrier: a molecule that helps to pass electrons along in creating ATP and heat energy.

oxidative phosphorylation: use of the electron transport system in which $FADH_2$ and NAD are oxidized to generate ATP with oxygen acting as the final electron acceptor. (See Krebs citric acid cycle.)

p

pacemaker: a small mass of specialized cardiac muscle in higher vertebrate hearts that generates a sequence of impulses to govern the heartbeat.

pachytene: a condition occurring during early prophase of meiosis; the stage following synapsis in which the homologous chromsome threads shorten, thicken, and exchange segments.

pancreas: a gland, located behind the mammalian stomach, that secretes a variety of digestive enzymes into the duodenum.

panmictic population: a population in which mating occurs in a random manner; any sperm cell is likely to be combined with any egg cell in the population.

parasite: an organism that lives in or on an organism of another species and derives its food from the host.

parasympathetic nervous system: one of the two divisions of the autonomic nervous system; it has an inhibiting effect on tissues and organs counter to that of the sympathetic system.

parenchyma: in lower animals, a spongy mass of vacuolated cells filling spaces between muscles, epithelia, and viscera; in plants, the soft, thin-walled cells that are used largely for support and for food storage.

parthenogenesis: unisexual reproduction involving the production of young by females that are not fertilized by males.

passive transport: see diffusion.

pathogenic: causing a disease syndrome.

pecking order: see dominance hierarchy.

penis: the male copulatory organ through which sperm are deposited in the female vagina and through which urine is excreted.

peptide: a compound of two amino acid units joined end to end, which are held together by a peptide bond. (See polypeptide.)

periderm: the cuticle or cornified outer layer of skin.

peristalsis: rhythmic muscular contractions in the digestive tract of vertebrates, serving to mix and move the contents of the tract.

phagocyte: an amoeboid cell that engulfs foreign particles found in blood and other body fluids as well as in the tissues of the lymphatic system.

phagocytosis: the process of ingesting a foreign particle by a phagocyte.

phagosome: an ingested particle that fuses with a spherical lysosome.

phenotype: the observable characteristics of an organism that depend on genetic and environmental factors. (See genotype.)

phenylketonuria: a congenital disease involving faulty metabolism of phenylalanine, producing mental retardation. (Abbreviated: PKU)

phloem: one of the vascular tissues in plants; it consists of sieve tubes and companion cells and transports nutrients around the plant body.

photon: a unit or quantum of energy of visible light or any other electromagnetic radiation.

photosensitivity: sensitivity to light.

photosynthesis: a series of processes involving the transformation of carbon dioxide and water in living plants into carbohydrates, using solar energy absorbed by chlorophyll.

photosynthetic phosphorylation: a process by which light energy is used to convert ADP to ATP (by the addition of a phosphate group) and NADP is converted to NADPH.

phototropism: the influence of light upon growth; for example, the tendency of many organisms to grow toward or away from light.

pH scale: a scale, varying from 1 to 14, used to specify the acidic or basic character of a solution.

phylum: the major primary subdivision of the plant or animal kingdom; it consists of one or more related classes. (Plural: phyla)

pigment: a minute granule that occurs in many types of plant and animal cells; any colored substance; a compound that absorbs some wavelengths of visible light and reflects others.

pinocytosis: a process by which particles or ions on cell surfaces are absorbed, followed by an inward progression of the material in a droplet form until it is pinched off and incorporated into the cytoplasm.

pituitary gland: a small, oval endocrine gland attached to the base of the brain; it produces numerous hormones.

placenta: a mass of blood vessels and membranes in the uterus of pregnant mammals; it functions in the exchange of nutrients and waste products between mother and embryo (or fetus).

plasma: the colorless fluid portion of blood in which corpuscles are suspended.

plasma membrane: the outer layer, or living membrane, bounding the cytoplasm of a cell.

plastid: one of several kinds of small self-propagating bodies in plant cytoplasm.

platelet: a colorless enucleated cell fragment of cytoplasm in the circulating blood of all mammals; contains thromboplastin and is important in the clotting of blood.

point mutation: a single change in the base composition of DNA.

polar covalent bond: a bond between two nuclei that do not share electrons equally; this situation creates a dipole moment, which is due to the differing attractive forces of the atoms. (See covalent bond.)

polarization: an electrical potential or uneven distribution of chemical ions or charges; for example, the inside of the glial cell has a negative potential with respect to the outside.

polymer: the product that results when two or more simple molecules (monomers) of the same substance combine.

polymerization: the process by which a complex molecule of high molecular weight is formed by the union of a number of simpler molecules, which may or may not be alike.

polynucleotide: a nucleic acid composed of a linear sequence of nucleotides in which the sugars are linked through phosphate groups.

polypeptide: a polymer consisting of three or more linked amino acid units (peptides); the basic structural component of a protein. (See peptide.)

polyploidy: the state of having one or more complete extra sets of chromosomes.

polysaccharide: a carbohydrate consisting of two or more monosaccharide sugars; includes starch, cellulose, and glycogen.

polytene: a condition in which a chromosome fibril consists of a large number of parallel identical strands.

population: all individuals of the same species living in the same area.

population genetics: the study of the relative frequency of hereditary characters in large samples of whole populations of a species.

postsynaptic cell: a cell whose dendrites receive impulses across the synapse.

precursor: a starting molecule entering a catabolic or anabolic reaction; a molecule that will be changed to form some product.

primary consumer: an animal (herbivore) that obtains its supply of organic nutrients from the tissues of producers (green plants).

primary immune response: a sequence of steps in a response to any sort of foreign material; it involves recognition of the invader, mobilization of special cells, and destruction of invading cells.

primitive streak: in a human embryo, the main body axis occurring before the end of the first month.

procaryotic: lacking a true nucleus; said of cells that lack an organized, membrane-enclosed nucleus; examples are bacteria and blue-green algae.

producer: an organism that is able to synthesize organic compounds by capturing energy; for example, photosynthetic plants.

prophase: the initial stage of both mitosis and meiosis; the chromosomes and chromatids become discrete, the centriole divides, and the nuclear membrane disappears.

prostaglandins: a group of substances within the plasma membrane; they set off the synthesis of any enzyme that aids in the breakdown of ATP to cyclic AMP.

protective response: the behavioral response of an organism that helps to minimize damage from an injury; examples are blinking and limping.

protein: a complex organic compound (a polymer made up of amino acids) that is required of all life processes; contains carbon, hydrogen, oxygen, nitrogen, and other elements.

proton: an elementary particle having a positive charge equal in magnitude to the negative charge of an electron; a fundamental component of the atomic nucleus.

purine base: a type of bicyclic (two ring) nitrogenous base, such as adenine and guanine; constituent of DNA and RNA.

pyrimidine base: a type of monocyclic (one ring) nitrogenous base, such as thymine, uracil, and cytosine; constituent of DNA and RNA.

r

radioactivity: a property, exhibited by certain elements, of spontaneously emitting alpha or beta particles or gamma rays from the nucleus of an atom.

receptor: a cell, tissue, or organ of an animal that cooperates with the nervous system in detecting and responding to internal or external stimuli.

recessive character: a genetic character that is evident in the phenotype only when an individual is genetically homozygous (both alleles of a pair are recessive); otherwise, a dominant allele will mask the effect of a recessive allele.

recombination: the occurrence of new gene combinations in an organism that were not present in either parent—produced by crossing over during gametogenesis, random segregation of chromsomes, and chance recombination of chromosomes.

rectum: a short section at the end of the large intestine where waste material is stored prior to elimination.

recurrent nerve: a nerve that turns back in its course, such as the recurrent laryngeal nerve.

red blood cell: see erythrocyte.

reduction: an increase in the negative valence of an element occurring as a result of the gain of one or more electrons in a chemical reaction.

refractory period: a brief recovery period following a nerve impulse, during which the nerve cannot generate a new impulse (absolute refractory period) and then gradually returns to normal (relative refractory period).

regulatory gene: a gene whose products interact with other genes and cause their expression to be changed.

relative refractory period: see refractory period.

renal tubule: see nephron.

replication: in genetics, the duplication of hereditary material (complementary chromosomes) that takes place in meiosis.

reproductive isolation: prevention of interbreeding between two or more populations because of different breeding seasons, incompatibility of the reproductive organs in the two sexes, reproductive behavior, or geographical barriers.

respiratory system: all the organs of an animal associated with the absorption of oxygen and the release of carbon dioxide.

resting potential: the difference in electric potential of a fraction of a volt between the inside and outside of a cell when no nerve pulse is being transmitted.

ribonucleic acid: a nucleic acid containing the sugar ribose; a constituent of nucleoproteins found in the cytoplasm and chromatin, composed of long chains of phosphate and sugar molecules acting as carriers and mediators of genetic information. (Abbreviated: RNA)

ribonucleotide: a compound consisting of the sugar ribose, a purine or pyrimidine base, and a phosphate group; a component of RNA.

ribosomal RNA: a type of RNA molecule found in the ribosomes that functions in protein synthesis. (Abbreviated: rRNA)

ribosome: a minute, granular element of cytoplasm, composed of protein and RNA; thought to function in protein synthesis and usually associated with the endoplasmic reticulum.

RNA: see ribonucleic acid.

s

saliva: the digestive and lubricative juice secreted into the oral cavity of many vertebrates and invertebrates.

scavenger: an animal that feeds on dead animal material that it has not killed.

scrotum: a sac or pouch of skin in the pelvic region; contains the testicles in mammals.

sebaceous gland: an epidermal gland of mammals; projects into the dermis and secretes a fatty substance (sebum).

second filial generation: all of the offspring produced by the mating of two individuals of the first filial generation. (Symbol: F_2)

secondary consumer: an animal that feeds upon the tissues of primary consumers.

secondary immune response: a resistance to further disease acquired by an enlarged population of cells following an exposure to foreign material or to an infectious disease.

secondary polar body: a small, nonfunctional haploid cell resulting from a second meiotic division of a secondary oocyte.

segregation: a genetic principle; when a hybrid reproduces, half its cells transmit the dominant character of one parent and the other half, the recessive character of the other parent.

selective permeability: membranes that selectively allow certain sizes of molecules into an internal environment that differs from an external environment.

self fertilization: the union of an egg and a sperm produced by a single hermaphroditic animal.

semiconservative replication: the process by which one of the two strands of DNA in a given gene is copied by each daughter molecule.

sensitive period: the period during which imprinting can occur in the young stages of an organism's life.

sensitivity: the susceptibility of tissues or an organism to the antigen action of a serum or other substance.

sensory neuron: a neuron carrying stimuli away from a receptor and toward the central nervous system

sensory system: all the organs, tissues, and cells associated with the reception of stimuli.

sex chromosome: see X chromosome; Y chromosome.

sex-linked trait: a genetic character determined by genes located on one or more of the sex chromosomes.

sexual recombination: a recombination of genes resulting in offspring with different genotypes than either of the parents.

sickle cell anemia: a hereditary disease caused by a defective form of hemoglobin and characterized by sickle-shaped red blood cells.

sigmoid curve: a logistic or S-shaped population growth curve, showing initial slow growth, rapid intermediate growth, and later slow growth.

skeletal system: all of the supporting structures in an animal, such as bone and cartilage.

skeletal muscle tissue: the fibrous, or striated, muscle tissue that acts to support the structure of organisms; examples are chitin, cartilage, and bone.

smooth muscle tissue: nonstriated muscle tissue.

social organization: the organization of a population into different group roles, each with its own pattern of appropriate social behavior.

social releasers: the natural behavior of an organism that influences other animals in a social situation.

soma: a body; for example, the body of a cell. (Plural: somata)

somatic cells: diploid body cells as contrasted to sex cells.

somatic nervous system: the part of the vertebrate nervous system that is involved in the control of voluntary responses to stimuli.

somesthetic cortex: the part of the parietal lobe of the brain that receives impulses from the touch, heat, cold, and kinesthetic receptors.

somite: one of a series of segmentally arranged masses of tissues on either side of the neural tube of the embryo and giving rise to voluntary muscles, most of the skeleton, and the skin.

specialization: a specific morphological, anatomical, physiological, or behavioral feature that fits an organism or one of its parts for a particular function or habitat.

speciation: the process by which a new species is formed from a previously existing species through the mechanisms of evolution.

species: the major subdivision of a genus or subgenus; it is composed of related individuals that resemble each other and are able to breed among themselves but usually not with other species.

spermatid: one of four haploid cells resulting from the two meiotic divisions of a primary spermatocyte.

spermatocyte: a cell derived directly from a spermatogonium, which undergoes the first meiotic division and produces two secondary spermatocytes.

spermatogonium: a cell originating in a seminal tubule and dividing twice to form four primary spermatocytes. (Plural: spermatogonia)

spermatozoan: a male sperm cell with a locomotory flagellum. (Plural: spermatozoa)

sperm cell: see spermatozoan.

S-phase: a phase of DNA synthesis, lasting from several minutes to eight hours, following the G_1 period and preceding the G_2 period.

sphincter: a muscle having its fibers in a circular fashion around an opening so as to close that opening upon contraction.

spinal cord: a thick, longitudinal bundle of nerve fibers extending from the brain posteriorly along the dorsal side in vertebrates.

spindle: a structure formed in the cell nucleus during mitosis and meiosis; it consists of filaments that radiate from the poles into the cytoplasm and attach to chromosomes during mitosis.

spirillum: a corkscrew-shaped bacterium. (Plural: spirilla)

spleen: a highly vascular, glandular organ, situated in man at the cardiac end of the stomach; it functions in the formation of lymphocytes, in the destruction of worn-out erythrocytes, and as a reservoir for blood.

spore: the asexual reproductive element of lower organisms, such as bacteria, protozoans, or ferns and mosses; capable of developing directly into an adult.

stroma: the connective-tissue matrix of an organ; for example, the fine filmy framework of an erythrocyte or the embedding matrix of chloroplasts.

structural gene: a gene that codes for the production of a structural protein.

substrate: in biochemistry, any substance acted upon by an enzyme; in biology, the medium upon which an organism grows.

substrate-level phosphorylation: fermentation; the breakdown of carbohydrates, such as glucose, yielding energy without requiring oxygen. (See anaerobic respiration.)

sugar: a crystalline carbohydrate with either six, twelve, or eighteen carbon atoms in the molecules; plays an important role in plant and animal metabolism.

symbiotic theory: the theory that traces the cellular ancestry of modern organisms to more than one ancient organism.

sympathetic nervous system: one of the two divisions of the autonomic nervous system; it promotes energy expenditure by mobilizing the body's resources, and it counteracts the parasympathetic system.

synapse: the junction of the processes of one axon of a nerve cell with the dendrites of an adjoining nerve cell.

t

taxon: any grouping in the classifying of organisms to demonstrate evolutionary or structural relationships. (Plural: taxa)

territoriality: the specific area over which an animal establishes jurisdiction for feeding or breeding purposes or both.

tertiary consumer: a second-level carnivore who feeds mostly on secondary consumers or other carnivores.

tertiary structure: the folded, three-dimensional structure of a protein molecule.

testis: the male gonad in which spermatozoa are produced. (Plural: testes)

thalamus: a cluster of neurons and glial cells in the rearmost part of the forebrain; it is involved in hearing and vision.

thermodynamics, first law of: energy can be neither created or destroyed, but only changed from one form to another.

362

thermodynamics, second law of: all energy changes result in a reduction in the amount of free energy available.

threshold value: the specific stimulus intensity below which a given irritable tissue exhibits no response.

thrombocyte: see platelet.

thymine: a pyrimidine base; a component of deoxyribonucleic acid, involved in cellular metabolism.

thymus gland: an organ, lying behind the top of the breastbone, that is large during infancy but atrophies in adults; it produces cells (immunocytes) that function in immune responses.

thyroid gland: an endocrine gland peculiar to chordates that secretes the hormone thyroxin, which functions to regulate energy metabolism.

thyroid-stimulating hormone: a hormone produced in the pituitary that causes the thyroid glands to secrete the hormone thyroxin. (Abbreviated: TSH)

tissue: an aggregation of similar cells having the same functions.

telophase: the last stage in mitosis, characterized by formation of a new cell wall between the two nuclei.

tolerance range: the maximum and minimum conditions that a species can endure, such as temperature; such conditions define in broad outline the range of a population.

trachea: in humans, the tube descending from the larynx to the bronchi; in insects, a complex system of branched, air-filled respiratory tubules. (Plural: tracheae)

transcription: the enzymatic process by which the base sequence of chromosomal DNA is transferred to messenger RNA, which forms a complementary copy.

transduction: in genetics, the transfer of genetic information from one cell to another by a virus.

transfer RNA: a relatively small molecule of nucleic acid that acts as a carrier of specific amino acids during protein synthesis on ribosomes. (Abbreviated: tRNA)

translation: the biosynthetic process by which amino acid sequences in proteins are determined by base sequences in the RNA template on ribosomes.

transmitter substance: a chemical released from a neuron that induces changes in the membrane of a receiving cell, tending to depolarize the cell and produce an action potential.

triplet code: a sequence of three bases at a time that are required to specify one amino acid residue in a protein chain. (See codon.)

trophic level: one of several successive levels of nourishment in a pyramid of numbers, food web, or food chain; plant producers constitute the lowest level and dominant carnivores, the highest level.

u

ultraviolet light: radiation just beyond the visible violet end of the spectrum, having pronounced actinic and chemical properties. (Abbreviated: UV)

umbilical cord: a tough, stalklike projection from the ventral surface of the embryo of placental mammals; contains blood vessels and connects the fetus to the placenta; transmits oxygen and nourishment (via the fetal blood) from the mother to the fetus.

unit membrane: the triple-layered cell structure characteristic of all cell membranes.

uracil: a pyrimidine base; a component of ribonucleic acid involved in cellular metabolism.

urea: one of the chief products of protein metabolism in animals, formed in the liver and excreted by the kidneys in vertebrates; also found in fungi and in certain seed plants.

ureter: the duct connecting the kidney with the urinary bladder in higher vertebrates.

urethra: the duct that carries urine from the urinary bladder to the exterior in mammals.

urine: excretory product usually formed in the kidneys.

uterine cavity: location in which the morula is free for a short period before becoming implanted on the posterior wall.

uterus: the muscular chamber of the reproductive tract in female mammals in which the embryo and fetus develop.

v

vacuole: a large, fluid-filled, internal cavity found in plant and animal cells.

vagina: the part of the female reproductive system that acts as a receptacle for the penis during copulation in higher vetebrates.

vein: a blood vessel that carries blood toward the heart and away from the tissues and organs.

villus: a small protrusion, especially a protrusion from the surface of a membrane. (Plural: villi)

virulent strain: a strain of bacteria or virus that is infectious and damaging to the host organism.

virus: a submicroscopic, noncellular particle that consists of either DNA or RNA and a protein coat; it is parasitic and reproduces by means of synthetic processes in the host cell; it appears to be on the borderline between living and non-living matter.

w

wagging dance: the "dance" performed by a scout bee to communicate the direction of a food source to other bees in the hive.

white blood cell: see leucocyte.

x

X chromosome: a special sex-determining chromosome, not occurring in identical number or shape in both sexes; in humans, the female has two X chromosomes and the male has only one coupled with a Y chromosome.

x-ray diffraction: the process by which molecular structure (particularly that of crystals) can be determined by bouncing x-rays off planar surfaces.

xylem: one of the vascular tissues in plants; it conducts water from the roots upward and represents the wood of the plant.

y

Y chromosome: a sex-determining chromosome; in humans, the female has no Y chromosomes and the normal male has only one coupled with an X chromosome.

yolk: the protein and fat material in the ova of animals; serves as food for the developing embryo.

z

zygote: a fertilized ovum in plants or animals; a diploid cell resulting from the fusion of male and female gametes.

zygotene: the longitudinal pairing of homologous chromosomes during the prophase of the first meiotic division.

Index

a

abomasum, *220*
abortion, 184, 327
absolute refractory
  period, 260, *261*
absorption
  in digestive tract, 230
  of pigments, 74–75
  plasma membrane in, 101
absorption spectrum,
  chlorophyll, *73*
absorptive cell, *101*
absorptive heterotrophy, 20
acetic acid, 79
acetyl-coenzyme A, 79
acetylcholine (ACh), 264
ACTH (adrenocorticotrophic
  hormone), 243, 245
actin, 225, *226*
action potential, 260, *261*
active site, 173
active transport, *100*, 234
actual birth rate, 304
adaptation, 61
adaptive radiation, 11, *12, 13*
adenine, *49*, 114, 119, 120, 125
adenosine diphosphate. *See* ADP
adenosine monophosphate.
  See AMP
adenosine triphosphate. *See* ATP
adenyl cyclase, 244
ADH (antidiuretic hormone), 243
ADP (adenosine diphosphate),
  66–67, 72, 75, 79–81
adrenal cortex, *243*, 245
adrenal gland, *243*
adrenal medulla, 243
adrenocorticotrophic hormone
  (ACTH), 243
adsorption, 45
aerobic, 70
aerobic respiration, 76–81
afferent sensory neuron, *265*
African gazelle, 312
African sleeping sickness, *110*
afterbirth, 205
age-specific birth rate, 304
agglutination, 249
agriculture, 328, 333–335
Alaskan fur seal migration, *280*
albinism, 168
albumin, 179
alcohol, 44, 67–68
alga(e)
  asexual reproduction in, 191
  blue-green, 16–*17*,
    82–83, 90, 135
  brown, 18
  characteristics of, 18–*19*
  coenocyte formation, 217
  flagella in, 18, 109, *110*
  green, 18
  kelp, 18
  red, 18
  yellow-green
    (Xanthophyta), 18, *19*
algal bloom, 335, *336*
allele, 161–171
  albinism, 168
  crossing over of, 163, *164*
  genetic drift, loss in, 169
  in chromosome mapping, 165
  in gene pool, 166
  selection action of, 170–171
allergen, 252
allergy, 251–252
allium leaf epidermis, *221*
alveolus, *223*, 234
amino acid, 44, 173–177
  breakdown in liver, 232, 241
  catalysis of, 51
  deamination, 241
  excretory systems and, 241

in enzymes, 113
in proteins, 51, 113
polymerization of, 176–177
ribosome formation, 127, 176
RNA translation into, 115
transfer RNA and, 132
amino acid activating enzyme, 132
amino acid metabolism, 173
ammonia
  deamination, 241
  in early atmosphere, 41
  nitrogen in, 40
  urea formation, 241
amniocentesis, 181, *182*
amniotic sac, 206
amoeba(e), 18, *19*, 96
  enucleated action, 96
  nuclear transplant of, 96
  pseudopods of, *19*
AMP (adenosine
  monophosphate), 244
amphibian, *257*
amphioxus, 28
amygdala, 273
amyloplast, *104*
Anabaena, 16, *17, 82, 83*
anabolic process, 54
anaerobic, 70
anaphase
  of meiosis, 147, *148–149*
  of mitosis, *138–139*, 140
anaphylactic shock, 252
anemia, hemolytic, 252
anesthetic drugs, 99
angiosperm, *192*
angstrom, *88*
animal(s)
  avoidance response in, 282
  characteristics of, 24
  chemical signals in, 255
  cold-blooded, 252
  connective tissue, 220, *222*
  energy in, 54
  gas-exchange system,
    24, 232–234
  gonocyte production, 150
  morphogenesis in, 199, 201
  muscle tissue, *224–225*
  physiological systems, 24
  warm-blooded, 252
animal cell
  cytokinesis in, *142–143*
  lysosomes in, *103*
  vacuoles in, 103–104
Animalia, 24, *25–31*
  characteristics of, 24
  copulation, 195
  ingestion in, 24
  sexual reproduction in, 195, *201*
  phyla, *24–31*
  See also animal(s)
anisogamy, 194
annelid, 256
Annelida, 24, 25
annulus, 93
ant
  communication, 290–291
  harvest, 25
anterior pituitary gland, 242
Anthophyta, 24
anthrax bacteria, 16, *17*
antibody, 249–250
anticodon, 132
antidiuretic hormone (ADH), 234
antigen, *247*, 248–250
  antibody reaction with, *249–250*
  blood types and, 248
  inflammatory response, *247*
aorta, 238–239
aplysia, 258
Arachnida, 26, *27*
arginine, 125, 177
artery, 236–239
Arthrophyta, 22, 23
arthropod, *13*, 26
  chitin in skeletons, 26
  ganglia in, 256
  jointed appendages in, 26
Arthrospira (blue-green
  algae), 16, *17*
artiodactyl, *30*

Aschelminthes, 24, 25
aster, *141*
Asterias rubens, 283
asthma, 251–252
atmosphere
  formation of earth's, 41
  oxygen formation in, 68–69
atom, 33–41
  bonding, 38–41
  electron shell
    configuration, *36, 37*
  electrons in, 35
  neutrons in, 35
  nucleus, 35
  protons in, 35
atomic number, 35, *37, 38*
atomic shells, *36, 37*
atomic theory, *37*
ATP (adenosine
  triphosphate), 66–81
  ADP, degradation to, 67
  aerobic respiration,
    formation in, 78
  composition, 67
  electron transport chain,
    formation in, 80–81
  fermentation and, 67
  from ADP, 67
  from glucose, 80–81
  from glycolysis, 78–79
  from photosynthesis, 67
  hydrolysis in membranes, 143
  in metabolism, 70
  in mitochondria, 105, 107
  in mitotic apparatus, 140
  Krebs cycle, formation in, 80–81
  oxidative phosphorylation,
    formation in, 67
  substrate-level
    phosphorylation, 67
atrium, 238
atrioventricular node, 238
australopithecine, 324
autoimmune disease, 252
autosome, 161
  genetic material in, 180
  in Tay-Sachs disease, 183–184
autotroph, 16, 66, 69–70, 76,
  90, 215, 229
autotrophic nutrition, 69–70
Avery, O., 117, 175
Aves, 28
avirulent bacteria, 116
Avocet, *28*
avoidance response, 282, *283*
axon, 223–224, 256, 259, 260
  as nerve fiber, 223
  in ganglia, 256
  in myelin sheath, 223
  of neuron, 223

b

bacillus (bacteria), 16, 55, 67, 75,
  90, 92, 116–117,
  232, 252, 301
bacteria, 16, *17*
  anthrax, 16, *17*
  as decomposers, 16, *314, 315*
  as procaryotic cells, 90
  avirulent, 116
  bacillus, *90*
  coccus, *90*
  contents of, 83
  *Diplococcus
    pneumoniae*, 116–117
  DNA in, 117
  enzymes in, 83
  in colon, 232
  phagocytosis of, 100
  photosynthetic, 73–74
  shapes of, *90*
  spirillum, *90*
  virulent, 116
  vitamin synthesis in, 232
bacteriophage, 55, *56, 57*
Balbiani, E., 94
Balbiani ring, 142
Bambyx mori, 291
barnacle, 26
barn swallow, *280*
basal body, 109

NOTE: *Italic* page numbers refer to *figure illustrations* in the text.

NOTE: *Italic* page numbers refer to *figure illustrations* in the text.

t

T-2 bacteriophage, *57*
T-4 bacteriophage, *56*
T-5 bacteriophage, *55*
tapeworm, 24
Taraxacum officinale, 22, 23
Tatum, E., 175
Tay, W., 182
Tay-Sachs disease, 182–184
teeth, 231
telophase
  of meiosis, *144–145, 146,* 147
  of mitosis, *138–139,* 140, 142
temperature, 38, 41, 44,
  59, 252–253
temporal lobe, 267
territoriality (behavior), 295,
  296–297, 298–299
testis, 196, 199, 289
thalamus, 265, *266,* 269
thermodynamics, first law
  of, 54, 63
thermodynamics, second law
  of 53, 54, 64
threshold value, 260
thrombocyte. *See* platelet
thylakoid, *105*
thymine, *49,* 114, *118, 120–121*
thymus, 249, *250*
thyroid, 243
  See also TSH
thyroid stimulating hormone.
  See TSH
thyroxin, 243
tick, 284
Tinbergen, N., 284
tissue, animal, 218–219,
  220, 223–227
tissue, plant, 218, 220, *221, 222*
Tjio, J., 180
tobacco mosaic virus (TMV), *55*
tolerance, *309,* 311
tonsils, 248
toothshell, 26
tortoise, *8*
trachea, 206, 232
tracheophyte, 22
transcription, *125,* 142
  by RNA polymerase, *126,* 177
  of codons, 177
  of RNA, *121,* 126
  RNA from codons, 115
transduction, 187
transfer RNA (tRNA), *130–131,*
  132, 133, 176
translation
  code end points in, 127
  RNA into amino acids, 115, *121*
  steps in, *132–133*
transport cycle, 238–239
triglyceride, *47*
trilobite, *13*
triplet code, 127, 129, 177
Trisomy-21, 181

tritiated-thymidine labeling, *140*
trophic levels, 314
tropism, *58*
tropomyosin, 225
Trypanosoma gambiense, 110
trypanosome, 250
tryptophan synthetase, 129
tryptophane, 333
TSH (thyroid-stimulating
  hormone), 243, 245
tubeworm, 24, *25*
tubule, 196–197
tundra, Arctic,
tunicate, 28, *29*
twins, *209*

u

Uca minax, 26
Uexküll, J. von, 284
Ulothrix, 137, 193, 194
ultraviolet radiation, 44, 45,
  69, 71, 73
umbilical cord, 207
unconditioned stimulus, *258*
unicellular organism, 18, *110,*
  189, 215, 255
unit-membrane configuration, *98*
uranium, 35
uracil, *125, 127,* 176
urea, 241
ureter, 242
urethra, 242
uridine, 125
urine, 242, *288*
uterus, 199

v

vacuole, 103–104
vagina, 199
valine, 174
vascular plant, 22, *23, 70*
vascular tissue, 22, *70,* 197, *222*
vein, 235, 239
ventral nerve cord, 279
Venus flytrap, 22, *23*
vertebrate, 7, 256, *257, 264–271*
villi, 231
virulent bacteria, 116
virus, 55–56, 111
  See also individual viruses
vitamin synthesis, 102, 232
vitamin A, 332
voltage, 260–261
voluntary response, 264
voluntary striated muscle, *224*
Volvox, 191, 217

spontaneous generation, 7
sporangia, 20, *21*
spore, *145, 191, 252*
sporophyte, 22
sporozoan, 18
squid, 26, *27*
staphylococcus, *90*
Stahl, F., 122, *123*
stamen, *192*
star coral, *25, 219*
starch, *47, 104,* 231
starfish, *27, 192*
starling, *282*
steady state, 63–64
Stentor, 18, *19, 110*
steroid, 102, 243
stickleback (fish), 284, *285*
stimulus, *258, 277, 278, 283*
stomach, 220, 231
streptococcus, *90*
stroma, 104
structural gene, 179
Sturtevant, A., 161
suberin, *221*
substrate, 173, *174*
substrate-level
  phosphorylation, 67, *78, 79*
sugar, 44, 46, *47, 51, 72, 76,*
  114–115, 126, 127
sulfur, 44
sun, 41
"sun-compass" theory, *282*
Sutherland, E., 244
Sutton, W., 157
swallow, barn, *280*
symbiont, 92
symbiotic cell theory, 90, 92
sympathetic nervous system.
  See nervous system
symmetry, bilateral, 27, 256
symmetry, radial, 24
synapse, 261–262
synapsis, 146, *164*
synaptic cleft, *263*
synaptic junction, 261–262
synaptic knob, *263*
synaptic region, 256
synaptic vesicle, *262, 263*

w

wasp, 275–276
water, 77, 104
  balance in body, 241
  charges in, 39–40
  dehydration synthesis and, *47*
  in cells, 87
  in early atmosphere, 41
  photolysis of, *75*
Watson, J., 119
Watson-Crick model, *119*
wavelength (of light), 44, 72, 89
wax, plant-cell, 111
Weinberg, W., 167
Weismann, A., 189
Wells, I., 174
whale, gray, *280*
white blood cell, *181,* 248–249
Whittaker, R., 15
worm, 24, *25,* 250

x

Xanthophyta, 18, *19*
X-chromosome, 160, 161,
  180, *181,* 196
x-ray, 158, 175, *207*
x-ray diffraction, 119, *236*
xylem, *70*

y

Y-chromosome, 161, 180, *181,* 196
yeast, 20, 67, *191, 303*
yolk, 111, 198

z

zygospore, *195*
zygote, 135–136, 150,
  193, 194, 195
  as fertilized egg, 135
  from haploid gametes, 193
  in sexual reproduction, 193
  See also gamete; egg; sperm
zygotene, 146

NOTE: *Italic* page numbers refer to *figure illustrations* in the text.

Credits and Acknowledgments

Unit I/Chapter 1

2: Terry Lamb; 4: Patricia Peck; 6: Paul Slick after Dr. Wilfred Wilson; 7: The Bettmann Archive, Inc.; 8: (top) The Bettmann Archive, Inc., (bottom) Photo by George H. Harrison/ Grant Heilman; 9: (top) Doug Armstrong, (bottom) courtesy of the National Maritime Museum, England; 10: John Dawson after J.M. Savage, *Evolution*, Holt, Rinehart and Winston, 1969; 12: John Dawson; 13: (top) Doug Armstrong, (bottom) The American Museum of Natural History; 15: Doug Armstrong after R.H. Whittaker, *Science*, Vol. 163, pp. 151–157; 16: M. Woodbridge Williams; 17: (left to right, top to bottom) Ward's Natural Science Establishment, Inc., Lynwood Chace/National Audubon Society, Walter Dawn, Patrick Echlin, Carolina Biological Supply Co., Lillian Holdeman from Smith and Holdeman, *The Pathogenic Anaerobic Bacteria*, 1968, Charles C. Thomas, Publisher, Springfield, Illinois, and the Center for Disease Control, R.A. Packer, Iowa State University, courtesy of the New York Botanical Garden Library; 19: (left to right, top to bottom) Carl Zeiss, Inc., OMIKRON, Douglas P. Wilson, Eric V. Gravé, Grant Haist, Eric V. Gravé (5); 20: (top) Hugh Spencer, (bottom) Grant Haist; 21: (left to right, top to bottom) Philip Hyde, Allan Roberts, Grant Haist, Jack Dermid, Dennis Brokaw; 22: Walter Dawn; 23: (left to right, top to bottom) F.T. Haxo, Eric V. Gravé, Douglas P. Wilson, Dennis Brokaw, Grant Heilman, Grant Haist, Dennis Brokaw, Philip Hyde, Grant Heilman, Jack Dermid (2); 25: (left to right, top to bottom) James W. Dutcher, Paul Dayton, James W. Dutcher (2), Carolina Biological Supply Co., Eric V. Gravé Runk/Shoenberger (Grant Heilman) (2), Allan Roberts (2), William H. Amos, James W. Dutcher; 26: (left to right, top to bottom) OMIKRON, Allan Roberts, Douglas P. Wilson, M. Woodbridge Williams, James W. Dutcher, courtesy of Frank A. Brown, Jr. and L.N. Edmunds, Jr., Grant Heilman, Dennis Brokaw, Allan Roberts (2); 27: (left to right, top to bottom) Grant Haist, Richard F. Trump/Biophotography, M. Woodbridge Williams, William H. Amos, OMIKRON; 28: (left to right, top to bottom) Runk/Shoenberger (Grant Heilman), James W. Dutcher, Jack Dermid, San Diego Zoo Photo/Ron Garrison and F.D. Schmidt (3), James W. Dutcher, San Diego Zoo Photo/Ron Garrison and F.D. Schmidt; 29: (left) Dick Oden, (right column, top to bottom) James W. Dutcher, San Diego Zoo Photo/Ron Garrison and F.D. Schmidt, Jack Dermid, San Diego Zoo Photo/Ron Garrison and F.D. Schmidt; 30: (left to right, top to bottom) Dick Oden, San Diego Zoo Photo/Ron Garrison and F.D. Schmidt, Jack Dermid, San Diego Zoo Photo/Ron Garrison and F.D. Schmidt; 31: (left to right, top to bottom) San Diego Zoo Photo/Ron Garrison and F.D. Schmidt (4) Allan Roberts, John Oldenkamp/IBOL.

Chapter 2

32: John Brenneis for *Scientific American*; 34: Philip Kirkland; 35: Doug Armstrong; 36: Volt Technical; 37: Culver Pictures; 38–39: Doug Armstrong; 40: Paul Slick after Earl Frieden, "The Chemical Elements of Life." © 1972 by Scientific American, Inc. All rights reserved.; 42–43: Joe Garcia after *Adventures in Earth History*, selected, edited, and with introductions by Preston Cloud. W.H. Freeman and Company. © 1970.; 45: John Dawson after Robert Jastrow, *Red Giants and White Dwarfs*, Harper & Row, 1971.; 47–48: Paul Slick after Dr. Wilfred Wilson; 49–51: Doug Armstrong; 52: courtesy of S.W. Fox, Institute for Molecular and Cellular Evolution, University of Miami, Coral Gables, Florida; 53: Doug Armstrong; 54: Tom Lewis; 55: (top left) OMIKRON, (bottom left and right) courtesy of the Virus Laboratory/University of California, Berkeley; 56: (left) L.D. Simon and T.F. Anderson, The Institute for Cancer Research, (right) F.A. Eiserling, UCLA; 57: courtesy of A.K. Kleinschmidt, *Biochim. Biophys. Acta*, 88 (1964), p. 146 (fig. 2); 58: Tom Lewis; 60: (all but bottom right) courtesy of The American Museum of Natural History, (bottom right) David W. Miller.

Chapter 3

62: William MacDonald/IBOL; 64: The Granger Collection; 65: (top) Doug Armstrong, (bottom) William MacDonald/IBOL; 66: John Dawson; 67: Paul Slick after I.M. Lerner, *Heredity, Evolution, and Society*, W.H. Freeman and Company. © 1968; 68: Tom Lewis after Payson Stevens; 69: John Dawson; 70: Philip Hyde; 71: Doug Armstrong; 73: The Art Works; 75: Doug Armstrong; 76: The Art Works after A. Lehninger, *Biochemistry*, © 1970, Worth Publishers, Inc.; 77: Robert Kinyon/Millsap and Kinyon after Payson Stevens; 79: Doug Armstrong, redrawn by permission from A.L. Lehninger, *Bioenergetics*, © 1965, W.A. Benjamin; 80: John Dawson after Dr. Wilfred Wilson; 82: (top left) Ward's Natural Science Establishment, (bottom left) Lynwood Chace/National Audubon Society, (right) John Dawson; 83: William T. Hall/Electro-Nucleonics Laboratories, Inc.

Unit II/Chapter 4

84: Terry Lamb; 86: William MacDonald/IBOL; 88: (top) Tom Lewis after C. Swanson, *The Cell*, 3rd ed., Prentice-Hall, 1969; 89: (top) Doug Armstrong after W.A. Jensen and R.B. Park, *Cell Ultrastructure*, Wadsworth, 1967, (bottom) William Jensen; 90: (top and center) John Dawson; 91: Robert Kinyon/Millsap and Kinyon; 92: John Dawson after Dr. Wilfred Wilson; 94: (top left) Doug Armstrong, (bottom left) David M. Phillips, (bottom right) Daniel Branton from *Proceedings of the National Academy of Sciences*, 55:1048, 1966; 95: from D. Fawcett, *The Cell: An Atlas of Fine Structure*, W.B. Saunders Co., 1960; 97: Vincent T. Marchesi; 98–100: Tom Lewis; 101: (right) K.R. Porter; 102: David M. Phillips; 103: (top) K.R. Porter, (bottom) William A. Jensen; 104: William A. Jensen; 105: (left) Robert Kinyon/Millsap and Kinyon after Payson Stevens, (right, top to bottom) William A. Jensen, Myron Ledbetter, Garth Nicolson/Salk Institute; 106: (left, top to bottom) K.R. Porter, H. Fernandez-Moran, T. Oda, R.V. Blair, D.E. Green, *Journal of Cell Biology*, 22:63, 1964 (2), (right) Robert Kinyon/Millsap and Kinyon after Payson Stevens; 108: (top left) David M. Phillips, (bottom left) Myron C. Ledbetter, (right) David M. Phillips, 109: (top to bottom) David M. Phillips, Doug Armstrong, David M. Phillips; 111: (left) H.E. Hinton, (right) Grant Heilman.

Chapter 5

112: courtesy of the Upjohn Company, Kalamazoo, Michigan; 114: Doug Armstrong, text diagram courtesy of David Eisenberg, UCLA; 115: Jim Leragy, model of a DNA molecule/courtesy of the Xerox Corporation, text diagram courtesy of David Eisenberg, UCLA; 117: Tom Lewis redrawn by permission from J.D. Watson, *Molecular Biology of the Gene*, © 1965, W.A. Benjamin, Inc.; 118–119: Doug Armstrong; 120–121: John Dawson; 123: Tom Lewis after J.D. Watson, *Molecular Biology of the Gene*, © 1965, W.A. Benjamin Inc.; 124: (top) Wide World Photos, (bottom) courtesy Linda Viviers; 125: (left to right, top to bottom) John Cairns, David M. Prescott (2), Dr. Henry S. Slater; 126: (left) Doug Armstrong, redrawn by permission from J.D. Watson, *Molecular Biology of the Gene* © 1970, W.A. Benjamin, Inc.; 127: (left) Photo supplied by Dr. A. Rich, (right) Doug Armstrong; 129: Neill Cate; 130–131: John Dawson; 132: from CRM Productions, *The Cell: A Functioning Structure, Part II*.

Chapter 6

134: Masami Teraoka; 137: (left) Tom Lewis, (right) Tom Lewis after William Keeton, *Biological Science*, W.W. Norton and Co., © 1967; 138–139: illustrations by John Dawson; 140: J. Herbert Taylor; 141: (top) John Dawson, (bottom) Andrew S. Bajer; 142: courtesy of Ulrich Clever; 144–146: illustrations by John Dawson; 147: Dr. James Kezer, Department of Biology, University of Oregon; 150: John Dawson after B.I. Balinsky, *An Introduction to Embryology*, 1970, W.B. Saunders Co.

Chapter 7

152–154: Doug Armstrong; 155: courtesy of the Moravsky Museum; 156: Doug Armstrong; 159: (top left) Runk/Schoenberger (Grant Heilman), (bottom left) Roseann Litzinger, (right) John Dawson; 160: John Pierce; 162: (top) Ray Bravo, (bottom) Tom Lewis; 164: (top) Tom Lewis after William Keeton, *Biological Science*, W.W. Norton and Co., 1967, (bottom) The Art Works after Dr. Victor A. McKusick; 165: from C.B. Bridges, *Journal of Heredity*, 26:60–64, 1935; 168: Felicia Fry; 169: Paul Slick after Dr. Wilfred Wilson; 171: from the experiments of Dr. H.B.D. Kettlewell, University of Oxford.

Chapter 8

172: Joe Garcia; 174: (left) courtesy of the University of California at San Diego Medical School, (top right) Paul Slick after T. Friedmann and R. Roblin, "Gene Therapy for Human Genetic Disease," *Science*, Vol. 175, March 3, 1972, (bottom right) Doug Armstrong; 175: Paul Slick after B. Frieden, *Modern Topics in Biochemistry*, Macmillan, 1966; 176: John Dawson; 178: Neill Cate after J.D. Watson, *Molecular Biology of the Gene*, W.A. Benjamin, 1970; 180: Patricia Peck; 181: William D. Loughman, Biophysicist, Donner Laboratory, University of California; 182: (top left) John Dawson, (bottom left) courtesy of the Department of Neurosciences, University of California, San Diego, (right) T. Friedmann, "Nerve Cells and Behavior," © 1971 by Scientific American, Inc. All rights reserved; 183: Tom Lewis; 184: Paul Slick after Dr. Wilfred Wilson; 186: Felicia Fry.

Chapter 9

188: construction by Sheila Ordean, photography by Werner Kalber/PPS; 190: (left) John Dawson, (right) John Dawson after C. Singer, *A History of Biology*, 1959, by permission of Murnat Publications, Inc.; 191: (left) Dick Oden, (right: left to right and bottom) Smith/OMIKRON, Ward's Natural Science Establishment, Runk/Schoenberger (Grant Heilman); 192: (left) Douglas Roy, (right) M. Woodbridge Williams; 193: (left) James Endicott, (right) Chicago Biological Supply House; 194: Roseann Litzinger; 196–198: Ray Bravo; 200: Dick Oden; 202: John Dawson; 203: Doug Armstrong; 205: (right) courtesy of the Carnegie Institute of Washington; 206: (left) John Dawson; 208: (left) John Dawson, (right) photos by Elizabeth Wilcox; 209: (top) Doug Armstrong, (bottom) William MacDonald/IBOL.

Unit III/Chapter 10

212: Terry Lamb; 214: Karl Nicholason; 216: (top) Doug Armstrong after Dr. Wilfred Wilson, (bottom) Paul Slick after Dr. Wilfred Wilson; 217: J.T. Bonner; 218: (left) John Dawson, (right) James Endicott; 219: (left) James W. Dutcher, (right) George Lower/National Audubon Society; 220: James Endicott; 221: (left to right, top to bottom) Franklin Photo Agency, Carolina Biological Supply Co., Dr. E.G. Cutter, Carolina Biological Supply Co. (2); 223: (left) Tom Lewis, (right) Lester V. Bergman and Associates; 225: Paul Slick; 226: (top) Hugh E. Huxley, (bottom) Clara Franzini-Armstrong.

Chapter 11

228: Construction by Earnie and Helga Kollar, photography by Werner Kalber/PPS; 230: Millsap/Kinyon; 231: Patricia Peck; 233: Ray Bravo; 234: John Dawson; 235: (left) Millsap/Kinyon, (right) John Dawson; 236: (top) Walter Dawn; 237: John Dawson; 238: Tom Lewis; 239: (left) John Dawson, (right) John Dawson after H.S. Mayerson, "The Lymphatic System," © 1963 by Scientific American, Inc. All rights reserved; 240: John Dawson; 243: Millsap/Kinyon; 244: John Dawson after Dr. Wilfred Wilson; 247: John Dawson after R.S. Speirs, "How Cells Attack Antigens," © 1964 by Scientific American, Inc. All rights reserved; 249: (left) John Dawson, (right) Roseann Litzinger; 250: (left) Doug Armstrong, (right) Ron Estrine; 251: F.M. Burnet; 252: Eli O. Meltzer, M.D.

Chapter 12

254: Paul Slick; 256: John Dawson; 257: Douglas Roy; 258: (left) John Dawson after E. Kandal, "Nerve Cells and Behavior," © 1970 by Scientific American, Inc. All rights reserved., (right) courtesy of Dr. James McConnell, *Worm Runner's Digest*; 259: (left) Lester Bergman and Associates; 260: Sir John C. Eccles; 261: (top) Millsap/Kinyon; 262: (left) John Dawson, (right) UPI Compix; 263: (left) S.L. Palay, (right) Doug Armstrong; 264: Dick Oden; 265: after William Keeton, *Biological Science*, W.W. Norton, 1967; 266: Dick Oden; 267: after W. Penfield and L. Roberts, *Speech and Brain Mechanisms*, Princeton University Press, 1959; 268–272: Dick Oden.

Chapter 13

274: Dick Oden; 276: (left) John Dawson, (right) Steve McCarroll/IBOL; 277: Paul Slick; 279: (left) John Dawson, (right) Doug Armstrong; 280: from *So Excellent A Fishe*, by Archie Carr, © 1957 by Archie Carr, reprinted by permission of Doubleday and Co.; 281: (top to bottom) Tony Florio/National Audubon Society, John H. Gerard/National Audubon Society, George Leavens; 282: Neill Cate; 283: (left) from K.D. Roeder, *Behavior*, 10: 300–304, 1962, (right) Douglas P. Wilson; 284: R.F. Head/National Audubon Society; 285: (left) Thomas McEvoy, *Life Magazine*, © Time, Inc., (right) after N. Tinbergen, *The Study of Instinct*, The Clarendon Press, 1951; 296: (left) Doug Armstrong, (right) J. Woodland Hastings; 287: Doug Armstrong after K.S. Rawson, from "Photoperiodism and Related Phenomena in Plants and Animals," *Pub. No. 55*, © 1959 by the American Association for the Advancement of Science; 288: (top) Doug Armstrong, (bottom) The Art Works; 289: Doug Armstrong; 290: Tom Lewis; 291: John Dawson; 292–293: Tom Lewis; 294: (left) courtesy of A.M. Guhl, (right) Grant Haist; 296–297: John Dawson; 298: Grant Heilman.

Chapter 14

300: Dick Oden; 303: Doug Armstrong; 304: Bernard Wolff/Photo Researchers; 306: Doug Armstrong after Kormondy, *Concepts of Ecology*, Prentice-Hall, 1969; 307: Doug Armstrong after W.C. Allee, et. al., *Principles of Animal Ecology*, W.B. Saunders Co., 1949; 308: San Diego Zoo Photo/Ron Garrison and F.D. Schmidt; 309: Doug Armstrong; 340: © Walt Disney Productions; 311: Doug Armstrong; 312: (top) Doug Armstrong after G.F. Gause, *The Struggle for Existence*, Williams and Wilkins, 1934, (bottom) Leonard Lee Rue III; 313: (left) Doug Armstrong after R.L. Smith, *Ecology and Field Biology*, Harper & Row, 1966, (right) Jack Dermid; 314–318: John Dawson; 320: Doug Armstrong, 321: John Dawson.

Chapter 15

322: Darrell Millsap/Millsap and Kinyon; 324: Doug Armstrong; 325: (top left) Neill Cate after *Population Bulletin*, Vol. 18, No. 1, (top right) Robert Kinyon/Millsap and Kinyon, (bottom) John Dawson; 329: Dan Morrill Photography; 330: illustration by Tom Lewis, photography by Werner Kalber/PPS; 331: (top) Neill Cate, (bottom) Neill Cate after R.R. Doane, *World Balance Sheet*, Harper & Row, 1957; 332: Doug Armstrong after J. Bonner, "The Next Ninety Years," from R.P. Schuster, (ed.), *The Next 90 Years*, California Institute of Technology; 333: P.W. Hay/National Audubon Society; 334: John Dawson; 335: Neill Cate; 336: (top) photography by Allan Roberts; (bottom) Doug Armstrong.